The Modern Business Data Analyst

Dominik Jung

The Modern Business Data Analyst

A Case Study Introduction into Business Data Analytics with CRISP-DM and R

Springer

Dominik Jung
Software & Digital Business Group
TU Darmstadt, Darmstadt, Germany

ISBN 978-3-031-59906-4 ISBN 978-3-031-59907-1 (eBook)
https://doi.org/10.1007/978-3-031-59907-1

This Springer imprint is published by the registered company Springer Nature Switzerland AG
The registered company address is: Gewerbestrasse 11, 6330 Cham, Switzerland

To Felix and Alexander

Preface

Solving Business Problems with Business Data Analytics

"Sorry, but I cannot tell fortunes!", is my disappointing answer if people hear that I am an expert for *business data analytics*. They assume that my expertise in this field equates to crystal ball-like abilities or an uncanny knack for foreseeing. Some people ask me about my predictions on general topics like stock market trends, the next vote results, or specific things like the development of an obscure cryptocurrency or the future soccer results of their favorite hometown team.

Business data analytics isn't about divination or fortune telling. It's a methodical process, a toolset honed to derive insights from business data and make informed decisions. It's not a crystal ball. It's a sophisticated compass, guiding businesses through the maze of information to better understand the present and make calculated steps toward the future. Business data analytics isn't about magic - it's about leveraging information to drive strategic action.

Let me illustrate it with a short story of my life: During my time as a research assistant at the university, there was the 2018 World Cup in Russia. Because soccer is a big thing in Germany, the whole country and thus also our institute was in soccer fever. Some colleagues even took leave to watch certain games (and enjoy the celebration afterwards). Even during work, we often had a livestream running in our office to follow the current games. And so, it came about that my colleagues organized a betting game. Everyone had to participate, including me. The problem was that I'm probably one of the few Germans who isn't interested in soccer at all. So, I was faced with the challenge of predicting things I knew nothing about - classic everyday life for a data scientist.

The good thing in this case was that there was no problem to build up a database. Just a short Google search brought up a lot of freely available data and information. And besides the target was relatively clearly defined because there are only `3` meaningful outcomes in a soccer game in a world championship (`win`, `lose` or `draw`) and likely results (`1:0` or `1:1` and unlikely like `13:0`).

With that in mind, I built a model and bet exclusively on the most probable winner and one of the most likely results. I also occasionally initiated conversations about football to pretend that I was seriously into it. Although I still couldn't follow the soccer conversations with my colleagues, it took only a few match days until I got into the top tier with my method. And to my own surprise, I even made it to 1st place with my model in the end and won the betting game. And that without knowing the relevant players, backgrounds, or strategies.

A few years later I finished my PhD and left university to go to industry. And as it goes, there was again a soccer world championship coming up. As every championship, my new soccer-loving colleagues organized a betting game. Confident of victory, I took part and even bet a small sum of money. I dug out my old procedures and blindly bet on the results from the model (error 1). As I was otherwise very busy, I immediately bet on all available matches (error 2).

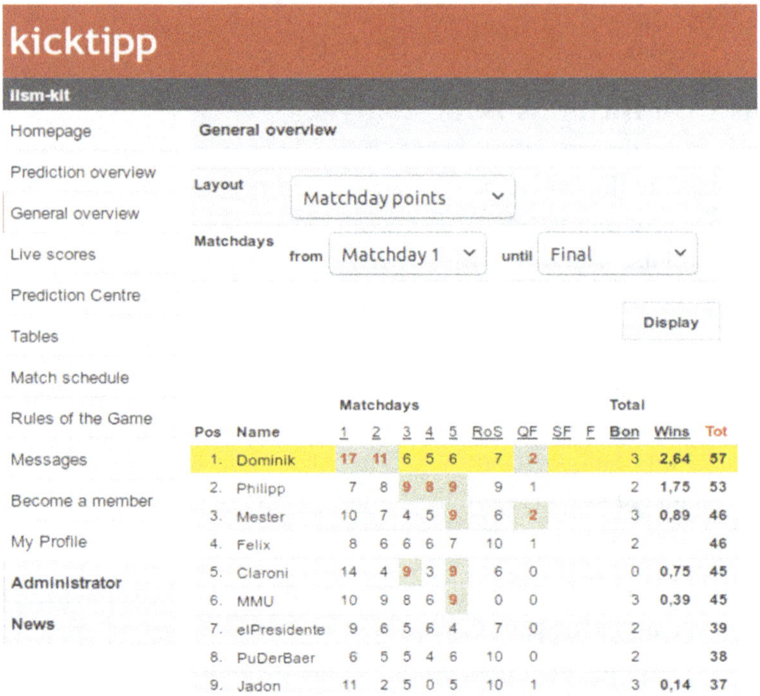

Figure 0-1 On my way to my glorious win against nearly 30 soccer-friendly colleagues.

When I looked at the table again a few weeks later (error 3), I was shocked! Almost all my results were wrong. I had almost always bet on the loser. When I looked at my model, it was clear that I had used it the wrong way (error 4). The betting game had progressed to the point where I had no chance of catching up, so I made one of the last places. And the cherry on top was that my boss forced me to bake a cake for the team as a punishment.

This anecdote illustrates very well what business data analytics is all about: It is about preparing a smart decision and then making it. You must not just blindly apply any code from your own archive or the internet (error 1). You have to crawl into the data and describe and understand the problem on the basis of the data. It is important to challenge the problem and its approaches regularly (error 2). Especially, at the beginning of a project a regular exchange with the stakeholders and the clear definition of a project plan and metrics are essential (error 3). These enable a clean evaluation and estimation of the quality of the data product (error 4).

As you can see, the reality of business data analytics is far from fortune-telling. It's not about gazing into a crystal ball but rather about deciphering complex data landscapes to extract meaningful insights. It's the art of transforming raw information into actionable strategies. And even if you proceed methodically and carefully, success is not guaranteed. Some days the models work out, and some days they don't because the project is too ambitious. And sometimes the models predict the right thing but are poorly interpreted or configured by the users – so the project fails in the long term. But every time data and models can help us to make qualified decisions in domains, where we have no knowledge, if we use them right. And with the support of good data and well-defined models most people can even make better decisions than many experts in specific application areas (like my soccer colleagues).

Unfortunately, most analytics books primarily focus on predicting and building sophisticated data science models. They give you long pages with mathematical formulas but not a single line of code. But the story is not just about building models. It's about understanding the data and what it tells you about the problem and how to solve it with analytics. Analytics in industrial practice is about making smart decisions and supporting a decision-maker in difficult situations. However, this is not always as easy as it seems in my anecdote. There might be many pitfalls like "bad" data or data quality, difficult or unclear target variables or missing documentation. And many other challenges you can't even imagine yet. But do not worry, we will learn methods and workarounds to handle these issues!

In this book we will focus on prediction models not as an end in itself but in order to make smart business decisions. Business data analytics without considering the users of an analytics solution is near to worthless. A dashboard to manage KPIs with bad usability, will not be used by the management and hence, be worthless. Or sometimes an excellent model that is not understandable is worse than a somehow average model that has weaker performance but can be well explained and easily understood by the users.

Hence, this book is particularly about the intersection of business data analytics and data science to taggle the following questions:

- How to plan and setup a business data analytics project
- How to work with R as a programming language to conduct your project
- How to understand and work with your business data
- How to use analytics models to predict and understand your business
- How to build tools that a manager can use to make well-informed decisions, based on facts and not on feelings

R for Business Data Analytics

You probably ask yourself, why should I learn R to make business data analytics? Why can I not use Microsoft Excel, Python or C++? The answer is: I would not call R my favorite programming language. There are other languages like Python that I find better designed and more beginner friendly. I also teach other programming languages like Python at university and see many benefits in it. But R on the other hand was built by statisticians as environment for statistics and business data analytics, which makes it the perfect tool to wrangle and dive deep into business data. And will help you to generate and deploy your analytics models easily.

In the R community many free and useful packages exist to boost your business data analytics pipelines: For instance, making good-looking visualizations that have publication-quality is a difficult undertaking in most analytics tools like Python or Matlab. And I can tell you that R and its `ggplot2` package is probably the most advanced tool and best solution to make professional visualizations to support your analysis and communicate your results. From my time at university, I know that even if you started in another language like Python and plan now to publish your results or want to design posters or infographics for big journals and newspapers you will end up using R for this specific task.

Dashboards with `shiny`, or reports with `rmarkdown` are state-of-the-art solutions for your business data analytics workflow. The most relevant basics can be learned in 1-2 hours of work and you can produce first results for yourself in little time. R allows you to communicate your models and findings in an interactive webapp easily without changing or migrating to another analytics tool.

Every time I want to quickly dive deep into data and understand and model something statistical, I end up using R. With R you can produce higher quality outcomes with less effort compared to other programming languages like Java, Skala and in particular Python, if you want to make business data analytics. And that is what we plan to do. R is the tool that is designed to do business data analytics and statistics. And hence, that makes it my first choice to make business data analytics.

However, the goal of this book is not that you become a kind of religious R-fanatic. I know there is a healthy debate raging over the best language for data science, artificial intelligence, data mining, data engineering, etc. Many people believe Python, Go, Java, etc. is the better language for handling all these kinds of problems in every domain. We call these people cognitive miser. Or as Abraham Maslow said: "If all you have is a hammer, everything looks like a nail". The goal of this book is to motivate you to go further in analytics and continuously learn new things. And this includes to learn further programming languages after reading this book.

How to Read this Book

Besides my job as data scientist, I am a longtime lecturer for data science and business data analytics at different universities and business schools in Germany. And I noticed that most students struggle that most concepts and algorithms of analytics and data science are described in academic language in thick books, sophisticated papers, or very abstract written blogs and websites with many errors. And hence, most people get wrongly the impression that they must get higher academic degrees in mathematics to figure this stuff out.

Thus, this book assumes that you have no academic degree or any experience with business data analytics or data science. Its goal is to illustrate and explain you the key concepts of business data analytics from scratch and to give you all the professional tools you need to start your journey as business data analytics professional. Business data analytics is art and science. There are many decisions where you have to rely on your intuition and cleverness. No scientific paper can teach you that. All you need is an open mind, willingness to puzzle and think mathematically, and a computer. And yes, this book!

We will cover a large number of different business data analytics techniques and best-practices. Rather than discussing the mathematical and theoretical foundations in detail, I will give you an intuitive understanding of the main concepts by illustrating you concrete practical examples. For that purpose, we focus on the application of the most relevant concepts and using mostly ready-to-go R frameworks and packages to get them working easily:

- `dstools` is a lightweighted package of different functions for professional data science development. It implements many useful algorithms and helper functions, which makes it a useful companion for data scientists and business data analysts.

- `tidyverse` is a popular collection of many data science algorithms and methods encapsulated in different packages (e.g., `dplyr`, `tidyr` or `ggplot2`). It's definitely the most relevant toolset for data science professionals and beginners. I recommend installing the `tidyverse`-collection instead of the many packages individually.

All these R packages are very powerful and provide an overwhelming number of functions and functionalities. The majority of packages we use in this book are part of the `tidyverse`-collection. The benefit is that these packages share a common way to do analytics and this will help you to get familiar with the topic.

Besides, we will use some other packages in this book that are not listed here. However, I made sure that these packages are compatible and follow the same principles. These specific packages are used for very specific tasks in this book (like web crawling) and therefore are not essential like the packages listed here.

You should also know, that it is not necessary that you know all R packages in detail or have already experiences in R or other programming languages. Nevertheless, coding experience will be helpful. If you are a Python programmer and you are familiar with some popular packages like `NumPy`, `Pandas`, and `Matplotlib` you will probably see many connections and similarities in the `dplyr`, `tidyr` and `ggplot` packages.

While you can read and understand the concepts in this book without writing one single line of code, I strongly recommend that you experiment and work with the different code examples in this book. For that purpose, I provide my code and further material for this book online at the website of the fictive *Junglivet Whisky Company*[1]:

www. junglivet.com

On this website I gathered all the material for this book and provide further information about the fictive company. You can use this material to deepen your background about the company, which can help you in the business cases of this book.

Furthermore, you find the lecture slides of the related university course of this book. You can use them to deepen your knowledge after reading this book or to look up some concepts, if you want another explanation about them or to learn more about their background.

Alternatively, you can also download the materials at my private git repository at:

https://github.com/dominikjung42/BusinessAnalyticsBook

If you are unsecure about git, or if you don't know how to use it yet and want to become familiar with it, the git tutorial documentation (*www.git-scm.com/docs/gittutorial*) is a great place to start. However, you don't need git skills in this book, but it will be definitely helpful in your future career as a business data analytics expert. After reading this book, you could continue your analytics journey by taking a deeper look at git and code versioning.

---

[1] *Similarities with people alive or other authorities are purely accidental and not intended.*

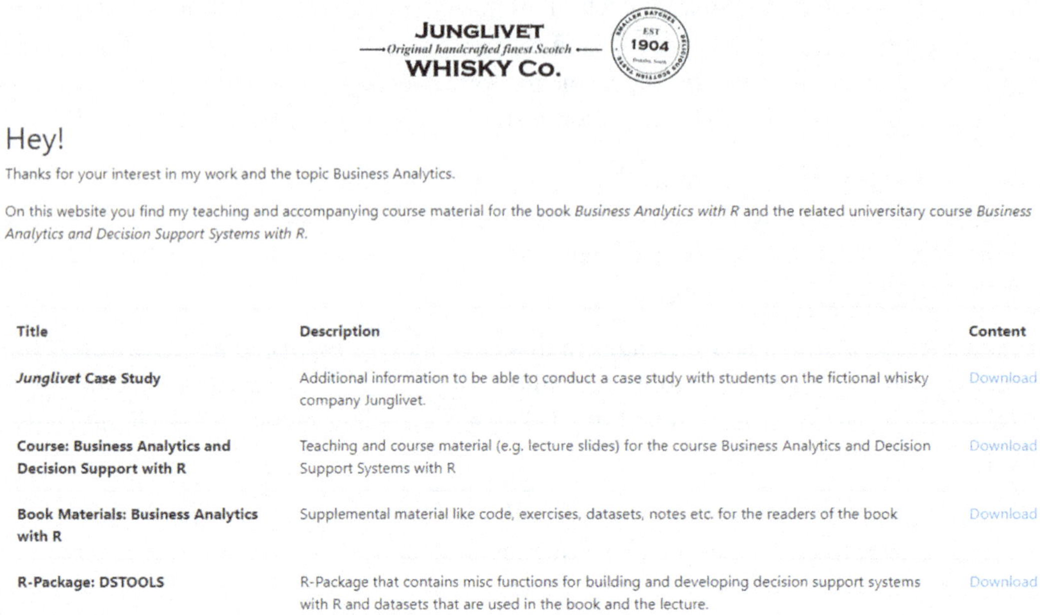
JUNGLIVET
Original handcrafted finest Scotch
WHISKY Co.

EST 1904

Hey!

Thanks for your interest in my work and the topic Business Analytics.

On this website you find my teaching and accompanying course material for the book *Business Analytics with R* and the related universitary course *Business Analytics and Decision Support Systems with R.*

Title	Description	Content
***Junglivet* Case Study**	Additional information to be able to conduct a case study with students on the fictional whisky company Junglivet.	Download
Course: Business Analytics and Decision Support with R	Teaching and course material (e.g. lecture slides) for the course Business Analytics and Decision Support Systems with R	Download
Book Materials: Business Analytics with R	Supplemental material like code, exercises, datasets, notes etc. for the readers of the book	Download
R-Package: DSTOOLS	R-Package that contains misc functions for building and developing decision support systems with R and datasets that are used in the book and the lecture.	Download

Figure 0-2 On the website of the Junglivet Whisky Company (www.junglivet.com) you can find further information related to the cases in this book.

Also, if you have never used R, in the chapter *Business Data Analytics Toolbox: R and RStudio* I will guide you through the installation process. You will learn the basic concepts of programming with R for business data analytics. The tutorials are for R beginners but also contain some information that might be also interesting for intermediate R programmers, so take at least a look at it before you decide to skip the chapter.

Throughout the book, we will learn more about business data analytics by solving case studies. You will be chief data scientist or chief business data analyst of the fictional *Junglivet Whisky Company*. By applying concepts and algorithms of this book you can test your skills and gradually leading the company along the path to success.

During this journey we will take a deeper look at all the different phases of analytics projects: *Business Understanding*, *Business Data Understanding*, *Business Data Preparation*, *Modeling*, *Evaluation* and *Deployment* (see the next chapter *How Business Data Analytics Experts Work* for more details). Sometimes we will focus on data processing and coding, sometimes we will look at the data itself and visualize it, and sometimes we will discuss concepts on a more general level.

In consequence, the book is organized in seven chapters that represent an introduction, the six different phases of typical business data analytics projects (whereby *Evaluation* and *Deployment* are combined in one chapter *Business Data Products*) and a last chapter to help you get started with real business cases.

As a first outlook, this book covers the following specific business data analytics topics and questions:

- What is business data analytics? And what are common business data analytics problems?
- What are the different phases and steps of business data analytics projects?
- Why business and data understanding are crucial for the success of your projects?
- How to handle, clean, and prepare your data for business data analytics?
- How to select and reduce your data to make it tidy?
- What are the most common analytics algorithms like regressions, k-nearest neighbors, decision trees, random forests, association analysis and ensemble methods?
- How to fit, optimized and train your analytics models?
- How to develop analytics solutions and data products for your users like the management or the domain experts?
- And finally, how to solve real business problems and case studies with business data analytics?

Conventions Used in this Book

The following typographical conventions are used in this book:

- *Italic,* indicates references like chapter names, URLs, email addresses and filenames.
- `Constant width`, indicates program code, as well as within paragraphs to refer to elements of the code such as variable or function names.

This symbol indicates other relevant things that are not in the text. If you see it, this box or element signifies a tip or important comment.

Code Examples

Further supplemental material (code examples, datasets, exercises, etc.) is available for download at: *www.junglivet.com*

Acknowledgments

I am also incredibly thankful to all reviewers and friends that helped me to make this book happen. They took themselves time to review my book in their free time besides their full-time jobs, family commitments and many other things they gave up to read my manuscript. In particular, I would like to thank Dr. Kevin Laubis for his feedback and for motivating me continuously for the project, Dr. Simon Behrendt for the comprehensive corrections and Dominik Raab for his numerous suggestions for improvement and the many entertaining discussions. Special thanks also to Prof. Dr. Peter Buxmann and Dr. Timo Sturm from University Darmstadt, with whom I was always able to discuss the application of machine learning in business data analytics and business informatics. I would also like to thank my former and actual colleagues at Karlsruhe Institute of Technology and Porsche AG, it was thanks to you that the idea for the project was born!

Above all, I want to thank my beloved wife, Trucy. She encouraged me to start and to continue working on this book. She is and was definitely one of my toughest critics, increasing the quality of this book.

Thank you all.

Contents

1 Introduction

1.1 Motivation

Congratulations to your new job! You are hired as business data analyst at the *Junglivet Whisky Company*. Through the book, we'll be learning more about business data analytics by building up analytics and reporting solutions for the different stakeholders at a traditional whisky distillery. However, the underlying problems are motivated by real problems which I faced during my analytics life-time. For educational purpose, I anonymized and adjusted the scenarios and details in this book.

We will start with simple straight-forward business problems and increase complexity step by step in each case study. Furthermore, we will implement basic algorithms and applications and even full decision-support systems from scratch. This will give you a strong understanding from business data analytics theory to application. And after working with this book, you will be ready to apply for a real business data analytics job in industry.

I would also like to point out again that you can download all the material we use such as code examples, datasets, notes, etc. from the course website at:

www.junglivet.com

So, you don't have to type all the coding examples by hand, just download the files and run them on your machine. However, I strongly recommend that you try out the exercises by yourself before looking at the solutions in this book or on the repository.

Additionally, I strongly recommend to install my R-package `dstools`. The name stands for "data science tools" and means that this package contains many functions that are useful for business data analytics and data science in general. Besides, it also contains many useful materials for this book. For instance, all the datasets of the case studies are included in the package so that you can load them easily with `data(<name of the dataset>)` in your R environment. You can solve all the exercises and tasks without the packages, but it will make your life much easier if you use it. And the good news: It is free to use!

If you decide to use it, you can install the newest version directly from GitHub with:

```
# Install the newest version from GitHub
install.packages("devtools")
devtools::install_github("dominikjung42/dstools")
```

> Don't worry if you have problems to understand the R code in this book right now. In chapter 2, we will take you through an R crash course, including how to install and manage such packages in R. For the moment, it's enough if you try to get a general understanding of what we are doing in the code.

D. Jung, *The Modern Business Data Analyst*, https://doi.org/10.1007/978-3-031-59907-1_1

1.2 Whisky Quality Problems in the Junglivet Company

Are you ready to go for your first analytics jobs in the *Junglivet Whisky Company*? Let us start with an introductory example. The purpose is to give you a first impression of the business data analytics process and the central concepts and problems we will discuss in the book. You can recreate it directly on your computer if you already setup R. But it will not be necessary if you haven't, we will setup R and the integrated development environment (IDE) RStudio together in the next chapter. So, best would be if you just read the code examples and try to understand them. Then we setup R and RStudio together in the next chapter and you can try to solve the business case by your own after getting a better understanding of everything. If you continue with this book, we will also learn other and better techniques than used in this introduction to analyze analytics problems. So, feel free to come back to this example any time later in your reading.

In summary, for the moment it's enough if you read it and try to understand the code examples on a general level. The purpose is to show you an analytics workflow, to tease you for the upcoming chapters and not how to write code. I want to motivate you and illustrate the process of a business analyst. Please keep this in mind when reading the following scenario. Don't worry, there will be only some minimal code examples given to help you to illustrate a typical analytics project.

1.2.1 Welcome to the Junglivet Whisky Company

The company sits in a charming old building some five-minutes-walk from the center of the next village. Your office and the production hall seem to be next to each other. When you arrive at the building with the cab, a mid-aged lady waves to you cheerfully with both hands. "Cheers, welcome to the Junglivet factory!" says the lady as you leave the car. "Glad you are here darling! My name is Miss Gleck. And because there is a lot of work for you to do, it is best if you report to Mr. Gumble at the distillery right away. There are considerable problems. I will see to it that your suitcases are brought to your office in the meantime. Let's go! What are you waiting for?"

Somewhat confused, you leave the office and go next door to the production hall. A friendly older gentleman in a workman's outfit greets you from afar. He has a very strong handshake and introduces himself as Mr. Gumble. He is from the distillery and responsible for the production team. You make some small talk while Mr. Gumble is full of questions about you and your analytics background. And while you talk, you recognize that until now the company made most decisions based on feelings, intuition and as a consequence has now serious financial problems. Thus, Mr. Gumble and his colleagues are very excited to have you in the team, and hope that you can give new drive by bringing your analytics knowledge aboard.

For now, he needs your help with some problems in the production line. He received complains about the quality of the whisky and tries to find the reason for that. Mr. Gumble gives you an USB-stick with the production lane data of the last two weeks and asks you to analyze it (which is quite uncommon, because in reality people don't hand you the data you need and you have to search for it).

He says that the dataset should contain all the data that you need (which makes the situation even more suspicious – you never get "all" the data you need in the first try in your whole career as business data analyst). As you plug in the USB-stick in the local computer you see that it consists of an excel sheet from the production lane containing 8 columns. An excerpt of the data is illustrated in Table 1-1.

Table 1-1. Excerpt from the dataset of the production lane (productionlog_sample.xlsx).

DAY	MONTH	MANUFACTURER	PRODUCT	SHIFT	COLOR	MALTING	TASTING
1	4	Leonard	Junglivet	Morning	0.27	Inhouse	895
1	4	Carlson	Junglivet Premium	Evening	0.27	Burns Best Ltd.	879
2	4	Leonard	Junglivet	Morning	0.28	Inhouse	938
...	...	...	...	...	...	...	...

Each row represents information of the shift and the final quality of the whisky production. For example, the first row shows all the information of the morning shift's Junglivet production (column SHIFT and PRODUCT) of the first of April (DAY and MONTH). The responsible brew master was Mr. Leonard (MANUFACTURER) and the malting was produced inhouse (MALTING). During production the color of the whisky was measured (COLOR) and after the production a quality assessment has been conducted. The final score in the quality check before delivery is 895 (TASTING), which is a quite good value (1000 is best says Mr. Gumble).

In summary the variables or features in the dataset are:

- DAY: The day of the production
- MONTH: The month of the production
- MANUFACTURER: Name of the leading brew master
- PRODUCT: Product type like Junglivet or Junglivet premium
- SHIFT: Indicates morning or evening shift
- COLOR: Color indicator on a scale between 0 and 1
- MALTING: Origin of the used malting products
- TASTING: Quality indicator based on a whisky tasting on a scale between 0 and 1000

If you want to learn more about the different features you can also download the additional material about whisky production on the *junglivet.com* website. But for the moment, let us continue helping Mr. Gumble!

1.2.2 Step 1: Business and Data Understanding

Unfortunately, you don't have your own laptop with your analytics toolbox with you, but you install Base R on the local computer of Mr. Gumble to take a quick look at the data. If you have R and RStudio installed and are familiar with it, you can follow the analysis by downloading the data from the book's website and open it in RStudio.

To start with your analysis, you have to load a couple of packages in your R environment (if you are not familiar with that, we will learn that later in chapter 2). While R has only some basic functions, it can be easily extended by loading further packages. In this case you decide that you will work with an excel sheet and probably some simple data processing and visualization packages. For that purpose, you install the following packages on your new laptop:

```
install.packages("devtools")
devtools::install_github("dominikjung42/dstools")
install.packages("dplyr")
install.packages("ggplot2")
install.packages("readxl")
```

And then load the freshly installed packages with the R-Bases `library()` function:

```
library("dstools")
library("dplyr")
library("ggplot2")
library("readxl")
```

Now that we have prepared everything for our analysis, we want to load the data into our R environment. For that purpose, you put the excel sheet from Mr. Gumble in the same folder as your R analytics project. If you are interested to follow this analysis you can download the file on the book's website, but if this is your first time reading this section it would be better if you just continue reading and come back to the chapter later.

After the dataset is in your local folder you load the production data directly form terminal with the following command in your R environment:

```
dataset = read_excel("productionlog_sample.xlsx")
```

Then you open the logfiles from the whisky production in your R environment. You want to check if your dataset is loaded correctly. You take a first visual look at the top lines of your dataset by calling the `head()` function:

```
> head(dataset)
DAY MONTH MANUFACTURER PRODUCT           SHIFT  COLOR MALTING          TASTING
1      4 Leonard    Junglivet         Morning 0.27  Inhouse              895
1      4 Carlson   Junglivet Premium Evening 0.27  Burns Best Ltd.      879
2      4 Leonard    Junglivet         Morning 0.28  Inhouse              938
2      4 Carlson    Junglivet         Evening 0.32  Inhouse              900
3      4 Leonard    Junglivet         Morning 0.32  Matro Ltd.           917
3      4 Carlson    Junglivet         Evening 0.28  Inhouse              900
```

If we have code examples that contain code and results / output of your code, I use the ">" symbol. The symbol means that this code is entered in the console of your RStudio IDE. Other lines in the code example without the symbol represent the R output.

As Mr. Gumble said, your dataset consists of `8` columns (good news – in most real cases your data description will not fit to the data you receive). As in the example, a single column represents further information about the whisky production shift. For instance, the third column

"`MANUFACTURER`" contains information about the responsible producer of the shift or the last column about the quality of the final product. These is quite important information and we want to use it to investigate the reason for the quality reduction in the last weeks.

All the datasets of this book are also available in the `dstools` package (online available at *https://github.com/dominikjung42/dstools*). Instead of loading them manually in your R environment you can just run `data("name of the dataset")` if you have the package installed. For instance, the following command loads the current dataset of this tasks from the `dstools` package:

```
my_data = data("productionlog_sample")
```

Before you can start further investigating the problem, you decide to take a look at the consistency of the dataset of Mr. Gumble. This is crucial, because, as a business data analytics expert, you know that data is often dirty. It can contain missing values or the wrong values. Or the format of the dataset might be broken or contain some logical errors (like wrongly named columns or other curiosities).

Thus, a natural next step is to investigate the consistency of your dataset and the related columns by calling the table view:

```
View(dataset)
```

A table pops-up in your RStudio containing your dataset in a tabulated form. You can see the results in Figure 1-1.

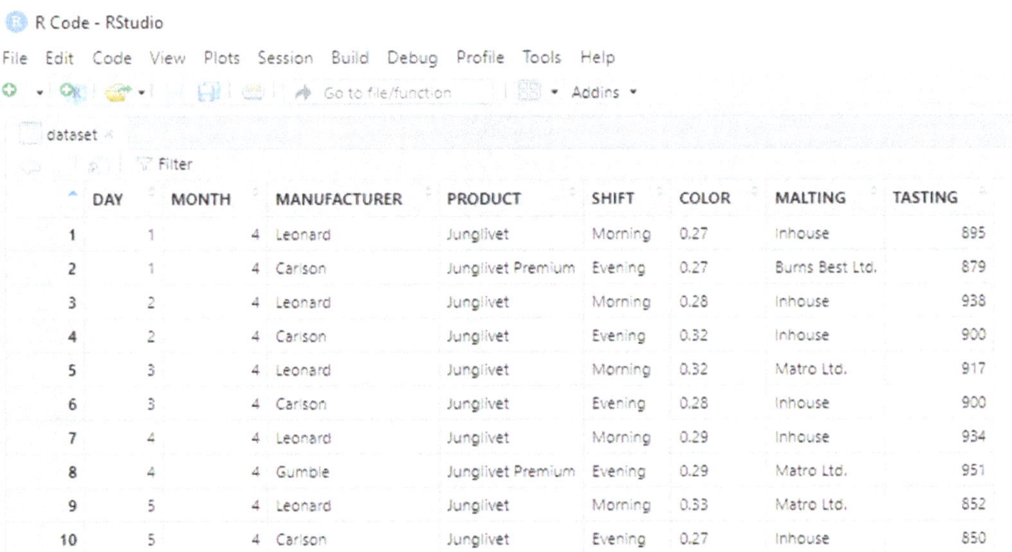

	DAY	MONTH	MANUFACTURER	PRODUCT	SHIFT	COLOR	MALTING	TASTING
1	1	4	Leonard	Junglivet	Morning	0.27	Inhouse	895
2	1	4	Carlson	Junglivet Premium	Evening	0.27	Burns Best Ltd.	879
3	2	4	Leonard	Junglivet	Morning	0.28	Inhouse	938
4	2	4	Carlson	Junglivet	Evening	0.32	Inhouse	900
5	3	4	Leonard	Junglivet	Morning	0.32	Matro Ltd.	917
6	3	4	Carlson	Junglivet	Evening	0.28	Inhouse	900
7	4	4	Leonard	Junglivet	Morning	0.29	Inhouse	934
8	4	4	Gumble	Junglivet Premium	Evening	0.29	Matro Ltd.	951
9	5	4	Leonard	Junglivet	Morning	0.33	Matro Ltd.	852
10	5	4	Carlson	Junglivet	Evening	0.27	Inhouse	850

Figure 1-1 Table with your dataset representing the different whisky production shifts and their output.

The table view of RStudio allows you to investigate your data in an excel-like format. You can see the rows with different IDs on the left. Besides there are different columns like `DAY` or

MANUFACTURER. You can sort your dataset by a specific column by clicking the column name or the small rectangle beside the name. Furthermore, the RStudio view has a Filter option. You can find it in the menu, right on top of the column names on the left. If you click on the button, you can filter the dataset to find specific rows. The magnifying glass in the right corner allows you to search for specific words or numbers in your dataset.

Figure 1-2 The filter function is very useful to look for specific values and check the quality of your dataset.

After eyeballing all the rows in the table, you see that row 11 has missing values. "Would have been too nice if the data quality would be perfect", you think to yourself. You decide to compute further descriptive statistics to check the quality of the dataset before you handle row 11.

A very common and useful function for this is the summary() function. You decide to run it by typing summary(dataset) in your R-console:

```
> summary(dataset)
      DAY            MONTH           MANUFACTURER         PRODUCT              SHIFT
 Min.   : 1.0   Min.    : 4     Length:21            Length:21            Length:21
 1st Qu.: 3.0   1st Qu.: 4     Class :character     Class :character     Class :character
 Median : 5.5   Median : 4     Mode  :character     Mode  :character     Mode  :character
 Mean   : 5.5   Mean    : 4
 3rd Qu.: 8.0   3rd Qu.: 4
 Max.   :10.0   Max.    : 4
 NA's   :1      NA's    :1

COLOR               MALTING             TASTING
Length:21           Length:21           Min.    : 822.0
Class :character    Class :character    1st Qu.: 875.0
Mode  :character                        Median : 925.5
                                        Mean    : 918.5
                                        3rd Qu.: 957.8
                                        Max.    : 999.0
                                        NA's    :1
```

As you can see R generated a short report of each column of your dataset. There is information if the column contains numeric or character values. And some more information like the different ranges, if the column is numeric.

Our missing row number 11 has also been detected by R (it is marked as NA). NA stands for "not available", which means in our case that there is one row that contains no values. Furthermore, the column or feature MONTH contains always the number 4. Features with only one value like MONTH are useless because they lack variability and provide no distinguishing information. Let us now have a look at these issues called data quality problems!

1.2.3 Step 2: Business Data Preparation

Therefore, you decide to start the next step of your analytics process, the business data preprocessing. Preprocessing means that you preprocess the data for the analysis and the modeling. This includes handling rows without values, rows with errors or remove variables from your dataset that are of no further interest.

You decide to remove row `11` and the column `MONTH` in the dataset. In a first step, you decide to start with the irrelevant column issue and remove it. Then you run `names()` to check if the column is removed and the dataset has the correct number of columns:

```
> dataset = subset(dataset, select = -c(MONTH))
> names(dataset)
[1] "DAY" "MANUFACTURER" "PRODUCT" "SHIFT" "COLOR" "MALTING" "TASTING"
```

Then you omit the row `11` with `NA` values to get a clean and usable dataset. You use the `na.omit()` command and to check if everything worked out correctly, you run `nrow()` to check the number of rows is reduced by one:

```
> dataset = na.omit(dataset)
> nrow(dataset)
20
```

For now, the dataset is fine and ready for the next step, so we can continue our analysis.

> Data quality problems are very common in typical analytics projects. And this process is very relevant and will consume most of your time. We will handle further concepts and algorithms for data cleaning and other major techniques for data preprocessing more detailed in the chapter *Business Data Preparation*. This dataset is an example dataset for the introduction and is hence very simple and has no complicated data quality problems.

1.2.4 Step 3: Business Data Analytics

In a next step, you want to investigate why there might be quality problems with the whisky production lines of Mr. Gumble. You decide to plot some specific features in the dataset to help Mr. Gumble to find some possible drivers of the production problems.

For that purpose, you run different visualization commands from base R on selected features. The features `MALTING`, `SHIFT` and `MANUFACTURER` seem to be interesting. Sometimes the existing raw materials in whisky production are not enough and the company has to buy more raw materials from suppliers. It could be that these differ in quality and have a negative impact on the whisky. This is represented in the feature `MALTING`. On the other hand, it could also be that the leading brew master was careless and not all production processes were carried out with the necessary accuracy, which is represented in the feature `MANUFACTURER`. Or that the employees have difficulties with concentration in the evening shifts (or the morning shifts), which is represented in the feature `SHIFT`.

Hence, you decide to investigate the relationship between `MANUFACTURER`, `SHIFT`, `MALTING` and `TASTING` even further. To investigate relationships between non-numeric, class features like `SHIFT` or `MANUFACTURER` and the numeric quality (`TASTING`) you use `boxplot()`. A boxplot is a simple but informative visualization that is used very often in business data analytics. In addition, you can often quickly see the influence of different categorical features in it.

Let us now take a first look at the supplier and their possible influence on the quality. You can do it with:

```
boxplot(TASTING ~ MALTING, data = dataset)
```

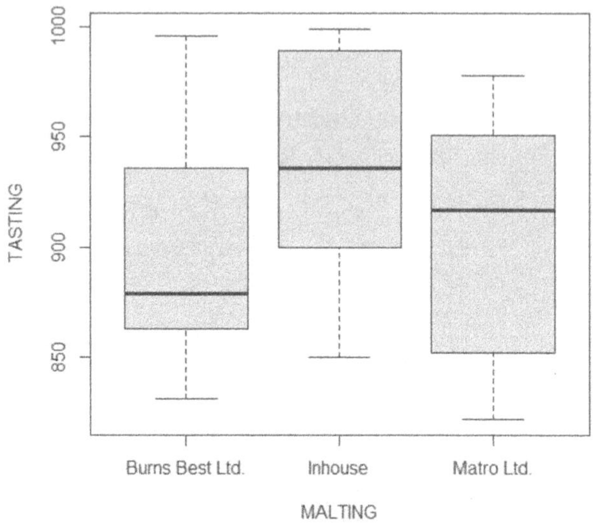

Figure 1-3. Boxplot visualization of the relationship between the classes "`Burns Best Ltd.`", "`Inhouse`", *and* "`Matro Ltd.`".

The result is a visualization with three boxes that you can see in Figure 1-3. The figure displays the distribution of the different supplier (represented by the column `MALTING`) of the *Junglivet Whisky Company* and their relationship on the quality in our whisky production dataset (represented by the column `TASTING`).

The box, which the name boxplot implies, shows in which range the middle 50% of all values are located. The bottom of the box is the point in the data where the bottom 25% have accumulated. The thick line in the middle of the box is the median. That means that up to this line 50% have accumulated. As we can see the Burns Best Ltd. Malting has the lowest median, indicating that it has wors quality. The top of the box marks the point where 75% of all values have accumulated.

We see that most whiskies with the supplier "Burns Best Ltd." have lower quality than if we use our own materials ("inhouse"). However, as a business data analyst you know that the plot gives only a short overview. It seems pretty clear that we have to further investigate the reason why "Burns Best Ltd." is associated with bad whisky. You might ask how can a business data analyst provide more evidence? A first step would be to compute further descriptive statistics to check your finding. Another step would be to run some statistical tests or models, which we will learn later in this book. For the moment, this seems to be enough for you because you know that Mr. Burns from Burns Best Ltd. has been condemned in the media for adulterated products in the last weeks.

To make a last quick check you decide to compute the mean whisky quality of your production with the bad supplier "`Burns Best Ltd.`". You can do that with:

```
> dataset_burns = subset(dataset, MALTING==c("Burns Best Ltd."))
> mean(dataset_burns$TASTING)
[1] 901
```

In the first line of code we make a new dataset named `dataset_burns`, which is a subset of our original dataset filtered by the supplier "`Burns Best Ldt.`". It contains only rows that have "Burns Best Ltd." as a supplier. The mean quality of whisky with this supplier is `901`.

To compare it with the whiskies that use inhouse materials you run:

```
> dataset_inhouse = subset(dataset, MALTING==c("Inhouse"))
> mean(dataset_inhouse$TASTING)
[1] 934.6
```

You were not surprised to see that the mean whisky quality is higher. You quickly close the laptop and decide to look at the other features `SHIFT` and `MANUFACTURER` later (which could be a very good exercise if you come back later to this chapter) and head for the office of Mr. Gumble. As you enter his office, you hear him discussing the whisky quality problems with Carl Leonard from the production team. You report your findings to Mr. Gumble and he is excited!

Mr. Gumble and Carl Leonard ask if you can also provide a possible indicator of the whisky quality that can be used in production. It would be very helpful if they could take a look at some easy measures like the whisky color during the production. And if the color has some deviations, they can stop the production instead of waiting of the final tasting.

What a day you think and go back to work. To investigate the relationship between numeric features like `COLOR` and quality you decide to use a quick scatterplot which can be made with `plot()`.

To investigate the relationship between `COLOR` and `TASTING` you run the following code, which puts `COLOR` on the x-axis and `TASTING` on the y-axis:

```
plot(x=dataset$COLOR, y=dataset$TASTING)
```

As you can see in the following Figure 1-4 there seems to be a relationship between the color of a whisky (`COLOR`) and the measured quality in the tasting (`TASTING`). The best color for *Junglivet* whisky is around `0.3` which indicates that color could be a possible indicator detecting quality issues already in the production.

After sharing your thoughts with Mr. Gumble (who is now completely enthusiastic about you), you realize, exhausted, that it is already really late. You go straight forward to your new appartement and find your bags and some papers with additional information about the company and a card with the weekly menu of the pub next door. You slip out of the building before anyone else can ask you for further help and go straight in the pub to get some food.

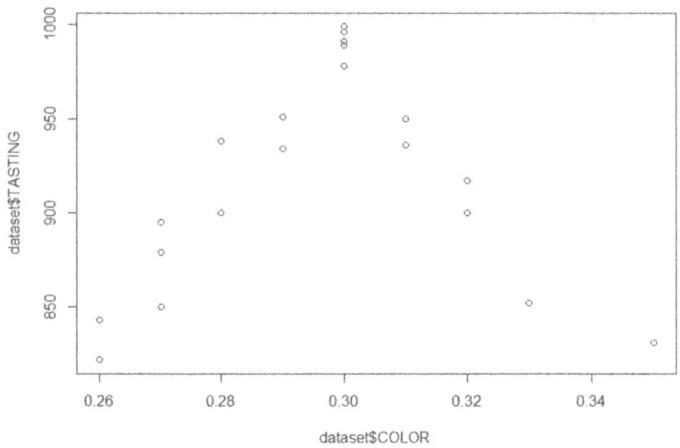

Figure 1-4. Basic plot of your dataset with the variables `TASTING` *and* `COLOR`.

"What an exciting first day, there is really a lot to do here" you think while taking a seat in the pub. You order some random stuff from the card and you doubt whether this was really the right decision to come here. While eating some horrible English food and drinking some glasses of 12-year-old *Junglivet* whisky to suppress the taste of the English food you start recapitulating your first day at the *Junglivet Whisky Company*.

You could help Mr. Gumble to solve his production problems, which will definitely increase future customer satisfaction. And your employees seem to be handsome and hard-working. However, if such simple problems have not been solved no wonder that the company got financial problems. After your fourth round of Junglivet whisky (the food doesn't seem so bad after all) you start to become optimistic. Based on some online reviews there is a huge number of customers loving the charm of the middle-class, family-controlled *Junglivet Whisky Company*. And besides all the problems in operational business, the whisky is delicious. Time to take some rest, there will be a lot of work for you to do!

The last chapter was an illustration of a small business data analytics case. We had a dataset, investigated and preprocessed it to derive further insights from it to support our business processes (whisky distillation). However, this was a quite simple business data analytics scenario. While, we detected the outlier in the analysis step, analytics experts normally check for outliers already in the data cleaning and do more advanced stuff in the analytics to detect patterns. In the following chapters of this book we will learn these sophisticated and automated techniques to detect relationships in your data instead of "fitting them by eye". Let's go!

1.3 The Business Data Analytics Mindset

Our first day at the *Junglivet Whisky Company* has illustrated that just running some R code will not make you a successful business data analyst. You must understand and clean your data. And if you present your first results to the management, there will always follow subsequent tasks and analyses. From my experience, I can tell you that a business data analyst needs a very specific mindset to be successful in practice.

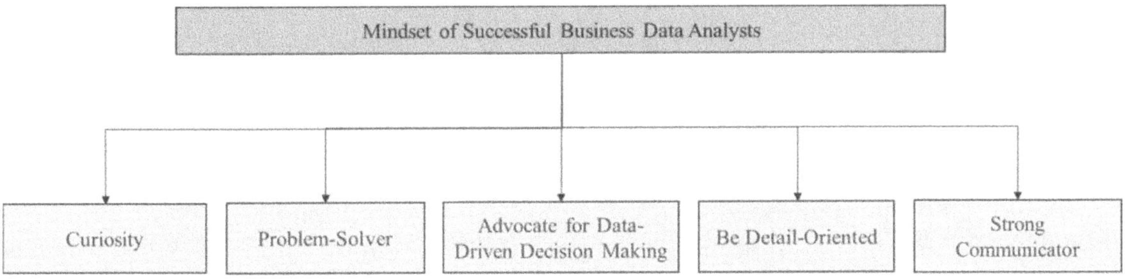

Figure 1-5. The key purpose of business data analytics is transforming data into actionable insights that guide business decisions.

First of all, you must see you as a kind of detective or investigator. You need an insatiable curiosity and inquisitiveness to explore business data in-depth. Ask yourself: How to handle and get your data for your project? What type of data is suited for your business case? Is the quality of the data sufficient to answer your questions? Is the dataset systematically biased or corrupted? How to handle missing values or outliers? How to use the data to support a business process? Rather than merely accepting numbers and figures at face value, you have to question the "why" and "how" behind the data patterns. This will lead you to meaningful connections and patterns that might go unnoticed by others.

Business data analysts also need a natural problem-solving mindset. They approach each data analysis challenge as an opportunity to provide solutions and make informed decisions. Armed with the methods and tools you will learn in this book, you have to break down complex problems into manageable parts, design statistical experiments, and conduct complex analyses to identify the drivers of a problem. There will be many challenges, especially if you work for no data-driven company but see them as chance to show the management potential for improvement and how business data analytics can contribute to the organization's success.

Third, business data analysts must see themselves as advocates of data-driven decision-making. They have to show that data is the most potent tool for informed decision-making. Rather than relying on gut feelings or intuition, you have to use and show how data can guide decisions. Build dashboards, setup reports and automate them, help others to use your findings. This data-first mindset is necessary to empower your colleagues and to influence key stakeholders in your organization with evidence-based insights.

In my humble opinion, in the world of data analysis, the devil is in the details. As a consequence, business data analysts have to understand the importance of accuracy in their work. They have to be meticulous and detail-oriented. Always pay close attention to business data quality, ensuring your business data is clean, consistent, and reliable. Have a keen eye for details allowing you to identify errors, outliers, or anomalies in the data early in your project, leading to more accurate and trustworthy results. And sometimes to better processes in your organization!

Fourth, business data analytics is useless if the insights and findings can and will not be used. If you want to provide the results of your business data analytics project like reports, visualizations, predictions, reports or dashboards to your users, your results should be designed always for non-experienced users. The insight or your analysis have to be communicated and provided in such a manner that your non-experienced users can use them effortless for decision-making. As a business data analyst, you must effectively communicate your findings to this non-technical audiences. Therefore, your results should represent the wording, working and cognitive capabilities of the end-users that are not firm with business data analytics.

In summary, the mindset of a business data analyst is the driving force behind turning data into actionable insights that propel businesses forward. Curiosity, problem-solving abilities, data-driven decision-making, attention to detail, and strong communication skills are key pillars of this transformative mindset of a data-driven organization. By embracing these characteristics, business data analysts can unlock the true power of data, revolutionizing how organizations make decisions, optimize processes, and gain a competitive advantage in the digital age.

1.4 How Business Data Analytics Experts Work

Since the beginning of business data analytics as a discipline, different process models have been proposed again and again as to what a successful analytics project should look like. There has been propositions from statistics-related organizations like the SEMMA framework from the SAS institute (SAS Institute, 2017). It organizes business data analytics as the process of sampling, exploring, modifying, modeling, and assessing. Other models come from data warehousing literature like the knowledge discovery process (Fayyad et al., 1996) that describe how unknown patterns can be identified in databases. Recent frameworks tackling new trends in software-development, like agile, are covered by Microsoft's TDSP - team data science process lifecycle (Microsoft Corporation, 2020), or focusing on specific subdomains like the DASC-PM – data science process model explicitly for data science in information systems (Schulz et al., 2020).

While analytics research is still proposing new guidelines and frameworks every year, there is one framework that has been established in business data analytics and data sciences since the beginning and most other frameworks build on it: The **cross-industry standard process for data-mining** (CRISP-DM). CRISP-DM has been developed by a consortium of over 200 analytics organizations in the nineties and published in 1999 to gather best practices across industries (Chapman et al., 1999). Since then it has become the most widely-used process standard for analytics and data science in general (Brown, 2015; Piatetsky, 2014).

In this book we will rely on the CRISP-DM methodology. If you are planning to work in an analytics team in industry your processes will be probably organized based on CRISP-DM. And if you understand the different phases and ideas behind them, you can easily translate them to the other frameworks. It's the most general and most widely used and will help you to organize your projects even if you work just for yourself.

The CRISP-DM framework divides analytic projects into six phases that are related to each other. The six phases are **business understanding**, **business data understanding**, **business data preparation**, **modeling**, **evaluation** and **deployment** and are illustrated in Figure 1-6.

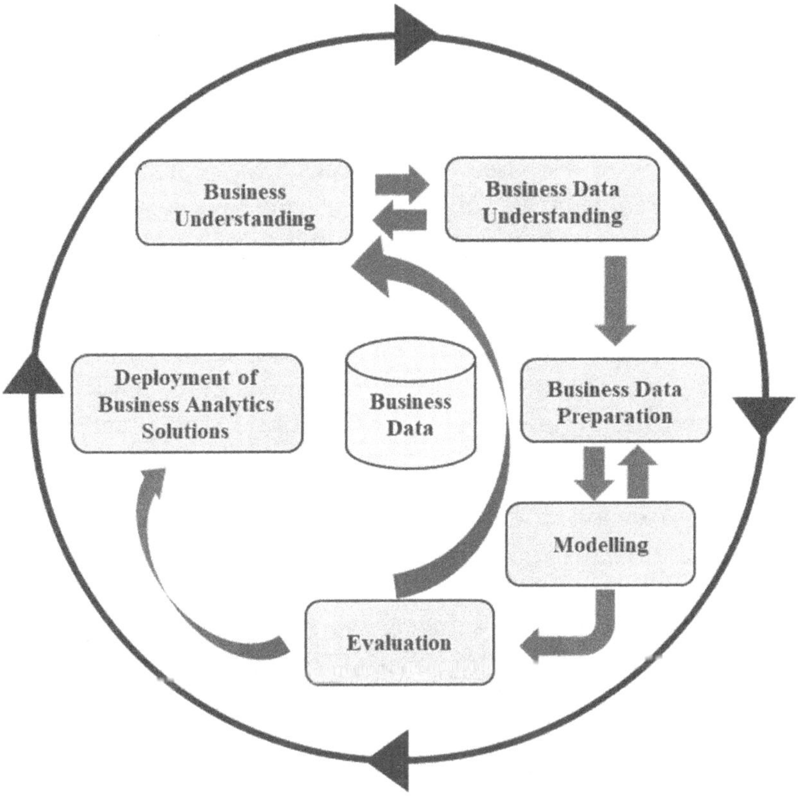

Figure 1-6. Business data analytics framework based on the cross-industry standard process for data-mining (CRISP-DM), adapted from the CRISP-DM 1.0 Guideline (Chapman et al., 1999).

The six phases build on each other but it is not required that you go through them step by step. In most real-live scenarios you will move back- and forth between them. In your analytics projects you will probably start with a business understanding workshop and setup the project management, then you will take a deeper look at the data basis. If you have a rough understanding of the data quality, you will go back to business understanding and update the project management timeline. Afterwards you can start with business data preparation. This illustrates that the phases depend on each other which is symbolized by the arrows in Figure 1-6 and explained more detail in Figure 1-8.

After you produced your first analytics results you can provide the results e.g., as a dashboard for your users. The new dashboard system can generate new insights and questions about the business problem, or your findings might influence your understanding of the initial problem, and hence the process starts again. This is symbolized by the outer circle.

The different steps or phases of a CRISP-DM based business data analytics process have different activities linked to them. The activities are common for every type of analytics projects and are necessary to avoid pitfalls and problems. It is important that you know them, and we will learn different methods how to actually do these activities in R (see Chapter 2). You will also find a checklist of the related activities for each phase at the end of each chapter. This way you can make sure you haven't forgotten anything!

Let us now look at the first phase. The main objective of the first phase **business understanding** or sometimes also called project understanding or domain understanding is to generate a general understanding of the objective of your business data analytics project in the team. Most clients will have a very abstract understanding of the potentials and limits of analytics. Or the risks or challenges of the projects have never been discussed. Some clients will probably have no technical understanding or unrealistic requirements. In this phase you have to taggle these pitfalls and identify the business problem you want to solve. Best-practice is to focus on the business problem with the highest impact (costs, revenue, etc.). Based on these specifications you organize your analytics project and try organizing and allocate manpower, time, hardware and budget over the whole project timeline. Your final project plan should be communicated as early as possible for every stakeholder. Find a shared consensus how to measure the success of the analytics project. This step is crucial for your project success and I will show you helpful methods in the next section.

The objective of the next phase is to overview the data sources related to the project. In particular, the data quality and availability are key success criteria. Do not forget that the best analytics system is useless if you cannot provide sufficient data for it. All activities of the **business data understanding** phase have the purpose to make sure that there is enough data and information about the data so that it can be used for analytics. Try to start your exploratory data analysis (EDA) with a small initial dataset and then go forward from there and investigate the other data sources. Check if the data quality is high enough to fulfill your requirements for the objective defined in the business understanding phase. If not, it is important to adjust the objective together with the stakeholder and discuss possibilities how to gather or generate the data that is needed. If that's not possible you have to cancel the project because there will be no usable solutions at the end.

If you got a rough understanding of the data, the data fields and data sources you have start the **business data preparation**. The data preparation step is the most time consuming one in every analytics project. Decide on which data you want to concentrate and how to transform the data to more general formats that can be used and shared easily in your team and between tools and systems. If you use raw business data you have to increase the data quality by removing corrupted data or replacing missing values. Do not forget that some specific data might require specific processing and anonymization steps due to national or international laws like the European general data protection regulation (Voigt & Von dem Bussche, 2017). Furthermore, redundant and unimportant data and features have to be removed or at least reduced as much as possible. Sometimes you have to generate new features or bootstrap your data because you have not enough of it. In addition, it is now also possible to carry out projects entirely on synthetic data.

Using your preprocessed data, you can finally start with **modeling**. In this phase, you apply business data analytics models and methods on your business data to generate insights to solve a specific analytics problem. Alternatively, you can use statistical methods to generate insights or simple models or decision rules that can also be used for decision support systems. Today exists a plethora of different methods and algorithms, the challenge is to find the right one for the right kind of problem.

In the next phase, the **evaluation** phase you compare the solution and performance and discuss the findings. In most cases you have to make many trade-offs like between understandability of your models and model complexity or performance (Allahyari & Lavesson, 2011), data or algorithmic modeling (Breiman, 2001), or generalization of your model and optimization (Geman et al., 1992). Furthermore, in most real-life projects you will face trade-off discussions about platforms, technical requirements and processes as a further complexity layer. This problem of finding the right trade-off between understandability and performance is also known as the modelers dilemma and illustrated in Figure 1-7.

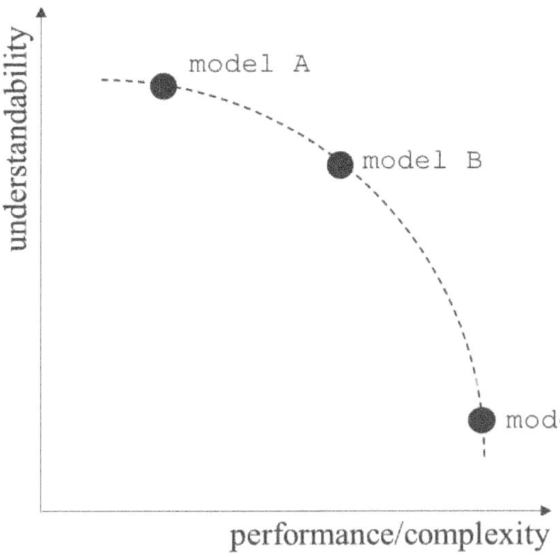

Figure 1-7 The trade-off between understandability and performance of your analytics solution.

In Figure 1-7 you see three models A, B and C. Model A has a very high understandability due to its low complexity. However, the performance is low compared to the other models. You can perhaps imagine it as a multiple linear regression. The other model B has a better performance but lower understandability, this might be a decision tree. The last model, model C has clearly the best performance but a very low understandability. This might be a random forest or other ensemble algorithm. In practice you must now decide if you prefer simple models that you can explain easily or models with better performance but are not easily explained. Sometimes there are laws or internal rules you must consider, and it depends on the type of prediction problem and how critical errors are. Consequently, you might decide for the linear regression model (model A) over the other models because it has a very simple structure and you can explain the influence of the variables easily.

After deciding for a model, it's time to consolidate your knowledge you have generated in your project so far. Knowledge is not just the model(s) from the modeling step, it's also the insights, visualizations, documentations, code, discussions, etc. And you have to evaluate how the knowledge and insights suit the objectives you defined at the beginning of your project in the business understanding. As a result of this discussion, you have to decide how you can actually give decision support for your clients and how your solution helps your clients with their business problems. Most projects result in the deployment of an app like a dashboard or a notebook. But this is not always necessary. In particular if you used simple analytics models it is often enough to communicate your results in a simple manner. For instance, you could write a report, generate a specific info plot or visualization with key findings, suggest a new KPI for the management, setup a script to automate your model, or just give a short review presentation with action points for the stakeholders. It all depends on the knowledge and the objective defined at the beginning.

If you decided to provide a system solution, which ranges from automated scripts or simple notebook reports to full information systems, you have to consider the activities of the **deployment** phase. This phase contains all activities to develop, deploy and maintain a decision support system.

1.5 Business Understanding and Project Setup

In order to add value as a business data analyst you need first to understand what your customer find valuable. Sometimes they will come to you and ask for specific statistics, plots or models. Don't be as foolish and just bring them what they asked for. Act as a consultant and try to understand why they want it. If you know the way, you can use your skills as business data analyst to show an easier or better way to get what they want.

Consequently, your first step in real-life analytics projects should be to understand the problem properly and to explore your data and possibilities carefully. You have to identify the core business problem and the general relationships in the field. In the next step, the project planning, we will use understanding to identify which analytics tasks are necessary to solve this problem. And finally, how the findings of your models can be transformed in a usable solution for decision support.

It is crucial to take enough time to get a profound business and problem understanding for yourself and the clients. Many analytic projects fail because of the inability of the stakeholder to achieve a shared business and problem understanding. For instance, the analytics team has no understanding of the technical terms of whisky production, while the product owner does not understand where the limitation of the model is, and how the performance is measured. Both teams cannot communicate about the technical problems and terms of the domain. They do not know how to help each other and will waste a lot of time with none-useful activities. Just because they never found a shared business problem understanding.

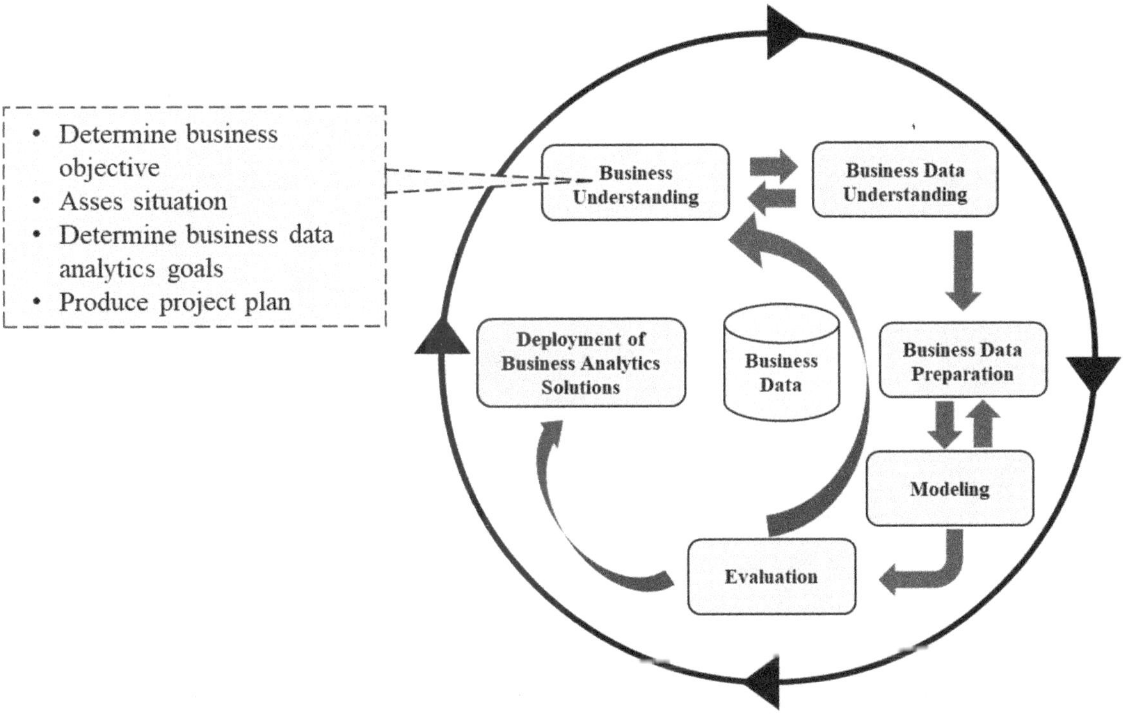

Figure 1-8 The generic tasks of the business understanding phase.

As illustrated in Figure 1-8, building on the CRISP-DM framework you should clarify the following aspects in the business understanding phase before you start with your project:

- **Determine business objective**: Gather background information about the business and business problem, define primary or business objectives and business success criteria
- **Assess the situation**: Make an inventory of resources, gather requirements, assumptions and constraints, evaluate risk and contingencies, and conduct a cost and benefit analysis
- **Determine project and analytics goals**: Define business data analytic goals, define business data analytics success criteria
- **Project planning**: Make a project plan and an initial assessment of tools and techniques

I have prepared a PowerPoint deck with many ready-to-use slides for you to use when you start your first real business data analysis project. You can download the file `CRISP-DM Ready-to-use Projectslides.pptx` from the course website. Feel free to use it!

1.5.1 Business Understanding Methods

Most of these points will require that you communicate a lot with your customers. You will probably prepare some PowerPoint slides or materials for that and will need some templates to start. But hey good news, to save you time, I prepared a PowerPoint slide deck with ready-to-go templates you can use. You will find it on the website of this book.

Besides fancy PowerPoint slides, you have to rely on established methods from decision theory to generate a better understanding of the problem space and gather the information you need. For that purpose, I will now give you a short overview of the most popular methods you can use in the business understanding phase illustrated in Figure 1-9:

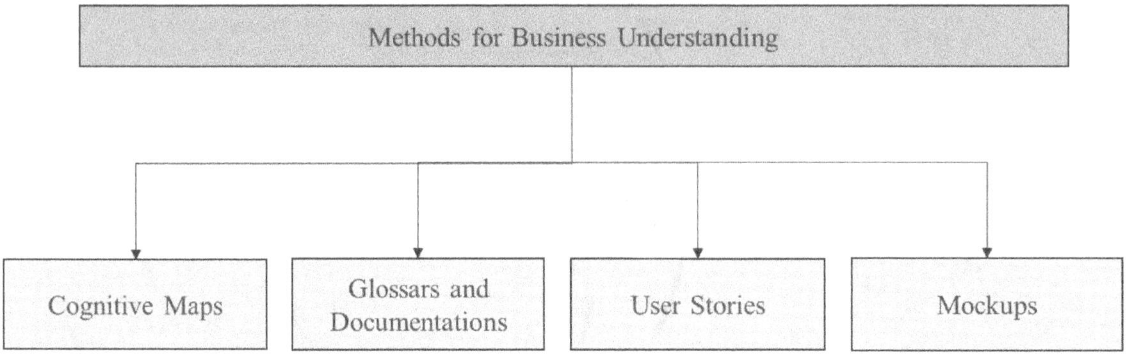

Figure 1-9 Most popular methods for business understanding.

You might be tempted to skip the list of to-dos and avoid discussing the full list of points with your customers, but the time in this phase is short compared to the time you will win during the whole project. I will promise you that time invested in business and problem understanding is a good investment and will save you huge amount of time in the subsequent steps.

1.5.2 Cognitive Maps

Cognitive maps are a powerful tool for business data analysts to better understand a particular subject, concept, or system in a business data analytics project. The cognitive maps reflect how people perceive and organize information in their minds, creating a mental framework that guides their thoughts, decisions, and actions related to that subject. This can help you to better understand their problem and how to help them. An example of a very simple cognitive map is illustrated in the following Figure 1-10.

I recommend that you build a cognitive map in a workshop that is part of the initial business understanding phase. Best it should be directly after the first kick-off event. Building cognitive maps in a workshop involves a structured and collaborative process to capture participants' collective knowledge and insights about a specific topic. Start by clearly defining the subject or topic for which you want to build a cognitive map. It could be a complex system, a problem, a customer journey, or any other concept that requires a deeper understanding.

Next, gather a diverse group of participants of the business domain where you plan to do your project. It is important that they have relevant knowledge and expertise related to the chosen topic. Do not invite people just for fun or because they like the cookies you offer in the workshop and ask you if they can join. Including the wrong individuals is not helpful, but including the right individuals from different backgrounds can offer unique perspectives and enrich the cognitive map. During the workshop, start with a brainstorming session, where participants share their thoughts, ideas, and understanding of the topic. Record all the input on a visual display, such as a whiteboard or a large sheet of paper.

After brainstorming, group similar ideas and concepts together. Identify key themes and categories that emerge from the participants' input. This helps organize the information and identify relationships between different elements. Then, transform the categorized information into a visual representation, such as a diagram or a mind map. The visual representation should illustrate the relationships between different concepts and how they fit into the larger context.

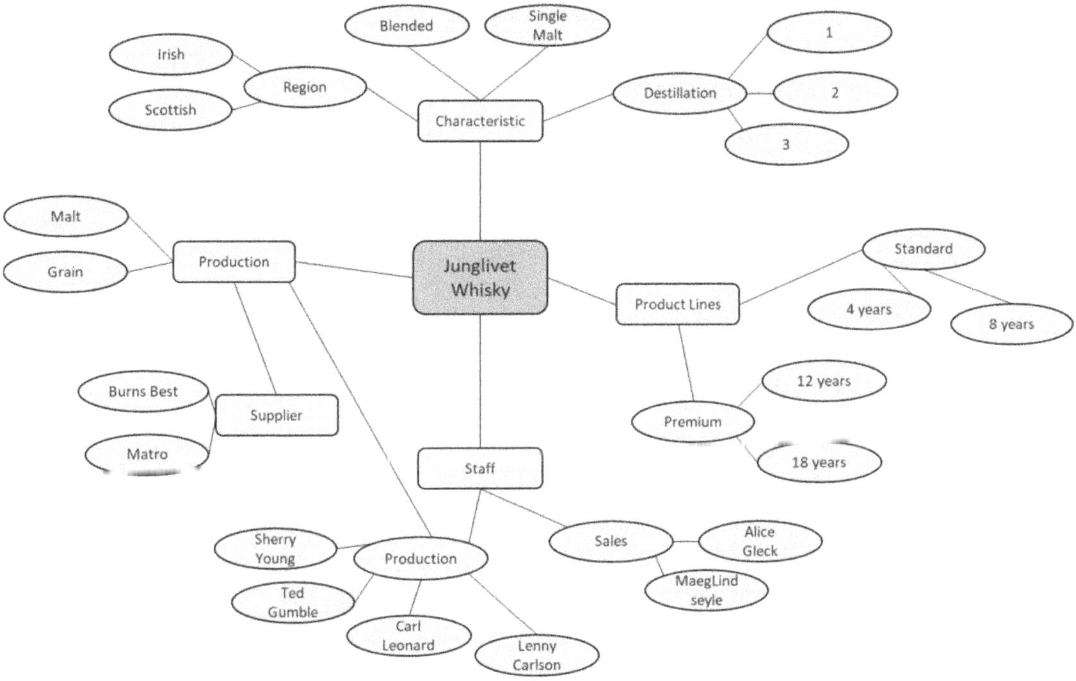

Figure 1-10 Example of a cognitive map after an initial workshop with the Junglivet team.

Review the cognitive map with the participants, seeking feedback and validation to ensure accuracy and completeness. Make necessary refinements based on the participants' input. Building cognitive maps may require multiple iterations, so consider holding follow-up workshops or sessions to further develop and refine the map based on additional insights and information.

Building cognitive maps in a workshop fosters a shared understanding among participants and promotes collective problem-solving. It helps uncover hidden insights, clarify complex concepts, and identify potential gaps or misconceptions. The collaborative nature of the process also encourages active engagement and participation, leading to a more comprehensive and accurate representation of the chosen topic. Utilizing collaborative online tools and platforms can facilitate remote workshops or capture inputs from a distributed team, enabling real-time collaboration and brainstorming.

1.5.3 Glossaries and Documentations

Another very helpful tool are glossaries. Glossaries used in various fields, serving as a compilation of specialized terms and their definitions. They play a crucial role in enhancing communication and understanding within specific domains, ensuring that everyone involved shares a common language. You can think of them as cheat sheets with the most relevant terms you should know.

Creating a glossary involves a systematic process of identifying key terms, their meanings, and contextual usage. It is essential to involve subject matter experts and stakeholders to capture a comprehensive list of relevant terms. Start by compiling the terms and their definitions, ensuring clarity and accuracy. Best practice is that your customers and other stakeholder provide a list with short explanations of the used technical terms in their everyday job. Furthermore, it can be helpful to show the analytics team the related business processes in real-life and write down all the questions and terms that are unclear.

To create an effective glossary, consider organizing terms alphabetically or categorically for easy navigation. Including cross-references and examples can further enhance users' understanding of the terms' applications.

Besides documentations are very useful for the business data analytics team to understand the background and concepts behind the problem. It will definitely help them later in the data understanding and modeling phase.

1.5.4 User Stories

The time of writing down long lists of requirements is gone. Today everything is agile and as a consequence, it might make sense to write down some user stories at the beginning of a business data analytics project. User stories are concise customer-centric descriptions of a feature or functionality of the analytics solution from the perspective of your end users. They serve as a fundamental component of agile development, helping development teams understand and prioritize user needs and requirements. In our case they will be helpful to better understand and prioritize the analytics solutions and products.

Creating user stories involves collaboration between product owners, stakeholders, and development teams. The process starts with identifying user roles and personas, representing different user types and their goals. Each user story typically follows a simple template: "As a [user role], I want [goal] so that [benefit]."

User stories focus on the "who", "what", and "why" of a feature, leaving the technical implementation details for later discussions. They provide context and clarity, enabling development teams to better understand the user's intent and deliver value.

In agile development, user stories are written on cards or digital tools, making them easily manageable and flexible for prioritization. They are part of the product backlog, and during sprint planning, development teams select user stories to be worked on during the iteration. You can decide if you want to work agile or more traditional. But I can recommend to write down some user stories to get a better understanding of the expectations of your users. User stories facilitate continuous communication between the stakeholders and you and your team which is in general a good thing.

If you are new to the topic agile development, I can recommend you to have a look at the "Scrum Guide". The guide is an excellent 16-page guideline by Ken Schwaber and Jeff Sutherland, the originators of Scrum. It's freely downloadable at: *https://www.scrum.org/resources/scrum-guide*.

1.5.5 Mockups

I also made very good experiences with mockups in my business analytics projects. In particular, if the requirements were very clear from the beginning and a specific system had to be designed.

You can imagine mockups as visual representations or prototypes of a product, often used in the early stages of design to demonstrate its layout, structure, and user interface (we will draw one in chapter 6, when we build a dashboard). In software engineering, they play a crucial role in the design and development process, providing stakeholders with a tangible preview of the final product.

Creating mockups is not difficult you can start with drawings, or sketches on whiteboards. Mockups are built relatively quickly and are an important part of business understanding. You can let your users work with pen and paper, on digital whiteboards, in PowerPoint or even create very high-quality mockups with special tools, depending on the project's needs.

Mockups help designers, developers, and stakeholders visualize the product's look and feel, ensuring that everyone involved shares a common understanding of the design direction. They act as a communication tool, facilitating discussions and feedback, and allowing for early validation of design decisions. Besides, mockups are also valuable for usability testing and user feedback. They allow designers to simulate user interactions, identify potential usability issues, and iterate on the design before actual development starts. This iterative approach helps save time and resources and leads to better user experiences.

I also recommend that you and your business data analytics team help the users to visualize potential solutions with mockup workshops or show cases of existing solutions from other projects. Most users have difficulties to understand the idea of a mockup and need your help. And last but not least, the discussions and conversations during the creation of a mockup are worth their weight in gold.

1.5.6 Business Objectives and Project Planning

After your first talks with your clients defining the business problem and scope, you might be tempted to start right away analyzing and building models. But with many years of experience, I must say, please resist this urge. I have seen so many business data analytics projects, where the clients and analysts had a project objective but didn't clarify how the quality of the results should be measured. As a consequence, the analysts couldn't hold the project plan because the clients "discovered" more and more requirements that are necessary until the solution was good enough. Please note, that could be also the other way round: the analysts could just provide a minimal solution, take the money and go. As a general rule, its good enough to clarify the evaluation criteria before you start working.

Consequently, it is necessary to use the first findings of the business understanding phase to clarify the background information about the business, the business problem and the business objectives. The main objective of you should be to thoroughly understand, from a business perspective, what your customer really wants to accomplish. As a consequence, the business objectives are clear definition of what has to be done in the business data analytics project. The objectives should provide a clear direction and purpose for the project, guiding decision-making and resource allocation. This might be to increase the revenues with a recommender system or to cluster your customers for a marketing campaign.

Related to the business objectives are the success criteria. They represent factors that determine a successful or valuable result of the project from a business perspective. This can be either specific and quantifiable, such as achieving a certain level of customer churn reduction, or more general and subjective, like providing meaningful insights into relationships. If the outcome is subjective, specify who will be responsible for making the judgment. Best practice is to combine them with key performance indicators.

You also will have to make an inventory of your project resources. Identify the project's available resources, encompassing personnel (business or domain experts, data experts, technical support, business data analysts), data sources (fixed extracts, access to live, warehoused, or operational data), computing resources (hardware platforms), and software tools (analytics tools, other relevant software).

Furthermore, you have to gather all project requirements, encompassing completion schedule, result comprehensibility, quality, and security, as well as legal considerations. Ensure that data usage is permitted. Identify the project's assumptions, which may pertain to both data and business aspects. Clearly list non-verifiable assumptions, especially if they impact result validity. Outline the project's constraints, which may involve resource availability and technological limitations, like dataset size for modeling.

In the next step, you have to identify potential risks or occurrences that could lead to project delays or failure. Provide corresponding contingency plans, outlining the actions to be taken in response to these risks or events.

In this context a cost and benefit analysis is important. Create a cost-benefit analysis for the project, which involves a detailed comparison of project costs with the potential business benefits in the event of successful completion. Whenever possible, use monetary measures for a more specific assessment, especially in commercial scenarios.

Building on the business objectives and success criteria, you will now have to define business data analytic goals and success criteria. For instance, the business objective could be "Enhance online sales to current customers", while the business data analytics objective could be "Forecast the number of whiskies a customer will purchase, considering their past three years of buying history, demographic details (age, salary, city, etc.), and item price". Outline the desired project outputs that align with accomplishing the business objectives. Define analytics success criteria in technical terms, such as achieving a specific level of predictive accuracy or generating a propensity-to-purchase profile with a designated degree of "lift". In cases where subjective terms are used, specify the individuals responsible for making the subjective judgments. In most cases there will be clear deliverables. Deliverables are tangible or intangible outputs produced in your business data analytics project, representing the generated knowledge and value provided to your users. In most cases they will be physical products, like analytics reports, visualizations or dashboards, or sometimes intangible outcomes, such as recommendations or general insights.

The deliverables serve as milestones to track project progress and ensure objectives are met. Each delivery plays a vital role in project management, providing clarity and accountability for team members and stakeholders.

Finally, it's time to finish your project plan. A typical project plan in business analytics consists at least of a written definition of the business objectives (you hopefully have now defined), a short project roadmap and a list with the project team or stakeholders. This is necessary to estimate whether the client's plans can be achieved at all with the available resources. And from my many years of experience I can assure you it gets even more important, because clients in general have often unrealistic expectations.

Following the CRISP-DM framework we know that there are different steps that build on each other and we have to estimate them in our project plan. However, this does not mean that we can just follow these steps without thinking. There are many decisions and consequences we have to consider. For instance, some analytic methods require a specific data format, platform or have expectations concerning the data. If you make a quick model evaluation on sample data and decide to use a regression model you have to go back, form the analytics step to the data transformation step and make sure that all your data is numeric. If your customers decide later, they don't want a regression model because accuracy is more important than understandability, you have to switch the model and hence the data preprocessing pipeline too. Therefore, make an initial tool evaluation early in the process since the selection of tools and techniques may influence the entire project.

1.6 Checklist

At the end of each chapter about CRISP-DM phases you will find a checklist. The checklist aims to serve as a tool for verifying the completion of tasks and outputs within each phase of the CRISP-DM methodology. It is adapted from the official CRISP-DM guideline (Chapman et al., 1999), and streamlined for this book. This checklist will ensure that you have nothing overlooked during the business analytics project. By detailing generic tasks and related outputs or deliverables for each phase, the checklist will help you maintain consistency to the CRISP-DM framework.

1. Phase: Business Understanding			
1.1	Determine business objectives	❯	• Background • Business objectives • Business success criteria
1.2	Assess situation	❯	• Inventory of resources and capabilities • Requirements, assumptions and constraints • Risks and contingencies • Terminology • Costs and benefits
1.3	Determine analytics goals	❯	• Business data analytics goals • Business data analytics success criteria
1.4	Produce project plan	❯	• Project plan: objectives, roadmap and team • Initial assessment of tools and techniques
2. Phase: Business Data Understanding			
3. Phase: Business Data Preparation			
4. Phase: Modeling			
5. Phase: Evaluation			
6. Phase: Deployment			

2 Business Data Analytics Toolbox: R and RStudio

For this chapter, please download and install R and RStudio on your local machine. We will use R and RStudio as a development environment and to get familiar with R. We start with some basic information about R programming, the RStudio interface, and move through how to write R code, import data like CSV files and the different structures of data in R. We learn data manipulation: how to deal with data frames and factors, how to add/remove rows and columns, how to calculate summary statistics from a data frame, and finally we go through a brief introduction of plotting.

> If you are not familiar with installing and setting up software like R and RStudio you should have a look at my bonus chapter "*Setup R and RStudio*" which you can download on the course website. In the tutorial I describe how to install R and RStudio software on your local machine and highlight common pitfalls during the installation process.

2.1 First Steps in RStudio to Run R Code

After you have installed your tools (R and RStudio) for your future as a successful business data analyst of the *Junglivet Whisky Company*, let's have a look at the big picture. For that purpose, please consider Figure 2-1.

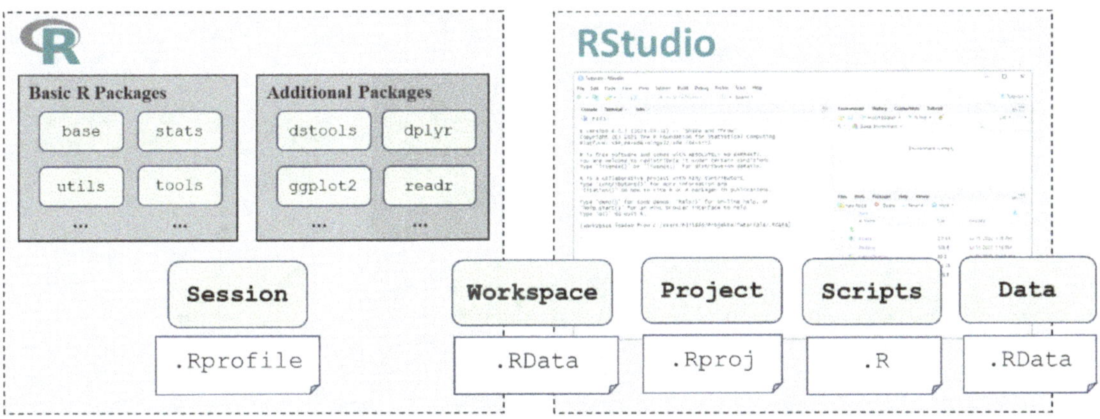

Figure 2-1 The R and RStudio universe of business data analytics.

We already know that R itself is a programming language for analytics and statistics. And RStudio is an IDE that allows us to use this programming language comfortably. In addition, R has a large number of its own packages that extend the functionalities of the language. However, it is also possible to install and load third-party packages.

A R package is a collection of functions, compiled code and sometimes even data or media files. The location where your packages are stored is called the R library. If you want to use specific functions in your R code, you can download suitable packages and install them. They will then be stored in your local R library. To actually use the package, you have to use the command `library(<package name>)` which makes the specific package available to you. This means you can then just call the package functions or use the data stored in the package.

D. Jung, *The Modern Business Data Analyst*, https://doi.org/10.1007/978-3-031-59907-1_2

If you launch R or RStudio, a R-session will be started automatically for you in the background. It starts when you start R or RStudio and ends when you close the program. During an R session, you can execute commands, load data, perform analyses, create visualizations, and interact with the R environment until you decide to terminate the session.

Furthermore, you can write scripts and generate or load data to solve your business data analytics problem. In real business analytics problems, there will many files related to your project. Best practice is to organize these files in R-projects. For instance, as business data analyst you want to organize the code for the *Junglivet Whisky Company* in projects in RStudio. This allows us to save and manage our scripts during the case studies of this book. But also helps to track and share the status of our work in the different workspaces. With that in mind, we have everything we need to start our work for the *Junglivet Whisky Company*. Let's go!

First of all, I want to show you the RStudio IDE and how to run R code in it. If you run RStudio for the first time you will see three panes. A big pane, the R console on the left. It contains a large block of text about your R version and a cursor at the end >. This is your R console where your future R code will run. On the right is a pane with a collection of tabs that focus on your current workspace and R environment, and a third pane on the bottom in the right that has tabs to browse and manage your project files.

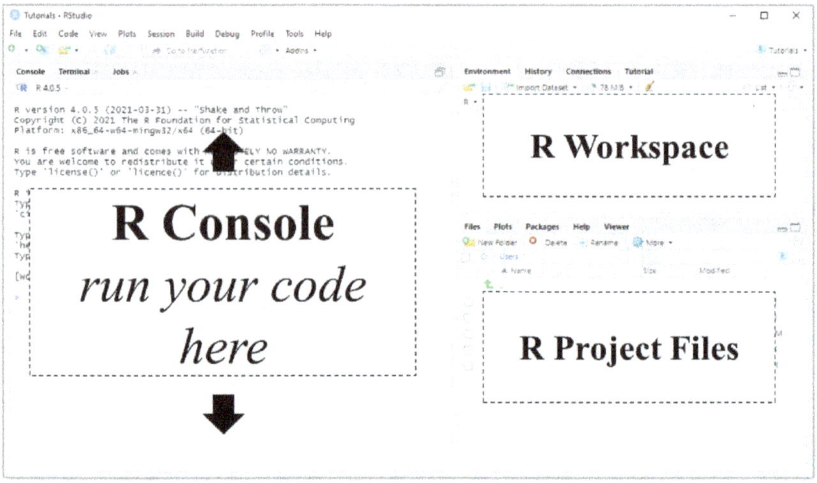

Figure 2-2 The different RStudio panes (console, workspace and project files).

To run your first R code, just click into the R console pane and type `1 + 1` and press enter.

The result should look like this:

```
> 1 + 1
[1] 2
```

Congratulation! You have run your first R code. R supports most mathematical operations and has a lot of build-in statical methods.

For instance, if you want to count the number of characters in the string `"Hello World"` just run:

```
> string = "Hello World"
> nchar(string)
[1] 11
```

R is based on the statistical language S, and hence also supports the variable assignment with an arrow "<- ". Consequently, the following code will work also:

```
> string_2 <- "Goodbye"
> nchar(string_2)
[1] 7
```

But before we continue to investigate the different functionalities of base R let us look at the other parts of your IDE. One very important pane is the workspace pane on the top right corner. If you did run the previous code example the character variables `"string"` and `"string_2"` have there appeared:

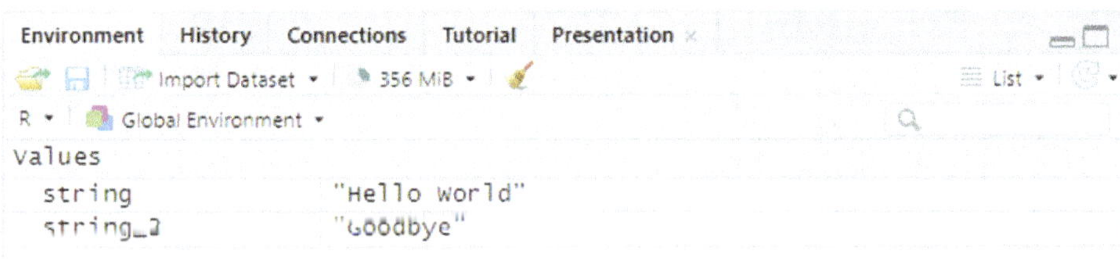

Figure 2-3 Your variables will appear in the environment tab of the workspace pane

The pane contains all your variables and functions of your current R workspace and possibilities to adjust and manage your R environment. If you want to get an overview of your current workspace variables you can look them up there (or just run the command `ls()` in your R console). You also can investigate, delete, or even import new variables in your workspace in this pane.

One advantages of an IDE like RStudio is that you can save your code in R scripts. The benefit is that you don't have to write the same code again and again. And that you can reuse and share your old code. You can edit these scripts with any text editor. RStudio has such an editor included that is specifically designed for programming with R. Specifically designed means it has many useful features like syntax highlighting and autocompletion to support you writing R code.

If you want to create a new R script just press on the small  icon in the navigation bar. Or click on the "File" menu, and then on "New File" and "R Script". A new pane with an empty file will open.

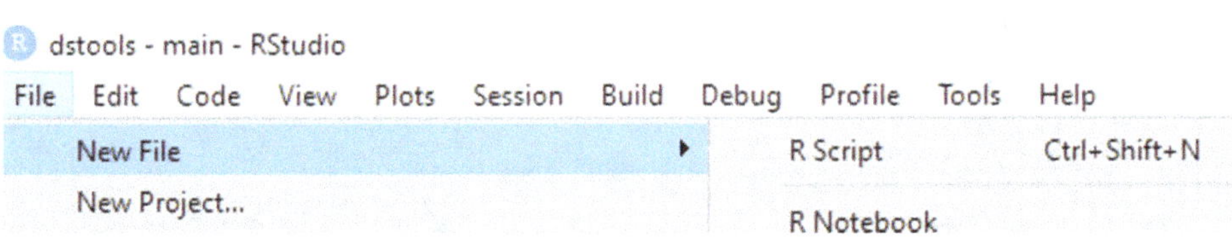

Figure 2-4 You can make a new file in the menu or with the shortcut `Ctrl+Shift+N`.

You can type the same command from above in this document. Then press on the Run button in the header or just press `Ctrl + A`, `Ctrl + enter` and your code will be transferred to the R console and executed automatically.

In the next step we save this file by pressing on the save icon . Or just by pressing `command + s`. This allows you to open this file later again and rerun your code. If you store the file in your current working directory it will appear in the last pane, the "R Project file" pane on the right bottom. You can think as this pane as kind of a file explorer.

2.2 Data Structures and Variables in R

Besides simple addition, R has many build-in functionalities you can use to make mathematical or statistical computations. But to actually use them you'll need a strong understanding of the basic data types and data structures.

R has five basic atomic classes:

- `Integer`: Whole numbers (e.g. `1,2,3`)
- `Numeric`: Real or double numbers (e.g. `1.0` or `1.2` or `10.0`)
- `Logical`: Boolean values (`TRUE`, `FALSE`)
- `Character`: Known as "strings" in other programming languages (e.g. `hello`)
- `Complex`: Complex numbers (e.g. `1+2i`)

To use that data types, R provides data structures like vectors, lists, data frames, tables and so on. To start, let us take a look at the most basic data structure - the vector.

A vector in R represents a sequence of elements of the same data type. Vectors can contain all available data types in R. One-element vectors can be created by assigning a single value to a variable, while multi-element vectors can be created using functions like `c()` to combine value sequences.

Vectors also play a fundamental role in many operations and computations within R. They can be of fixed or variable length and serve as building blocks for more complex data structures like matrices, arrays, lists, and data frames.

Based on the data types, R has the following modes of a vector:

- `Integer`: Whole numbers
- `Numeric`: Real or double numbers
- `Logical`: Boolean values
- `Character`: String values
- `Complex`: Complex values

Which allows you to declare one-element vectors vise-versa:

- `Integer: X = 1`
- `Numeric: X = 1.2345`
- `Complex: X = 1 + 2i`
- `Logical: X = TRUE`
- `Character: X = "One"`

In R exist many base functions to check the class (`class()`), mode (`mode()`) and type (`typeof()`) of your variable. But you can also check the different types explicitly with `is.numeric(), is.integer()`.

If your variable has the wrong type, you can also cast your variable vice-versa with `as.numeric(), as.integer()`:

```
> X = 101
> X = as.character(X)
> is.numeric(X)
[1] false

> is.charcter(X)
[1] true
```

We have the following R classes, modes and types we will use in the subsequent sections of this book:

Table 2-1 Overview of the different classes, modes and types in R

Example	**Class**	**Mode**	**Type of**
`1L`	`integer`	`numeric`	`integer`
`1`	`numeric`	`numeric`	`double`
`1 +2i`	`complex`	`complex`	`complex`
`TRUE, FALSE`	`logical`	`logical`	`logical`
`"good bye"`	`character`	`character`	`character`

As we discussed before, you can also create a vector of multiple values (multi-element vector). For that purpose we have to concatenate distinct values with the function `c()`:

```
> my_vector = c(2,1,3,4)
> funky_vector = c(TRUE, TRUE, FALSE)
```

Besides, we can also combine multiple vectors to a new vector containing the values of the subvectors:

```
> really_funky_vector = c(my_vector, funky_vector)
> really_funky_vector
[1] 2 1 3 4 1 1 0
```

However, you always must separate each element by a comma. Furthermore, you can also comment your code with the # command. Everything behind the # is ignored:

```
> my_vector = c(2,1,3,4) # write your comment here
> my_vector
[1] 2 1 3 4
```

To check if your vector has the right class you can check that with the command `class()` of the vector. Because, if you start to mix them, R will implicitly coerce them:

```
> X = c(TRUE, 2, 3, 4)
> X
[1] 1 2 3 4
```

Besides your normal variables, there are some special values in R. `NULL` is the null object, which has a length of zero and allows you to delete variables by assigning them to `NULL`.

```
> my_vector = NULL #deletes the poor guy
> my_vector
NULL
```

Furthermore, there are three values that indicate missing or corrupted values: `NA`, `Inf` and `NaN`. `NA` is the value used to represent missing values and stands for "not available". While `Inf/-Inf` indicates positive/negative infinite. `NaN` indicates that the value is "Not a Number". Please note that this is different from `NA`.

I will illustrate that with the following examples:

```
> 1/0
[1] Inf

> 0/0
[1] NaN
```

In a nutshell, data structures like vectors can be used to store data (storage mode). The values in a vector must all be of the same data type, either all integer, real, complex, characters, or logical.

In a next step, we can now use our vector to make some R magic with it. For instance, you can also name the different components of a vectors in R:

```
> my_funky_vector2 = c(A=1, B=2.3, C=100)
> my_funky_vector2
    A     B     C
  1.0   2.3 100.0
```

You can check the length of a vector with the `length()` function:

```
> length(my_funky_vector2)
[1] 3
```

If vectors of different length are combined, R reuses the previous values:

```
> c(1,2,3,4) + c(1,2)
[1] 2 4 4 6
```

Or you can make some basic vector arithmetic. The related functions are illustrated in Table 2-2.

Table 2-2 Basic arithmetic functions in R

`+`	Addition		`^`	Raising to power
`-`	Subtraction		`sqrt()`	Square root
`*`	Multiplication		`%%`	Modulo
`/`	Division		`%/%`	Integer division

Besides vectors there are other data structures in R, which can be differentiated by their dimensionality and the number of data types they can store. The whole overview of all the available data structures is illustrated in the following Figure 2-5.

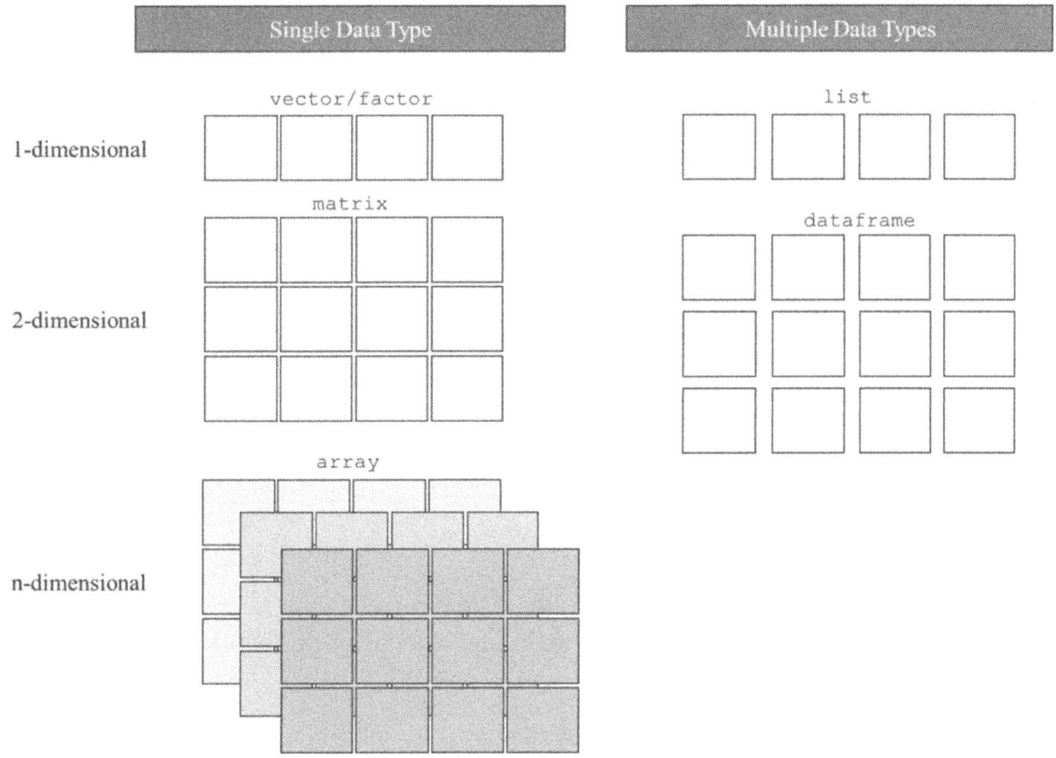

Figure 2-5 Basic data structures in R to store the different data types.

I will illustrate these different data structures in the following part of this chapter. First, we will take a look at vectors. You can find them in Figure 2-5 in the first row on the left.

You already know vectors, which are one dimensional data structures that store values in a list-like fashion. An alternative to vectors are factors. You can think of factors as kind of special vectors that represent categorical data ("integers with labels"). If you want to generate a variable that contains categorial information like the doneness of your steak, you can do it like that:

```
my_factor = factor(c("medium", "rare", "rare"), levels = c("blue",
"rare", "medium", "medium rare", "well-done"))
```

As you can see, compared to vectors, they can only contain pre-defined values. Thus, factors include the value and the level of the categorical data:

```
[1] medium rare rare
Levels: blue rare medium medium rare well-done
```

The different levels can be ordered (like the different steak temperature levels), or unordered (like gender: male, female, else). Sometimes they are better than integers, because the levels of the factors are self-describing.

Another helpful feature is that you can compute a frequency table with `table()` to get an overview of your variable:

```
> table(my_factor)
my_factor
       blue         rare        medium medium rare    well-done
          0            2             1           0            0
```

The next data structure on our list (see on our overview on Figure 2-5) is a matrix. A matrix is a 2-dimensional combination of vectors, or a rectangular array of data elements arranged in rows and columns. You know matrices probably from the math lectures in school. In R you can define and use them in the following manner:

```
> my_matrix = matrix(
      + c(1, 2, 1, 1, 2, 2), # the elements
      + nrow=3, # number of rows
      + ncol=2, # number of columns
      + byrow= TRUE) # fill matrix by rows
> my_matrix
[,1] [,2]
[1,] 1 2
[2,] 1 1
[3,] 2 2
```

If you want to compute multidimensional data like a time series of matrices, the data structure of your choice is an array. An array is a kind multidimensional vector comparable to multiple matrices that are linked together.

You can setup an array in R in the following manner:

```
> my_array = array(1:24, dim=c(3,4,2))
```

```
> my_array
, , 1
[,1] [,2] [,3] [,4]
[1,] 1 4 7 10
[2,] 2 5 8 11
[3,] 3 6 9 12
, , 2
[,1] [,2] [,3] [,4]
[1,] 13 16 19 22
[2,] 14 17 20 23
[3,] 15 18 21 24
3 rows
4 columns
2 dimensions
```

These data structures are quite useful, but what do you do if you have different data types in a dataset and want to handle them in one single data structure. For instance, if you made a survey and have different variables of different data types in your columns like age (`numeric`), name (`character`) etc., don't worry. The good thing in R is, if you want you can also mix different data types in a 1-dimensional data structure with a list. A list is a generic object that can contain different objects (like a "more flexible" vector). Let us take a look at the following example to see how a list works:

```
>    funny_dog_names=    c("Winnie,    the    Poodle",    "Bark    Twain",
"OzzyPawsborne")
> random_numbers= c(4,8,15,16,23,42)
> coin_flips= c(TRUE, FALSE, FALSE, TRUE, TRUE)
> stuff = list(funny_dog_names, random_numbers, coin_flips, 7)
> stuff
[[1]]
      [1] "Winnie, the Poodle" "Bark Twain" "OzzyPawsborne"
[[2]]
      [1] 4 8 15 16 23 42
[[3]]
      [1] TRUE FALSE FALSE TRUE TRUE
[[4]]
      [1] 7
```

As you can see our list `stuff` contains other things like another list with dog names or numbers or even a single number seven. Pretty cool, right?

If we come back to our survey example, common practice is to store such data in a data frame which you can image as a list with different sub vectors (de facto a data frame is a list of different vectors of equal length). Alternatively, I always tell my students that you can also imagine it as a matrix with different data types per column.

The result is a tabular representation of your different data types and variables.

```
my_data_frame = data.frame(Numbers=c(0,1,2), Booleans=c(TRUE, FALSE,
TRUE))
> my_data_frame
Numbers Booleans
1 0 TRUE
2 1 FALSE
3 2 TRUE
```

In business data analytics data frames will be our most common data structure. If we store our data into it and use it for analytics modeling, business analysts assume that the data in data frames is in tidy form (Wickham, 2014).

This means:

- each variable is a column
- each observation is a row
- each type of observational unit forms a table.

If the data has multiple values stored into one column or an observation is stored in multiple rows we have to preprocess and restructure our dataset. We will come back to that in *chapter 3 - Business Data Understanding*.

An extension of the data frame is the `tibble` data frame from Hadley Wickham. Tibbles are an enhanced version of the traditional data frame and are part of the `tidyverse` ecosystem, designed to work seamlessly with other `tidyverse` packages. Tibble data frames have several advantages over standard data frames. They display data more cleanly, showing only a limited number of rows and columns when printed, making it easier to work with large datasets. Tibbles also have stricter data checking, making it less likely for common data-related mistakes to occur. Additionally, they have a more consistent behavior across different operations, ensuring a more predictable and user-friendly data manipulation experience.

You already met one in the introduction, where you printed the dataset:

```
> head(dataset)
# A tibble: 6 x 8
    DAY MONTH MANUFACTURER PRODUCT             SHIFT   COLOR MALTING            TASTING
  <dbl> <dbl> <chr>          <chr>              <chr>   <chr> <chr>                <dbl>
1   1      4 Leonard     Junglivet           Morning 0.27   Inhouse                895
2   1      4 Carlson     Junglivet Premium Evening 0.27   Burns Best Ltd.        879
3...
```

You can see that the dataset you used is a `tibble`. In most cases `data frames` and `tibbles` will behave the same.

But if it should be necessary you can also convert them vice-versa with `as.data.frame()` or as `as_tibble()`:

```
> dataset_new = as.data.frame(dataset)
> head(dataset_new) #df
    DAY MONTH MANUFACTURER        PRODUCT     SHIFT     COLOR    MALTING     TASTING
1  1      4    Leonard           Junglivet  Morning  0.27     Inhouse           895
2  1      4    Carlson Junglivet   Premium  Evening  0.27   Burns Best Ltd.     879
3...
```

2.3 Work With Data in R

After illustrating how to define a dataset with values, you are now interested in, how you can access the specific values in your data objects and how to investigate them in more detail. This is a common practice to check the data quality of your data or to better understand if the data is usable for your analytics project.

Common questions you want to answer during data understanding are

- How does my data look like?
- Which data types does it contain?
- How is it structured?

R and RStudio has many functions and tools to help you to answer these questions. You can use base R and the console or the user-interface of your RStudio for that. Let us now take a look at this in more detail.

If you want to take a look at your data in the console, just move your cursor in the terminal in the bottom left and type the following commands:

```
dataset = read_excel("productionlog_sample.xlsx")
#       or       if       dstools       is       installed       just       run
dstools::data("productionlog_sample")
View(productionlog_sample)
```

If you run these commands a new tab should pop-up and show the structure and contents of the data frame. Make sure that you use the right path to the file in the `read_excel()` command. In the code example above, I assume that the file `productionlog_sample.xlsx` is in your R working directory. Alternatively, you can also go to your variable overview on the right and click on the dataset `productionlog_sample`. The same tab with the same content should appear.

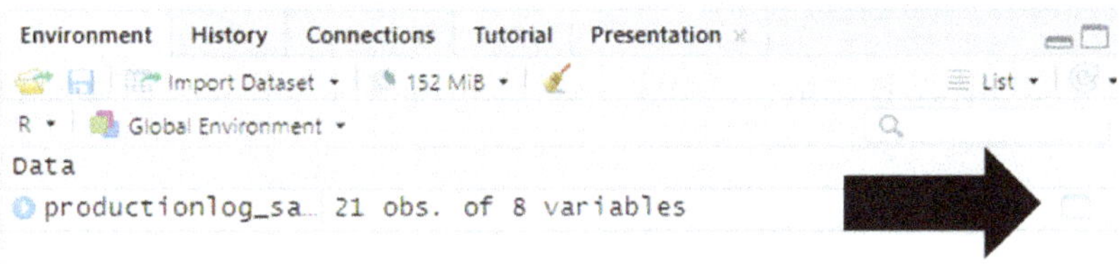

Figure 2-6 Press the small table button in the environment in RStudio and the data will pop-up in tabular form.

As you can see in the new tab, the data is presented in a tabular form with the individual observations on the rows and the different variables like "`LOCATION`" or "`Distillery`" on the columns.

Another relevant information is the data type of the different variables. In our case we would be interested if the ratings are a factor or normal numeric values or if the coordinates are numeric or characters. As you can see the `str()` command helps us to answer these questions.

```
> str(whisky_collection)
'data.frame':     40 obs. of  14 variables:
 $ NAME: chr  "An Cnoc" "Ardbeg" "Auchentoshan" "Ballantine's" ...
 $ DISTILLERY: chr  "Knockdhu" "Ardbeg distillery" "Auchentoshan
distillery" "Pernod Ricard" ...
 $ LOCATION: chr  "Scotland" "Scotland" "Scotland" "Scotland" ...
 $ TYPE: chr  "Single Malt" "Single Malt" "Single Malt" "Blended" ...
 $ REGION: chr  "Highland" "Islay" "Lowland" "Lowland" ...
 $ FOUNDATION: num  1894 1815 1823 1957 1779 ...
 $ COORDINATES: chr  "57.78497402327632, -1.6207674406702595"
"55.6406114241932, -6.107878725864291" "55.922066708629245, -
4.437533995096446" "55.9659724419173, -4.568164585896366" ...
 $ WIKIPEDIA: chr  "https://en.wikipedia.org/wiki/Knockdhu"
"https://en.wikipedia.org/wiki/Ardbeg_distillery"
"https://en.wikipedia.org/wiki/Auchentoshan_distillery"
"https://en.wikipedia.org/wiki/Ballantine%27s" ...
 $ RATING: num  2 4 4 2 3 4 3 3 3 2 ...
 $ REVIEWS: num  9 9 6 8 8 10 8 8 8 7 ...
 $ CRITIQUES: num  90 91 91 81 88 95 90 89 91 89 ...
 $ SMOKNESS: num  -1 5 -2 0 3 3 -4 0 4 3 ...
 $ RICHNESS: num  -3 -5 4 2 1 0 1 0 -4 3 ...
 $ PRICE: num  32 46 25 23 35 2700 75 36 45 19 ...
```

The log in the console shows us structural information of our dataset. For instance we can see the data structure, which is a "`data.frame`" in our case. How many observations (rows) and variables (features) exists. In our case we have `40` observations, which means we have `40` different whiskies in this sample, and `14` variables storing further information like name or rating of the whisky.

As before we can also use the GUI of RStudio instead of the terminal to investigate our whiskies. In this case go to the environment tab on the right top of your editor, click on the small blue arrow before the variable "`whisky_collection`". Then the arrow should expand a detail view, which shows the same information as our console log.

As you have seen, RStudio and the R terminal go hand in hand and you can use both to investigate your data.

The functions we use so far are termed "diagnostic functions". Diagnostic because they help us to diagnose data problems like a doctor does when you come to him with a problem. R has many of such diagnostic functions that you can use.

Here is a short overview of the most important once:

- `str()`: Internal structure of the R object
- `names()`: Names of the elements within the object
- `class()`: Internal class of the object
- `mode()`: Storage mode of the object
- `length()`: Dimension of the object
- `unique()`: Unique values of the object
- `sessionInfo()`: Version information about R and loaded packages

After you checked the structure and content of the dataset of your interest, we want now access the different values stored in this dataset for deeper analysis. Now I want to help you to do so and to access the data in your data objects like data frames, lists or vectors.

Common practice is that you retrieve values in a vector by declaring an index inside a square bracket "[]" operator.

```
> my_vector = c(10, 20, 30, 40, 50, 60, 70, 80, 90, 100)
> my_vector [1]
10
```

This kind of indexing has the following rules. If the index exists, the value is returned. If the index is negative, this value is not returned and if the index does not exist, the return value is "`NA`".

Indices can be numbers, logicals, and even names which can increase the readability of your code. Hence, I recommend clearly to use names, if possible, instead of numbers. Furthermore, numbers can change due to reordering of the data structure but the names mostly remain.

To illustrate the different ways and possibilities to index values from your data objects we will make a new sample vector `my_vector` with numbers from `10` to `100`. We could do that as we did before with declaring the vector with "`my_vector = c(10, 20, 30, 40, …)`". But this time we want to make that a bit smarter and use the `seq()` function for that. Writing so many numbers is quite embarrassing. And as a business data analyst we want to avoid such work.

The `seq()` function needs a start value and an end value. Furthermore, we can give a step size (default is `1`) to define the values between the start and the end value. To get the same value as we did manually, we use the `seq()` function in the following manner:

```
my_vector = c(seq(10, 100, by=10))
```

This generates a new vector with the numbers `10, 20, 30, 40, 50, 60, 70, 80, 90` and `100`. Using a function for that gives us more flexibility, if we want to add further values easily or want to expand our vector later.

If we want to access the distinct values, we can do that as before with the brackets:

```
> my_vector[1]
10
> my_vector[2]
20
```

If we want to access a sequence of values in our vector we can use the `seq()` functions we have learnt some sentences before:

```
> my_vector [seq(1, 3)]
10 20 30
```

Instead of the sequence you can also use other vectors, if the required values are in no sequence, but require some work:

```
> my_vector[c(1,3,2)]
10 30 20
```

You can also except values with the „-" operator:

```
> my_vector[-1] # all values except the first
20  30  40  50  60  70  80  90 100
```

You can also index by Boolean values:

```
> my_vector[c(TRUE, FALSE, FALSE, TRUE, FALSE, FALSE, FALSE, FALSE,
FALSE, FALSE)]
10 40
```

Or by logical comparison:

```
> my_vector[my_vector >= 90]
90 100
```

Or if you have a very good day, you can set names for each element and then index them by name:

```
> names(my_vector) = c("Holger", "Samantha", "Bernd", "Anna", "Fred",
"Nicky", "Louis", "Jennifer", "Jack", "Zoe")
vector["Zoe"]
100
```

To complete this overview, I want to let you know that lists can be indexed the same way. For instance, you can name the elements of the list we setup in the previous example. After initializing the list, you can access the elements later by name:

```
> beauty_contest = list(dogs=funnyDogNames, rating=coinFlips)
> beauty_contest$dogs
[1] "Winnie, the Poodle" "Bark Twain" "OzzyPawsborne"
> beauty_contest[[1]]
[1] "Winnie, the Poodle" "Bark Twain" "OzzyPawsborne"
```

Now, let us come to our multidimensional data objects. You retrieve values in a matrix by declaring a 2-dimensional index inside a square bracket "`[]`" operator. For instance, in a 2-dimensional matrix, indexing works as `your_matrix[<row_index>, <column_index>]`. While for arrays with more dimensions, it extends to indices specified for each dimension as `your_array[<first_index>, <second_index>, <third_index> ,…]`. This indexing scheme allows us to access and manipulate specific elements, rows, columns, or slices within these multidimensional structures, enabling precise data extraction and manipulation:

```
> my_matrix = matrix(1:9, nrow = 3)
> my_matrix
[,1] [,2] [,3]
[1,]    1    4    7
[2,]    2    5    8
[3,]    3    6    9

> my_matrix[1, 2] # index the 4 in the first row, second column
4
```

The multidimensional sibling of the matrix is the data frame. Data frames share similarities but offer more flexibility. In contrast to matrices data frames can contain columns of different data types (numeric, character, logical, etc.), making them versatile for handling heterogeneous data. You can think of a data frame as a spreadsheet-like structure, where columns can represent variables or attributes, and rows represent observations or instances. They are widely used for data manipulation, analysis, and are especially prevalent in handling real-world datasets. Especially, when versatility in data types and operations is necessary. The indexing of data frames is comparable to matrices, which means you can subset them like this:

```
my_dataframe = data.frame(x = c(1,1,2), y = c(2,1,2))
my_dataframe[my_dataframe$x== 2, ]
> x y
3 2 2
```

Or select columns:

```
my_dataframe$x  # or my_dataframe[,"x"]
> x
1 1
2 1
3 2
```

2.4 Write Functions and Logic in R

Now we are ready to start writing our first script. Sometimes, the built-in functions in R might not cover all the tasks you need to perform. When you find yourself needing more complex computations beyond basic arithmetic operations, you'll want to explore control structures or even create your own functions. Or you simply want to avoid repetitions. Therefore, the rule of thumb is: If you have to copy code, make a function out of it.

Control structures like loops (e.g., for, while) and conditional statements (e.g., if-else, switch) enable you to execute R code conditionally or repetitively based on specified criteria. They provide the flexibility to handle various scenarios and perform complex operations.

Furthermore, writing your own functions allows you to encapsulate specific tasks or calculations into reusable blocks of code. This modularity helps in organizing and simplifying your R-scripts, as well as promoting code reusability and readability. You can define your own function like this:

```
my_function= function(arg1,…)
{
expression1
…
}
```

And you can call this function as follows:

```
my_function(arg1…)
```

A function takes one or more inputs (or none), known as arguments, which can be optional or mandatory based on how the function is designed. These arguments serve as the parameters that the function operates on or uses to perform specific tasks. Arguments can be of various types, such as scalars, vectors, data frames, or even other functions.

Within the function's body, these arguments are manipulated, processed, or used to produce desired outputs. Functions can return results, which might be a single value, a vector, a data frame, or any other R object. The returned output can then be stored in variables, printed, or used as input for other functions or operations.

Understanding how to define and utilize functions effectively is a crucial aspect of programming in R. Here is a simple example of a function multiplying the input parameter by 2:

```
multiply = function(x){
x*2
}

multiply(1) # run function
```

Sometimes it might be necessary to expand your functions with the help of other control structures like if and else commands or loops as illustrated in the following Table 2-3.

Table 2-3 Control flow structures in R

Control Structure	R Example	Description

if	if(a>b){
	  print("a>b")
	}
		If the condition is TRUE *the expression is computed*

ifelse	ifelse(a>b, "Yo", "No")	*If the condition is TRUE, the first value is returned, the second otherwise*

if-else	if(a>b){
	  print("1")
	}else{
	  print("2")
	}
		Choose between two expressions depending on condition

if-else-then	if(a>b){
	  print("1")
	}else if(a>c){
	  print("2")
	}else{
	  print("3")
	}
		Choose between multiple expressions depending on condition

switch	foo = switch(
	  expression,
	  case_1,
	  ...,
	  case_n
	)
		Combining multiple conditions, if TRUE the related expression is computed

for	for(i in 1:3){
	  print("ho")
	}
		Iterate a specific number of times with index i

repeat	repeat {
	  statement
	}
		Executes the same code until a break condition breaks the loop

while	while(cond){
	  doing
	}
		Executes the same code as long as the condition is fulfilled

2.5 Expand your Code with External R Packages

In the previous section we have seen how to build our own functions. However, this is not always necessary. R is open source and has its own ecosystem the “R-chitecture”. That means that various contributors provide many packages for the most statistical problems and areas for free. Their R packages encapsulate the most relevant functions and dataset of a specific topic, and you can use them in a single line of code. In most R projects, the most scripts start with loading packages in your library and expanding base R by loading further packages that will be used.

For instance, if you want to use the methods and datasets I prepared for this book, you have to install the `dstools` package. Or if you want to work with an API (see chapter 6), you will need some API packages.

Let us now have a short look at the R-achitecture in Figure 2-7.

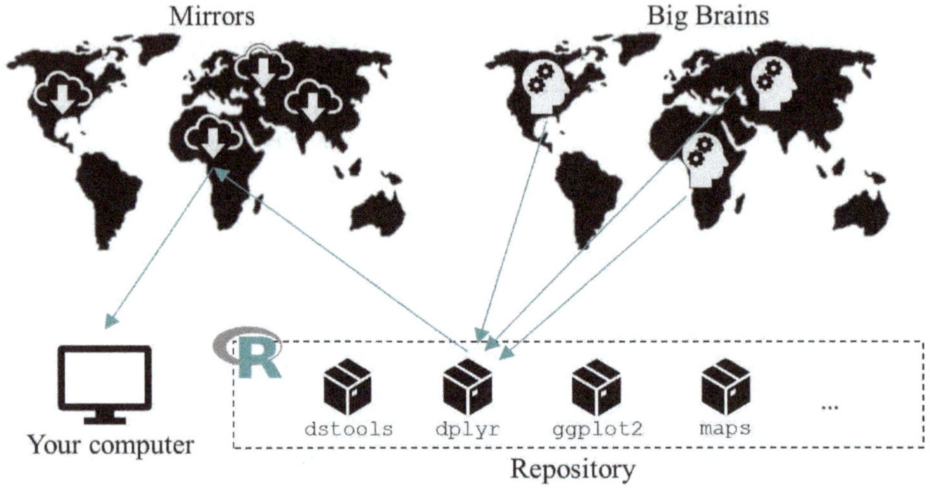

Figure 2-7 The "R-chitecture" adopted from Andy Field (Field et al., 2012).

The essential part of the R ecosystem are packages. Packages are the primary extension mechanism for R (and Python). They are the standard to share your functions, datasets, and documentation with colleagues or other business data analysts. From a technical perspective you can think of a package as a set of structured folders and files.

In Figure 2-7 the authors of the packages are illustrated as "Big Brains" because they put a lot of effort in the package developing process and provide their work for free to you. A bundled package is a package that's been compressed into a single file for download. By convention, package bundles in R use the extension ".tar.gz", but you can store them also as .zip if you want. These bundled packages are then stored in a repository.

In R there exist three relevant repositories or sources of packages:

- **CRAN**: R's central software repository, supported by the R-Foundation
- **Bioconductor**: A repository for analysis of genomic and biological data
- **Github**: Today it is more and more popular to store your package code and provide them in version-controlled directories

Some companies also offer their own repository where you can download packages, but this is very specific, and I recommend that you contact your IT coordinator in such a case.

If you want to install packages and use it like the `tidyverse` from CRAN you can just run:

```
install.packages("tidyverse")
```

Or if you want to install multiple packages:

```
install.packages(c("tidyverse", "devtools"))
```

If you work with other scripts and the command `library()` raises the following error message:
`there is no package called '…', you have to install them first, then run library() once again`

You have to install the missing package to use the package that is missing.

If you want to install a package from github, or a current release of a package that is not available on `CRAN`, you run the `install_github(<developer>/<packagename>)` and install it from the developer's repository. In our case the developer's name is "`dominikjung42`" (that's me!) and the name of the package is "`dstools`".

```
library(devtools)
install_github("dominikjung42/dstools")
```

But make sure that you have installed `devtools` from `CRAN` first to use this functionality.

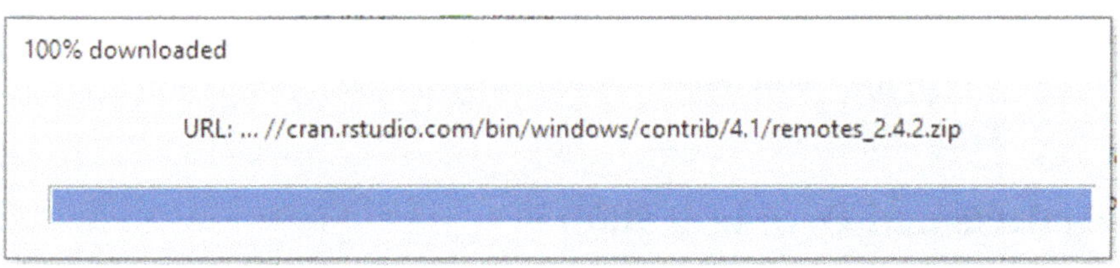

Figure 2-8 If you download packages in RStudio the following message will appear and show you the state of the installation process.

After downloading the packages with CRAN or from github, we'll be able to use them in our R scripts or RMarkdown files by loading them in your local library. Package C are just directories containing installed packages. When a package is requested by R, R searches the different library directories to find the installed package.

You can load packages from your active library in your R environment with:

```
library()
```

For instance, you can load `tidyverse` and `devtools` with:

```
library(tidyverse)
library(devtools)
```

If you are using RStudio as an IDE then your code is checked and you will get a warning if you are missing the required packages for the code (see Figure 2-9). Just press on "Install" in the warning and the missing packages will be installed for you!

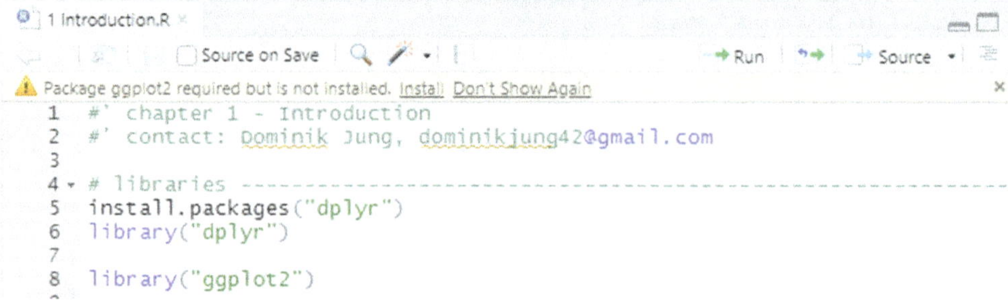

Figure 2-9 If you try to load not-installed packages, RStudio prompting you to install the required packages.

If your R library path is not the system-level library path, installing packages for the first time from the RStudio interface can cause the following error:

```
Error in unzip(zipname, exdir = dest):
      cannot open file 'C:/Users/...": Permission denied
Calls: <Anonymous> ... .install.winbinary ->
unpackPkzZip -> .zip.unpack -> unzip
Execution halted
```

You can run the install command manually (`packages.install("ggplot2")`) to solve this problem. Alternatively, you have to run RStudio as administrator and change the library path.

2.6 Manage your Projects and Data in RStudio

If we think back at our first business case from the introduction, you will remember the following code section loading our whisky data in your RStudio:

```
dataset = read_excel("productionlog_sample.xlsx")
```

The following code loads the local file `productionlog_sample.xlsx` in your R environment. This procedure is the straight-forward way if you want to import a file from this book (you can download it on the git repository of this book).

Alternatively, you can use the RStudio user-interface for that. Both ways are suitable and work vise-versa. If you decide to load the dataset with RStudio go to the "Files" tab in the right bottom corner. Right-click on your dataset and a pop-up with the options "View File" and "Import Dataset" will appear. Click on the latter one. In the next seconds a wizard for data upload will appear and guide you through the import process. In our case it's enough to just press "Import" without further specifications.

The wizard is quite suitable if you are not sure how to import the specific filetype in your project. Furthermore, the wizard generates the R-code to load the file, so you can save this code and use it directly in your script the next time you want to load this file. My suggestion is, if you know how the dataset looks like and are familiar with the data type you can import it directly in your code and save some time by avoiding the wizard. Later in this book we will further discuss the RStudio wizard to import data.

Nevertheless, we have a problem. If you close your RStudio and start a new analytics project, your data or working status might be lost. But good news: RStudio has so called “projects” to handle this problem. RStudio allows you to manage all your files and data with “projects”. The benefit is that you can throw all your stuff in one folder and have directly access without browsing your files. Furthermore, you can save your current session, and reload it easily. You can find an example project based on CRISP-DM on the git repository (look for “sample project structure”).

Congratulations! So far you have survived the R introduction. In the last sections of this chapter, I would like to quickly provide you with some additional information.

For the first time, you'll encounter the "Useful R functions" section. You will find this kind of subchapter in every chapter. I have put there important and useful functions and best practices that you don't need for an introduction, but that will surely help you in the future in your business data analyst life. If you need some time to learn and recapitulate the chapter than it’s nothing wrong about to skip this subchapter and come back in some days to it.

2.7 Further Beginner Resources

Finally, this was my quick and sharp introduction into programming with R. However, R programming is no rocket science and there is no shortage of R tutorials. If you feel unsecure or want to deepen your basic R skills, I can recommend that you take a look at the following tutorials. But don’t worry, we will sharpen your R and analytics skills with many case studies and exercises in this book in the upcoming chapters. 😊

- **RStudio 's Online Learning Resources**: RStudio provides a variety of free online resources, including interactive tutorials, cheat sheets, and articles, designed for beginners to learn R and data science concepts: *www.rstudio.com/resources/training*
- **Coursera**: Coursera hosts various R courses taught by top universities and instructors. Many of these courses are free to audit and they cover a wide range of topics for beginners: *www.coursera.com*
- **YouTube**: Many R experts and data science educators have created YouTube channels dedicated to R tutorials. You can find a wide range of video tutorials covering different aspects of R programming: *www.youtube.com*
- **R-bloggers**: R-bloggers is a popular blog aggregator that compiles posts from various R bloggers. It is a great resource to discover tutorials, tips, and tricks shared by the R community: *www.r-bloggers.com/*

Furthermore, this book is only the first step in a long journey in R coding. There will be definitely situations you will soon face problems that are not answered in this book. In the following section I want to give you some recommendations where to find help.

- **Google**: The best practice to find a solution to a coding problem is to use Google. Copy the error message or type your question in combination with "R" to get some results. Chances are very high that you get something useful. In the past Google could not handle single characters like "R" in search queries but since some years Google could fix this issue.
- **Stackoverflow**: Another relevant source for help is the website *www.stackoverflow.com*. You can also add "stackoverflow" as a keyword to your google query to get results from this website directly. There is a whole section for R programming, where you should find a solution to most of your problems. If you cannot find a solution to your problem, you can also ask a question describing your problem. It is recommended that you add some code examples to reproduce your error or to better understand your problem.
- **GitHub**: If you have some problems with an R package or get error messages using the package you can also ask for help on GitHub. Most R packages are hosted on GitHub and if there are issues with the current version you will find there the latest discussion and bug fixes. You can also ask the authors of the package directly if you have some questions.
- **Chatgpt**: And last but not least services building on large language models like Chatgpt (*www.chat.openai.com*) also have shown incredibly good performance to write R code and help you to understand R code. Just paste your R code and ask ChatGPT to explain it or ask it directly how to write a specific code snippet.

2.8 Useful R Functions for Everyday R Programming

Good to see that you decided to stay! Let us have now a look at the first best practice: loading data for this book. In the whole course there are two possibilities: Download the data and import it as you learnt in this chapter or load it directly from the `dstools` package. Which means instead of downloading the file `productionlog_sample.xlsx` and then running:

```
dataset = read_excel("productionlog_sample.xlsx")
```

You can alternatively just load the library `dstools` and run:

```
dataset = data("productionlog_sample")
```

This works for all files that are used in this book and will help you to make sure that there are no errors in the data import.

Another helpful best practice is that if you work with external data from the internet that you put the download process directly in the code so that your future you or your colleagues can easily replicate your analysis and run your code directly.

Therefore, you can create a folder with data with the following code:

```
dir.create("data_raw", showWarnings=FALSE)
download.file(url=
"https://github.com/dominikjung42/BusinessAnalyticsBook/tree/main/data",
             destfile="productionlog_sample.xlsx", mode="wb")
```

Sometimes you might want to delete your variables or datasets. Then you can do that with `rm()`:

```
rm(dataset)
```

Perhaps, you remember the function `paste()`. We used it in the subchapter to print variables in a control structure like a for loop. Sometimes you might want to print multiple variables and if you do it with `print()` then it will be not what you expect:

```
> a = 1
> b = 2
> print(a,b)
[1] 1
```

Because if you want to "print" multiple variables you have to paste them together with `paste()`:

```
> paste(a,b)
[1] "1 2"
```

Furthermore, we skipped `Date` objects and formats in the introduction. That's because working with them is a bit of a mess in R and I did not want to confuse you at the beginning. Working with date objects in R involves using the built-in date and time classes, as well as various date-related functions from the base R or other packages.

You can create `Date` objects with:

```
my_date = as.Date("2022-11-22")
```

The `as.Date()` function allows a variety of input formats through the `format` argument. The default format is a four-digit year, followed by a month, then a day, separated by dashes. But you can also use it to convert a character or numeric string to a date object with a specific format. If you want to generate a date with a specific format, a format string has to be composed using the elements shown in Table 2-4.

Table 2-4 Options for the Date format

Parameter	Value
`%d`	Day of the month (decimal)
`%m`	Month (decimal)
`%b`	Month (abbreviated)
`%B`	Month (full English name)
`%y`	Year (2 digit)
`%Y`	Year (4 digit)

The following examples show some ways that this can be used:

```
> as.Date("8/17/2023",format="%m/%d/%Y")
[1] "2023-08-17"

> as.Date("April 20, 1989",format="%B %d, %Y")
[1] "1989-04-20"
```

Furthermore, you can extract components of a date object like the year or month with `weekdays()`, `months()`, `days()` or `quarters()` etc.:

```
> my_date = as.Date("2022-11-22")
> weekdays(my_date)
[1] "Thursday"
```

2.9 R Beginner Exercises

At the end of each chapter, you will find an exercise case study with a business problem you can solve to train and recapitulate your knowledge of the content of the previous chapter. Because you recently started your R career, I decided to replace the case in this chapter with some basic R exercises so you can sharpen your basic R skills before you have to start with a whole business case exercise. Use the `onlineshop` dataset from the website for this exercise.

Lindsey Maegle from the marketing team contacts you, because she needs some help to understand the sales in the online shop. Can you help her with the following tasks?

1. Download the `onlineshop` dataset from the website of this book and load it in your R environment.
2. How many columns and rows has the dataset?
3. When was the first order in the online shop?
4. What is the average turnover per purchase? What is the highest and what the lowest turnover?
5. Which customer buys most frequently? And how many different customers exist?
6. Lindesy is interested in the shopping behavior of some specific users. Can you help her and provide a subset with the logs of `ron_swanson76`, `horst_lüning` and `dorothy_parker`.
7. What is the average turnover of each of these users?
8. Who is the oldest customer?

Good job! Now it is a good point to grab a cup of tea, some cookies and take a short break!

3 Business Data Understanding

3.1 Introduction

Now that you have the basic knowledge of programming and the mindset, let's turn to the raw material of business data analytics: the business data. Business data is all data that is generated in business operations and that we use for our questions and projects. This encompasses various types of information such as statistical data like the whisky quality of different production batches, unprocessed analytical data from the online shop, customer feedback from the local shop, sales figures from the marketing team, and other diverse datasets like weather data or information about the soil in order to better estimate the quality of the ingredients. The next chapter "*Business Data Understanding*" is about exploring and understanding the different types of business data.

In this crucial phase, we delve into the process of comprehending the data landscape, its nuances, and its significance within the broader business context. By gaining a deep understanding of the business data, its sources, and potential limitations, we pave the way for informed analysis (*Modeling*). Equally essential is the preprocessing of this data (*Business Data Preparation*), where we cleanse, transform, and enrich it to ensure accuracy and readiness for analysis.

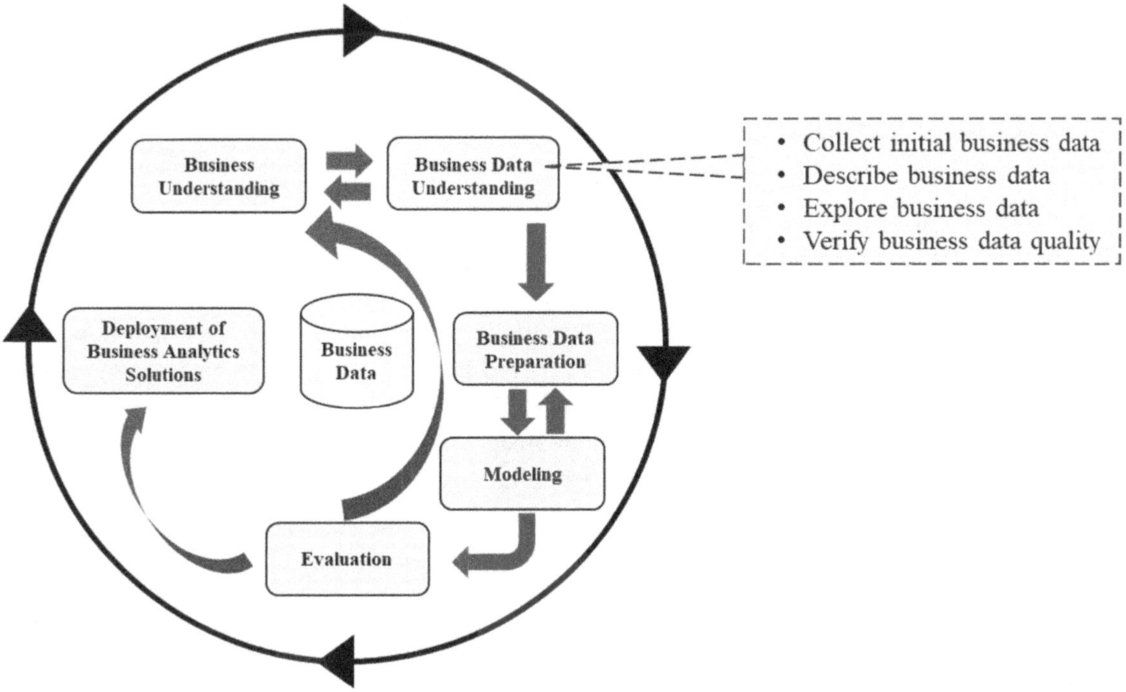

Figure 3-1 The generic tasks of the business understanding phase.

As illustrated in Figure 3-1, the next phase "Business Data Understanding" is about laying the foundation for the next phases by understanding our most relevant resource: business data. We want to collect the different sources and understand our data before we can continue in our

D. Jung, *The Modern Business Data Analyst*, https://doi.org/10.1007/978-3-031-59907-1_3

project. Therefore, you should clarify the following aspects in the business data understanding phase:

- **Collect initial business data**: Collect and write an initial data collection report
- **Describe business data**: Write a short description of the dataset to allow your team to understand the data also in the subsequent phases
- **Explore business data**: Make an exploratory data analysis and write a data exploration report
- **Verify business data quality**: Identify quality issues and report them

First, obtain the business data specified in the project resources, either by accessing it directly or collecting it as needed. This initial phase includes loading the whole or samples of the business data, which is required for understanding it. For instance, we have investigated a sample of the production data in the example of the introduction in chapter 1. This process may also involve the initial steps of data preparation. It's important to note that if multiple data sources are acquired, the issue of integration needs to be considered, either at this stage or later during the data preparation phase.

Second, we have to describe the initial, high-level characteristics of our business data and present the findings. Doing so we have two tools: descriptive statistics and data visualization and/or a combination of both. In this phase it can make sense to prepare a short summary report in form of a PowerPoint presentation, which provides an overview of the acquired data, including its structure, format, volume (e.g., record and field counts per table), field identities, and any noticeable surface features. Furthermore, you have to check if the acquired data aligns with the specified requirements, we identified in the business understanding and project planning.

Building on our first visualizations and descriptive statistics, the next step is about exploring our business data. This can be done through filtering and summarizing, further visualizations, and reporting methods. These methods encompass the examination of feature distributions (e.g., the target feature in a prediction task like the whisky quality in our introductory example in chapter 1), correlations between pairs or a limited number of features, outcomes from basic aggregations, characteristics of notable sub-groups within the data, and straightforward statistical assessments. These analyses serve the primary business data analytic objectives; they can also enhance or fine-tune the data description and quality reports and provide input for the subsequent transformation and other data preparation procedures necessary for further analysis.

Finally, conduct a comprehensive evaluation of the data's overall quality, encompassing inquiries into various aspects. These include assessing data completeness, ensuring it encompasses all the necessary cases. Scrutinize the correctness of the data, identifying any existing errors and gauging their prevalence. Additionally, identify the presence of missing values within the dataset. In this case, examine the representation of these missing values, their occurrence patterns, and their frequency within the dataset. By addressing these crucial questions, we gain a thorough understanding of the business data's reliability and potential limitations.

3.2 Business Data Manipulation with dplyr

However, before we turn to visualizing and exploring our business data, let's take a look how to manipulate your data that it suits your needs. Data manipulation is necessary so that we can deal with the data easily and quickly. Therefore, we make use of the most well-known package for this: `dplyr`.

The `dplyr` package provides lots of utilities for the most common data manipulation tasks. It is designed to work directly with the most important data structures that we learned about in chapter 2. It can be used to manipulate data in a variety of ways to streamline your code and business data understanding. Whereby many common tasks are optimized by writing it partly in C++ (don't worry this happens under the hood, we will keep programming R).

Another advantage, which may not be so obvious at the moment is that you have the possibility to work directly with data stored in an external database. As we will see in the next chapters, this has the advantage that the data can be natively managed in a relational database, queries can be performed on this database and only the results of the query are returned.

To test and work with the `dplyr` package, we load an example dataset from the *Junglivet Whisky Company* in our R environment:

```
whisky_collection = read_excel("whiskycollection.xlsx")
```

Or if the `dstools` package is installed and loaded you can load it directly to your R-environment and run:

```
whisky_collection = data("whiskycollection.xlsx")
```

The `whisky_collection` dataset is a small collection of about `40` outstanding and award-winning whiskies from around the world. In addition, it contains further information about the distillery, the place of production and the results of different whisky tastings (among other from myself). We will use this dataset in this chapter to learn how to visualize and explore a new dataset.

The dataset consists of the following variables or features:

- `NAME`: The name of the whisky
- `DISTILLERY`: Distiller of the specific whisky
- `LOCATION`: Production location of the whisky (mostly countries or regions)
- `TYPE`: Specification of the whisky type like e.g. `single malt` or `blended`
- `REGION`: Region of the whisky production (mostly relevant for Scotchs)
- `FOUNDATION`: Year of the first whisky production
- `COORDINATES`: Longitude and latitude values of the distillery
- `WIKIPEDIA`: Link to the related article of the English Wikipedia
- `RATING`: My personal rating of this whisky (I am open to discuss it, just write me an email if you think the rating is wrong)
- `REVIEWS`: The average rating of this whisky based on consumer reviews from many whisky online shops in 2023

- `CRITIQUES`: The average rating of this whisky based on reviews from professional critics until 2023
- `SMOKENESS`: How smoky vs. delicate it tastes, negative values implicate delicate
- `RICHNESS`: How rich vs. light it tastes, negative values implicate light
- `PRICE`: The average price level in Euro of the youngest 10/12 year or consumer version in the whisky exchange 2023

After you have loaded the dataset in your R environment, we want to investigate it in RStudio. Thus, we run the `View()` or the `head()` function and take a look at the dataset:

```
View(whisky_collection)
head(whisky_collection)
```

The dataset is an excerpt of a fictional whisky collection. Each row represents one whisky. The dataset also contains general character data like the name of the whisky or the distillery, but also categorial data like the type (`single malt`, `blended` etc.). Furthermore, there is numeric data like the year of the foundation and geographical data in numerical form like the coordinates of the distillery. This is a quite realistic dataset with mixed data types, we have to consider when we run our analyses. If you want to learn more about the dataset, which is part of the `dstool` package, you can load the package and run `?whisky_collection` in you console to open its help page.

But how do we now work with this dataset? How can we handle it, if we want to compute the mean rating of whiskies from Scotland vs. Ireland? Or if we are interested in the number of whiskies in each region?

The solution to many business data analysis problems follows always the same strategy. You break up a big problem into manageable distinct pieces, operate on each piece independently and then put all the pieces back together (Wickham, 2011).

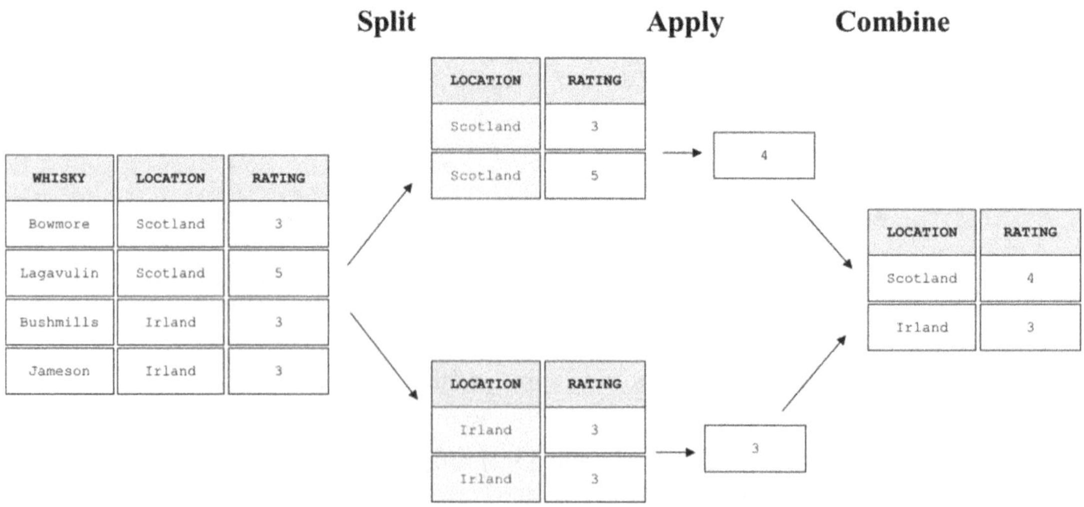

Figure 3-2 The split-apply-combine strategy to handle business data processing.

As you can see in Figure 3-2, the computation of the average rating of whiskies in Scotland vs. Ireland, requires splitting our dataset in two parts. One part containing whiskies from Irland and one part containing whiskies from Scotland. In the next step we apply a function (average computation) on each subset. Then we take these values and combine them together in a new dataset. In this new dataset we have all the information we wanted.

This strategy is called split-apply-combine strategy. And the `dplyr` package is built with that in mind to provide functions and methods to help you handle your data.

In a next step we are going to learn some of the most common functions you will need if you want to follow the split-apply-combine paradigm in your analyses:

- `select()`: Pick specific columns of your dataset by their names
- `filter()`: Pick rows on conditions based on their values
- `%>%` (aka pipe): We can use it to build data processing pipelines
- `mutate()`: Create new columns in your subset
- `group_by()`: Group data to run further functions on it
- `summarize()`: Create summary statistics on your grouped data
- `n()` and `count()`: Counts discrete values in a selected dataset

All of them work in the same manner or are compatible to each other and can be combined to manipulate your data for your analysis.

If you are reading this chapter for the first time and are not familiar with R you probably don't have the `dplyr` package and other upcoming packages installed. Which means you can't use their functions in your R console or RStudio. Hence, you have to install and activate them, like the `dplyr` package in a first step with:

```
install.packages("dplyr", dependencies = TRUE) #install, just run once
library(dplyr) #activate, run everytime
```

3.2.1 select()

The first important command in `dplyr` is `select()`. It allows us to select specific columns in our dataset and built a subset out of them. The first argument of the function is your data frame, and the subsequent arguments are the columns you want to keep.

To select only the names of the whiskies and their rating you can run:

```
select(whisky_collection, NAME, RATING)
```

To exclude certain columns you can put a "-" before the name of the column you want to exclude. For instance, to return all columns, except `WIKIPEDIA` and `RATING`, run:

```
select(whisky_collection, -WIKIPEDIA, -RATING)
```

Please note that the commands do not modify the inputs. If you want to save them or use them later you have to assign them to a (new) variable:

```
my_selection = select(whisky_collection, -WIKIPEDIA, -RATING)
```

If you want to assign them and print the results you can do that by putting the assignment in parentheses:

```
(my_selection = select(whisky_collection, -WIKIPEDIA, -RATING))
```

Moreover, `dplyr` has many helper functions which allow you to build more complex selections. The most popular allow to select columns based on matches like:

- `starts_with("WIKI")`: Selects columns that start with `WIKI`
- `ends_with("PEDIA")`: Selects columns that end with `PEDIA`
- `contains("KIPE")`: Matches columns that contain the characters "`KIPE`" in the name

You can use them inside the `select()` command, e.g. to select `SMOKENESS` and `RICHNESS`:

```
select(whisky_collection, contains("NESS"))
```

3.2.2 filter()

To choose certain rows instead of columns you use `filter()`. The `filter()` function allows you furthermore to define specific criterions to select the rows you are interested in. This selection can be done based on the values of your rows. For instance, to select all whiskies from Scotland you run:

```
filter(whisky_collection, LOCATION == "Scotland")
```

As you see, it works the same like the `select()` command from before. The first argument is the data frame and the second one is your condition(s). For that you can use the basic R notation you know from the previous chapter: > means bigger, < means smaller, >= means equal or bigger. If you want to filter multiple categorial variables like locations, you can use the `%in%` operator to search in a vector with your filter criteria:

```
filter(whisky_collection, LOCATION %in% c("Scotland", "USA"))
```

You can also combine different filter arguments with logical operators. You will have to use these conditions, if you want to select your data effectively in one command. I illustrated them in Figure 3-3.

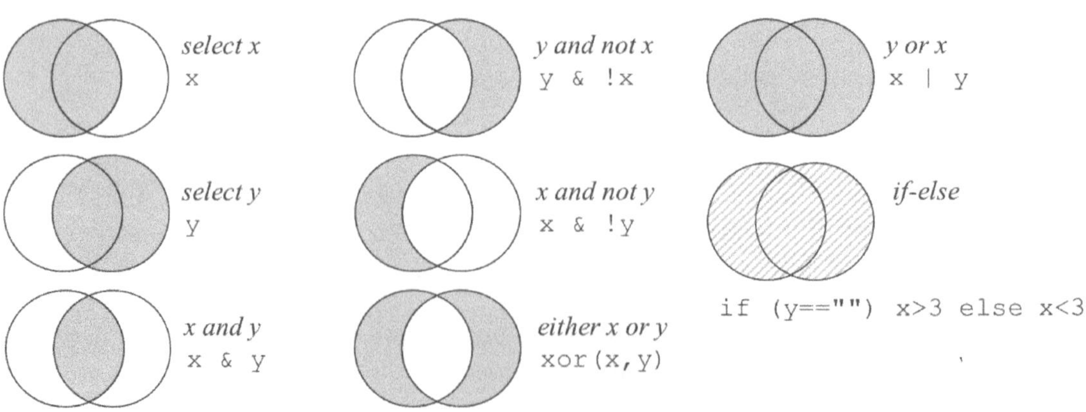

Figure 3-3 Available filter operators in `dplyr`.

For instance, to select whisky from Scotland or single malt American whisky you can do that in the following manner:

```
filter(whisky_collection, LOCATION == "Scotland" | (LOCATION == "USA"
& TYPE == "Single Malt"))
```

In natural language these commands can be translated as “Filter my dataset for all whiskies that are from Scotland or are single malt and are from the United States”.

> As many other programming languages R also has the `&&` and `||` command. They are reserved for conditional execution, and you should not use them here.

So good, so far. But as you might probably assume, if you start to use multiple `select()` and `filter()` together your code will get very messy. If you want to select the best whiskies from Scotland, you write probably something like that:

```
subset_scotland = filter(whisky_collection, LOCATION == "Scotland")
subset_scotland_red = select(subset_scotland, NAME, RATING)
subset_scotland_red_final = filter(subset_scotland_red, RATING >= 4)
```

In each step, you have to make a stop and think about a new name for your dataset. This is readable, but the best way to clutter up your workspace with lots of long named variables. If you have a big code project, this will become hard to keep track of. Furthermore, the code is not very messy, which makes it hard to debug – for you and for others.

3.2.3 Pipelines with %>%

For that purpose, `dplyr` allows you to pipeline your code instead. You can do that with the “`%>%`” command. In RStudio you can just press `Cmd + Shift + M` if you want to insert it. If you have a Mac, you press `Cmd + Shift + M`. The “`%>%`” is the so named “pipe” command, which stands for “use the output of the left for the command on the right”. I explain it to my students with the following analogy: $f(x)$ becomes “`x %>% f()`”. This is very useful when you need to do many things to the same dataset and you can use it to build sophisticated data pipelines with it.

Now our previous example becomes this:

```
whisky_collection %>%
  filter(LOCATION == "Scotland") %>%
  select(NAME, RATING) %>%
  filter(RATING >= 4)
```

As you see the code became much shorter and more readable. We use the pipe to send the `whisky_collection` data through the `filter()` to keep whiskies from Scotland. Because the pipe takes the output from the left and passes it directly, we do not need the first argument with the dataset any more. Then we select the columns `NAME` and `RATING`. Finally, we filter it again to select whiskies with a higher rating.

I suggest my students to think of the pipe operator like the word "then". In our example above, we took the `whisky_collection`, then we selected all the whiskies from Scotland then columns `NAME` and `RATING`, and then we filtered for high rated whiskies.

This example was very simple, but if you want to write more complex scripts this is very helpful. You do not have to use the pipe command, but I strongly recommend it because it will help you and others to find bugs much faster.

3.2.4 mutate()

Let us now compute the more complicated things. For example, what happened if we want to make new columns. Like a column `AGE` that returns the age of the distillery based on the year of foundation, or a new column `FAVORITE`, which represents whiskies you like. The `FAVORITE` column is based on the values in existing columns like `LOCATION` (Scotland) or `RATING` (if not from Scotland, it has to be rated with at least `4` points) as in our example before. For both new columns we have to use the `mutate()` function.

To create a new column `AGE` to compute the destillery's age in numbers we run:

```
whisky_collection_new = whisky_collection %>%
  mutate(AGE = 2023 - FOUNDATION)
```

This adds a new column at the end of the data frame with the name `AGE`. Run the `View()` command or the `head()` function to check if everything worked correctly.

To create a new column `FAVORITE` to mark whiskies from Scotland or single malt whiskies from the `USA` is a bit more complicated. For that purpose we have to use the `ifelse()` function from chapter 2.

The `ifelse()` function has the following structure:

```
ifelse(test, yes, no)
```

If you want to make a new feature that represents if a whisky comes from Scotland, you would run something like that:

```
whisky_collection_new = whisky_collection %>%
  mutate(FAVOURITE = ifelse(LOCATION == "Scotland", TRUE, FALSE))
```

To add a further check, namely if the whisky is not from Scotland, we will still mark it as favorite if it has at least `4` points in the rating. So we have to combine two `ifelse()` checks in our `mutate()` function.

As illustrated in Figure 3-4, we first have to check if the location is in Scotland. And then if whisky is distilled somewhere else, we have to check if it has a rating of at least `4` points.

We can do that in R by combining two `ifelse()` functions in the following manner:

```
whisky_collection_new = whisky_collection %>%
  mutate(FAVOURITE = ifelse(LOCATION == "Scotland", TRUE,
                            ifelse(RATING >= 4, TRUE, FALSE)))
```

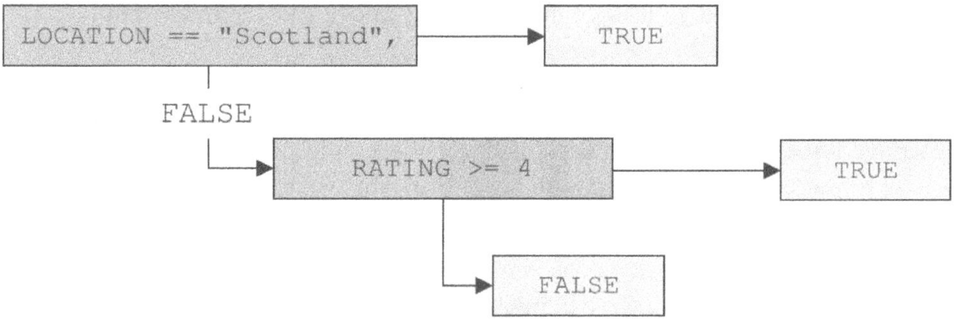

Figure 3-4 Illustration of our logic to compute our whisky favorites.

The next relevant function is `group_by()`. This function allows you to group different rows of your dataset by a value in one or multiple columns. This is very useful if you want to `mutate()` new columns based on subsets or subgroups in your dataset.

For instance, you want to compute the average rating of all the whiskies in each country (Scotland, USA and so on)

You can do that in the following manner:

```
whisky_collection_new = whisky_collection %>%
  group_by(LOCATION) %>%
  mutate(BENCHMARK = mean(RATING))
```

In a first step we group our whiskies based on the `LOCATION` into different groups (one group with Scottish whiskies, and one with American and so on). And then we compute for each of those group an average rating.

3.2.5 summarize()

If the average ratings of each country would be the results you are interested in your analysis, you could also run a `summarize()` in your pipeline. The `summarize()` function would reduce your dataset and summarize it to the results you are interested in:

```
whisky_collection_new = whisky_collection %>%
  group_by(LOCATION) %>%
  summarize(BENCHMARK = mean(RATING))
```

After building a new data frame `whisky_collection_new` we take a look at it by pasting its name in the R console in RStudio:

```
> whisky_collection_new
1 Canada        2
2 Irland        3.17
3 Japan         3.5
4 Scotland      3.39
5 Taiwan        4
6 USA           1.67
```

As you can see, the result is no more a big data frame with many rows and columns. The `summarize()` function has reduced your huge dataset to the values you are interested in.

3.2.6 n() and count()

Another very useful function is `n()`. This function counts the distinct elements in a group. We can use that to add the number of whiskies in each location to each our dataset from the example above. We will use a new column with the name `NUM` for that. Because the `summarize()` function excludes all columns that are not summarized, we have to add our new variable `NUM` there:

```
whisky_collection_new = whisky_collection %>%
  group_by(LOCATION) %>%
  summarize(BENCHMARK = mean(RATING), NUM = n())
```

If you want to count the distinct rows you can alternatively use `count()` for that. Which means that `whisky_collection %>% count(NAMES, LOCATION)` is roughly equivalent to our example from above.

3.2.7 across()

Finally, we will have a look at the `across()` function from the `dplyr` package, which can be used in R to apply a transformation to multiple `columns`.

Instead of writing complex mutate or summarize functions we can use `across()`.

For instance if we want to summarize some of the columns of the whisky collection, you would write a very long `summarize()` function like this:

```
summarize(
    AVG_RATING = mean(RATING),
    AVG_REVIEW = mean(REVIEWS),
    AVG_CRITIQUE = mean(CRITIQUES),
    AVG_PRICE = mean(PRICE),
    AVG_PRICE = mean(PRICE),
    AVG_PRICE = mean(PRICE)
  )
```

But instead you can use `across()` to write the following code to compute the mean:

```
whisky_characteristics = whisky_collection %>%
  select(LOCATION, RATING:PRICE) %>%
  group_by(LOCATION) %>%
  summarize(across(RATING:PRICE, mean)) %>%
  ungroup() %>%
  mutate_at(vars(-LOCATION), normalize)
```

Because this might be new for you let me explain it step by step. As usual we start by selecting specific columns from our dataset, including `LOCATION` and a range of columns from `RATING` to `PRICE`.

Next, we organize the data by grouping it based on the `LOCATION` column. This allows us to perform subsequent calculations within each unique location.

Within these grouped sections, we calculate the mean values for each numeric column (from `RATING` to `PRICE`). This step results in summarizing the arithmetic average for each column from `RATING` to `PRICE` and for each distinct location in the dataset

If our the summary calculations are completed, we remove the grouping structure with `ungroup()`, returning the dataset to its original ungrouped state.

Finally, we apply a function, named `normalize()`, to standardize the numeric columns excluding `LOCATION`. This step ensures that these numeric values are uniformly adjusted.

Last but not least let me highlight that you can use this technique also to apply your own function(s) on selected columns, for instance to increase each value by `1`:

```
df %>%
  mutate(across(c(LOCATION, RATING), function(x) x+1))
```

3.3 Business Data Visualization with ggplot2

R has been developed by statisticians and analysts. And therefore, one of its central tasks is to create graphics and visualizations and consequently R has a variety of possibilities and packages to create outstanding graphics. These range from packages for visualizing city maps like `Leaflet`, to using Javascript graphics packages via the wrapper `highcharter` or 3D graphics via `RGL`. And so, it is not surprising that numerous scientific journals, in which visualizations play a central role, create their graphics almost exclusively using R.

In general, it can be said that almost all of the known packages in R rely on two main graphic systems: graphics and grid. Both have their different strengths and weaknesses. And also differ in the way graphics are created and defined. Since many packages are based on one of the two systems, let's take a look at the most important package of each graphics system to better understand how we can create graphics in one way or another.

`Graphics` is the first and therefore oldest graphics system developed for R. It has a few years under its belt but creates simple elegant graphics. It is still the central plotting system of R and is installed and shipped with the basic version of R. It is loaded by default when we start R and is immediately available. It contains many well-known functions like `plot()` which we will look at in detail in a moment. On the other hand, it is a bit outdated and unfortunately does not meet some more modern requirements.

`Grid` is a more modern graphics system. It implements a different approach to creating graphics that is independent of the core graphics package. It is rather generic and for this reason we don't call functions from the grid package directly but use packages that do it for us. The best known of these is the `ggplot2` package. We will take a closer look at this package as well since it seems to become more and more the standard for graphics in analytics.

In summary, R has two main graphic systems in the back and endless packages building on these two systems. Base R methods building on `Graphics` provide most of the basic visualizations, but the configurations options are very limited. The modern system `grid` works differently and gets more and more popular. Hence, I will give you the base R code if it exists but will focus on the `ggplot2` package (and its brothers and sisters) using `grid`.

3.3.1 Basic Plotting with Base R

To visualize the relationship between two numeric variables a simple scatterplot (or just plot) is best practice. You can do that in base R with the command `plot()` and in `ggplot2` with the command `geom_point()`. We will now use our dataset `whisky_collection` and plot the two numeric variables `REVIEWS` and `CRITIQUES` to illustrate a possible relationship between user reviews and professional reviewers. We do this both in base R and `ggplot2` to understand the two graphic systems in detail.

In base R we can create a simple scatterplot with `REVIEWS` and `CRITIQUES` on both axes with:

```
plot(x=whisky_collection$REVIEWS, y=whisky_collection$CRITIQUES)
```

In the base R `plot()` function we always have to specify the x-axis and the y-axis. If we don't do this R will make a guess and the result is most often not what we want. In this case, I decided to put the `REVIEWS` on the x-axis and `CRITIQUES` on the y-axis.

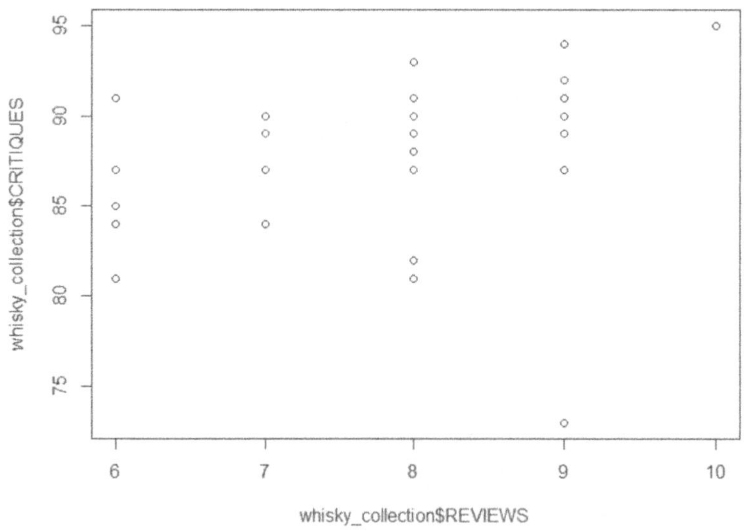

Figure 3-5 Scatterplot to illustrate the relationship between `CRITIQUES` and online `REVIEWS`.

The plot in Figure 3-5 shows that with each professional rating (`CRITIQUES`), the grade given by online customers in their review (`REVIEWS`) also increases. As business data analyst, we assume that there seems to be a linear positive relationship between the online reviews of a whisky and professional ratings. In other words, our plot supports the idea that online users give comparable ratings like whisky experts in a professional tasting.

You decide to communicate the finding to your new colleagues from the *Junglivet Whisky Company*. But before you can show them the plot, we will enrich our plot with some further information. The current axis labels are the name of the variables, that is fine, but the Junglivet management will be probably a bit confused to get cryptic labels like "`whisky_collection$CRITIQUES`" on a visualization. Perhaps they plan to publish this visualization in the company's newsletter, therefore it would be helpful to have some more speaking labels. We also want to adjust the range of the axes and add a more speaking title like "Relationship between critic and user ratings" to our plot.

```
plot(x=whisky_collection$REVIEWS,
     y=whisky_collection$CRITIQUES,
     xlab="average online review rating",
     ylab="average critic score",
     xlim=c(5, 10),
     ylim=c(70, 100),
     main="Relationship between critic and user ratings")
```

If you run this command, you will get a new plot in the plots pane of RStudio, which is illustrated in Figure 3-6. The plot looks now a bit more sophisticated and is ready for your presentation.

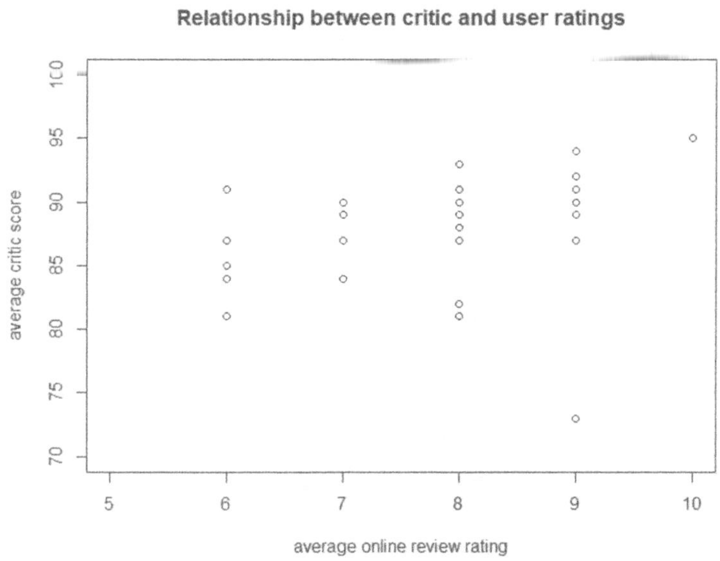

Figure 3-6 Relationship between `CRITIQUES` and user `RATINGs` as scatterplot.

As you have seen, base R allows you to make fast simple plots. However, I have to admit that the structure of the commands may vary from one problem to the next. If you want to make another type of plot you will have to learn a completely new function with different parameters. Do not worry, we will learn them in this chapter but to be fair it is difficult to know what function or parameters have to be varied in order to make changes like the label of the x-axis. Furthermore, it's nearly impossible to change the background to a grid for better readability or create complicated overlays. Thus, let us now make the same plot with `ggplot2` which solves these shortcomings.

3.3.2 Advanced Plotting with ggplot2

The `ggplot2` package is part of the `tidyverse` package (collection), which gives you a lot of possibilities for data visualization. Using `ggplot2`, we can specify different parts of the plot, and combine them together using the "+" operator. The operator has some similarity to the "`%>%`" pipe operator we have met in the previous section.

Illustrated in Figure 3-7, in `ggplot2`, all visualizations follow a consistent underlying grammar of graphics (Wickham, 2010; Wilkinson et al., 2005). That means that they consist of independently specified building blocks we have to combine to create any kind of plot.

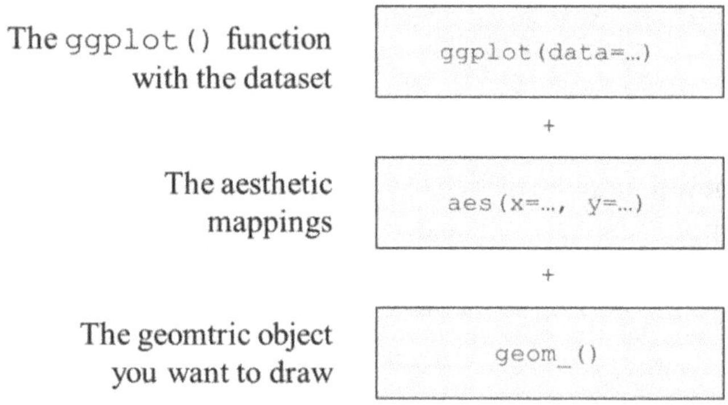

Figure 3-7 The structure of ggplot2 plots consisting of data, aesthetic mapping, and geometric objects.

The most relevant building blocks of a `ggplot2` visualization include:

- **data**: The source and sample of your dataset that you want to visualize
- **aesthetic mappings**: How your data is mapped to the geometric objects like color, x or y values
- **geometric objects**: The graphical visualization of your data like a point, line or area

> You can get an overview of available geometric objects in `ggplot2` running the following code in your R environment:
> `help.search("geom_", package = "ggplot2")`

To generate a plot we have to specify the data and the data source we want to use in this plot with `ggplot(data=...)`. Then in the next building block we specify the aesthetic mapping with `aes()`. Aesthetic mappings describe how variables in the data are mapped to the properties of your geometric objects, which we have to define in the last block. I recommend my students to think of geometric mapping as the act of drawing of a plot and geometric objects as the type of drawing like line or point. In this case we want to draw points with `geom_point()`. This will produce a scatterplot comparable with the one we generated in the step before with base R.

```
library(ggplot2)
ggplot(data = whisky_collection) +
  aes(x = REVIEWS, y = CRITIQUES) +
```

```
  geom_point()
```

In the first step of the code snippet, we load the package `ggplot2`. If the loading fails, you probably didn't install it correctly together with the `tidyverse` package in the previous chapter. In the next step, we specify our plot with the three building blocks data, aesthetics and geometric mapping. Our geometric object is a point, because we want a scatterplot. The properties of a point are x-axis and y-axis. Therefore, we specify that R uses the values from the variable `REVIEWS` as value for the x-axis and the values from `CRITIQUES` as value for the y-axis in the coordinate system. All this is done in the aesthetic mapping `aes()`. The final scatterplot is illustrated in Figure 3-8.

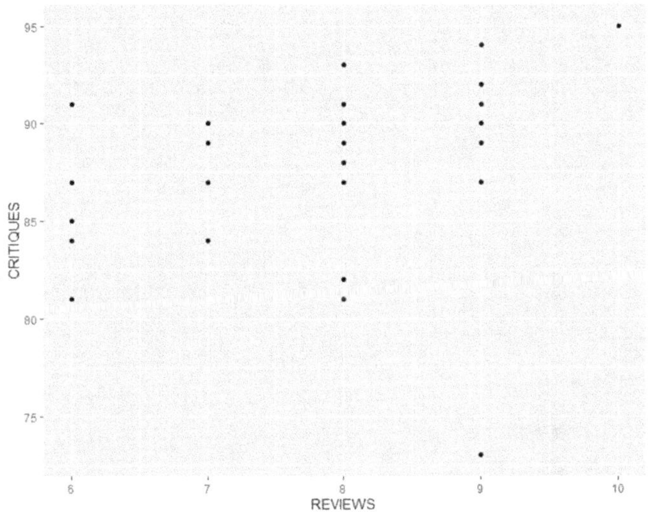

Figure 3-8 The same scatterplot from before with `ggplot2`.

In a next step, we also want to prepare our new plot for presentation at the Junglivet management meeting. Therefore, we have to improve the design and information a bit.

As before, let us add two more concrete labels on the x-axis and the y-axis of our plot. Furthermore, we can highlight the country of each whisky by using the `LOCATION` as a color parameter. We also set a new design which is more suitable to the presentation we plan:

```
ggplot(data = whisky_collection) +
  aes(x = REVIEWS, y = CRITIQUES, color=LOCATION) +
  geom_point(size=3) +
  theme_bw() +
  labs(x = "Average online review rating",
       y = "Average critic score") +
  ggtitle("Relationship between critic and user ratings")
```

If you read the code above carefully, you see some comparable lines of code like in the base R plot. But also, there are some differences and new code elements like `theme_bw()`. Instead of specifying many parameters in one big `plot()` function, we add design specifications step by step by adding one building block after another. The result of the code is illustrated in Figure 3-9.

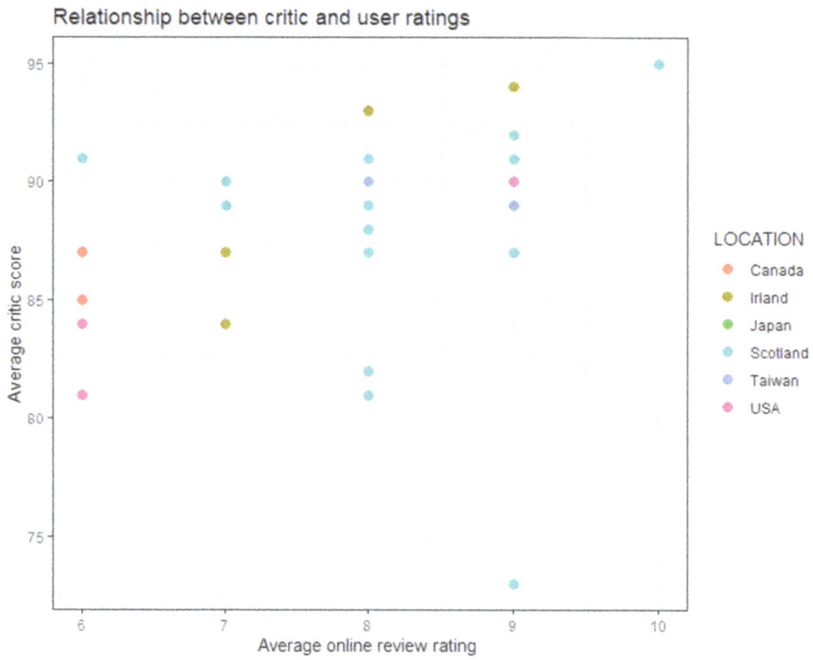

Figure 3-9 Enhanced `ggplot2` plot with further characteristics like legend or colors.

The plot in Figure 3-9 has now many differences compared to Figure 3-8. Its design is clearer, the data points are bigger and we even were able to add further information (`LOCATION`) to our plot, which is pretty awesome! Most of the changes are done by further building blocks. In `ggplot2` exists many such further building blocks that are made to enhance your plot. The most relevant are:

- **coordinate systems**: adjust or choose a specific coordinate system
- **faceting**: create separate subplots for subsets of data
- **position adjustments**: apply minor tweaks to the position of elements within a layer
- **scales**: further specify how the geometric mapping should happen
- **statistical transformations**: give further statistical specifications
- **themes**: if you want to change the general design based on different design themes

The default coordinate system is the grey colored one like in Figure 3-8. Then, if you want to change it or choose between many other styles and backgrounds. For instance, you can change to a map system to map geographical projects with `coord_map()` or to a cartesian coordinate system with `coord_cartesian()`.

Sometimes you want to compare a categorial variable and their numeric values. Or you want to generate different plots based on one column in your dataset. Facet provide the possibility to do so. You can define a subset as the levels of a single grouping variable with `facet_wrap()` or a subset as the crossing of two grouping variables with `facet_grid()`.

All layers have automatic position adjustments that resolve overlapping. You can change that default by using position adjustments to apply minor tweaks to the position of elements within a layer. For instance, you can dodge overlapping objects side-to-side with position_dodge() or add a small jitter with position_jitter().

When mapping a variable to a shape with aes(shape = TYPE), we only tell ggplot2 to use the different values of the variable TYPE ,to select a shape for each data point. But we can not specify how that should happen, like which shape should be used. If we want to specify what color, shape, size, etc. should be used, we have to modify the corresponding scale. We can do that with a series of functions of the format scale_<aesthetic>_<type>. If you want to change the shape, you can do that with e.g. scale_shape_manual(values=c(3, 16, 17)), which selects the icons with the ids 3, 16 and 17 as a scale.

Scatterplots and most simple plot types do not require statistical transformations. Each point is plotted at x and y coordinates equal to the value in the dataset. But other plots, such as boxplots or histograms require statistical transformations. For instance, you can change the size of a bin in a histogram to 100 with geom_histogram(stat="bin", binwidth=100). We will come back to that in the section about histograms, later in this book.

3.3.3 Methods of Data Visualization in R

Data visualization is the generation of plots and visuals that represents a specific view or underline an insight you found in your data. Therefore, data visualizations have to be as simple as possible without simplifying the key messages of your analysis. A good visualization might also guide your further analysis and show you key aspects of your data.

Visualizations are generated easily in R, but it's very hard to produce usable ones. And furthermore, there exists a panoply of different visualizations. How to select the right one for your use case?

For that purpose, I made a simple decision diagram for you to select the best visualization for your business data. The diagram summarized in Figure 3-10 is a recommendation from my experience and probably you will develop your own preferences in the future. Just use it as a starting point for your business data understanding, if you are new to business data analytics.

You do not have to try to make each of the graphics by your own straight away. I will show you how to make these different visualizations as well with base R, if you want a quick solution and how to make them with ggplot2, if you want a more sophisticated, nice-looking plot.

My recommendation here is that you go fast through this chapter and each visualization type with the overview in Figure 3-10 in mind. Get a rough understanding and overview of the different types but make no deep dive. Better go back to the example from the introduction and try to visualize the dataset by your own using some of the visualizations of the following chapter you are interested in. Then you can continue with the next chapters. For future projects like the upcoming case studies, you can always use the decision tree in Figure 3-10 to decide which visualizations fits best. This is the most suitable way to learn how to implement all these plots in R.

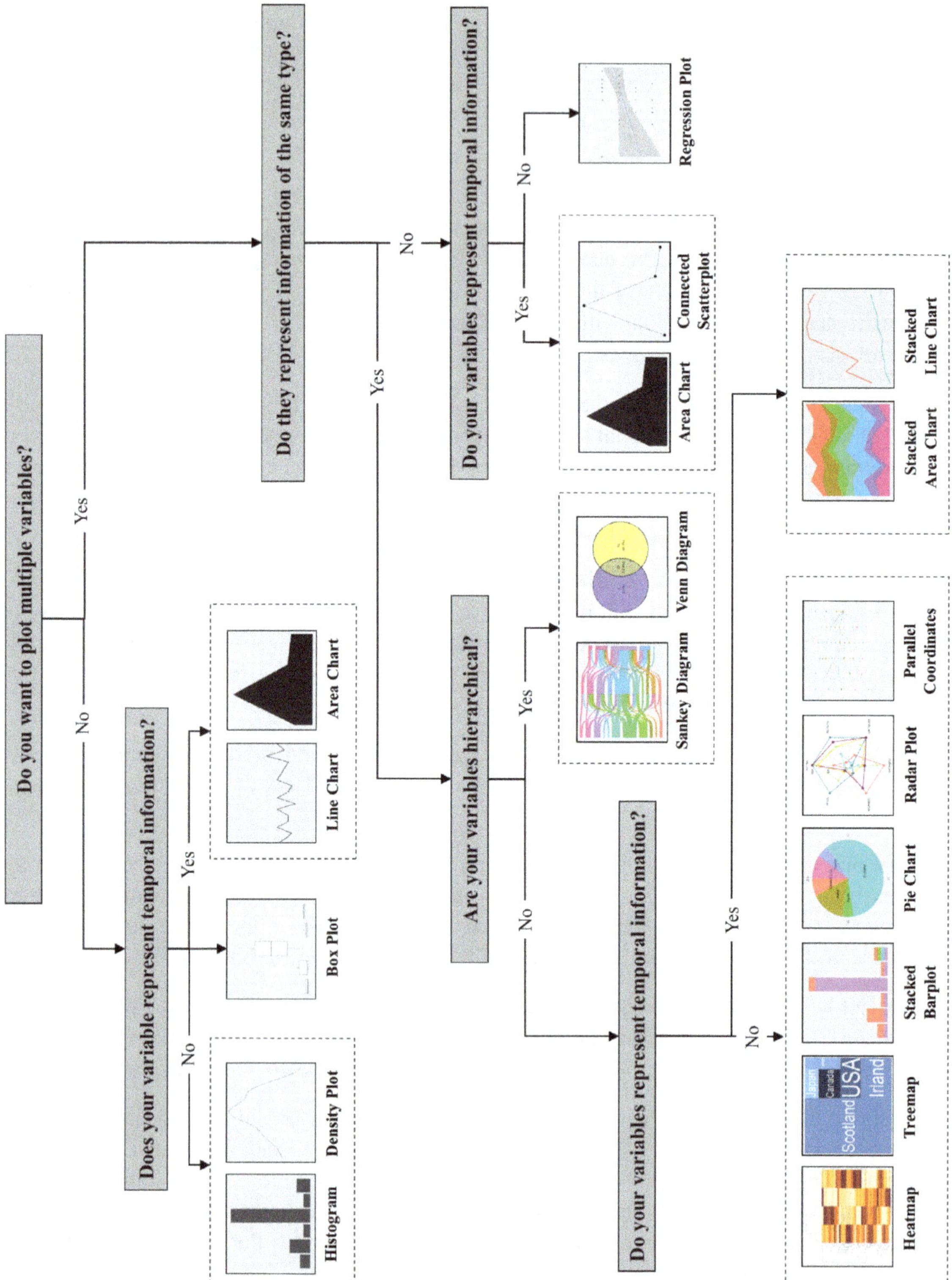

Figure 3-10 Decision tree with the most relevant graphic types in business data analytics.

Histogram

Histograms are a special type of bar plots, a kind of visualization with bars representing categorial values. While bar plots are often used to visualize the frequencies of categorial values, a histogram can be used to plot the frequency distribution of numerical variables. You can think of histograms as graphical representations of the frequency distribution of variables that are not ordered like classes or categories in bar plots.

In most cases you use numeric variables in your histograms. Thus, histograms classify data into different bins based on different ranges and count how many data points belong to those classes. The data is then plotted against the number of points in each class. Hence histograms show a frequency on the y-axis and the different classes on the x-axis. For that purpose, the range of the numerical variable is discretized into fixed bins. If the bins or intervals are of equal length, business analysts speak of equal-length bins.

A histogram is often used to conveniently illustrate the main features of data distributions. It is also a useful tool when you are dealing with data sets of more than 30 rows. For business analysts it can help to identify unusual observations (outliers) or gaps in the business data.

Let us now make a simple histogram of the expert ratings (CRITIQUES) in our whisky_collection with base R. This can be helpful to better understand the quality of the whisky collection in general.

```
hist(whisky_collection$CRITIQUES,
     xlab="Rating from critics",
     main="Whisky ratings from critics")
```

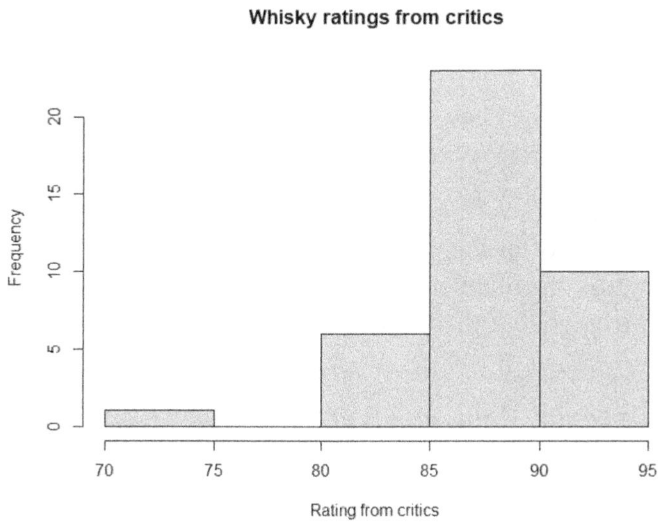

Figure 3-11 Whisky RATINGS from CRITIQUES as histogram with base R.

The plot immediately shows that most whisky ratings are between 86 and 90 points inclusive. About 25 whiskies in the collection are in this range, and another 10 have a better rating of 91 or more points. Thus, it can be concluded that the collection is really very exquisite.

In ggplot2, we can make a histogram with:

```
> ggplot(data = whisky_collection) +
> +    aes(x=CRITIQUES) +
> +    geom_histogram()
`stat_bin()` using `bins = 30`. Pick better value with `binwidth`.
```

The final histogram in Figure 3-12 is comparable to the base R version but has more bins. It also uses the label "count" instead of "Frequency" on the y-axis. Nevertheless, the message is the same and the plots are directly comparable.

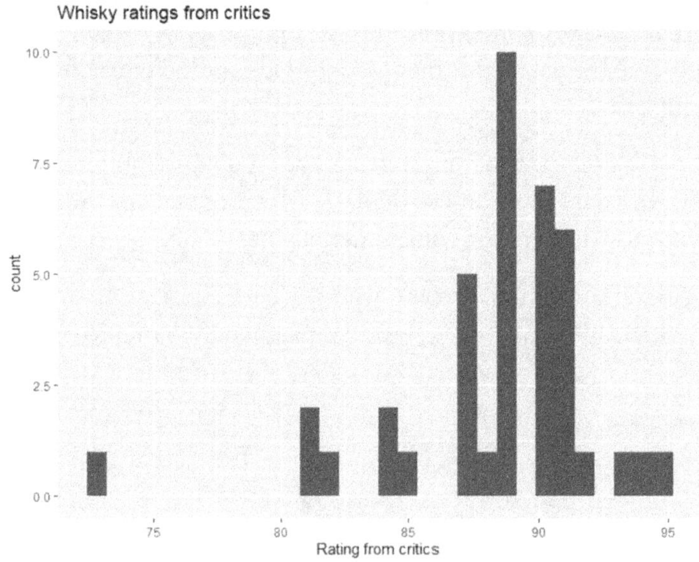

Figure 3-12 Whisky RATINGs from CRITIQUES as histogram with ggplot2.

You might already have noticed that if you run the code, R will show you a comment: `Pick better value with `binwidth``. This is a friendly reminder from `ggplot2` that you can change the bin size with the statistical transformation if you are unhappy with it.

But how to choose the right number of bins? So far there is no silver bullet to find the right number of bins. R will automatically determine the number of bins for your histogram based on Sturge's rule (Sturges, 1926), which says you choose the number of bins based on the size of your dataset n as:

$$Number\ of\ bins = \lceil log_2(n) + 1 \rceil$$

If you want to specify the number of bins manually you can do that using the `breaks` parameter of the `hist()` function or in `ggplot2` with a statistical transformation by setting a new `binwidth` in `geom_histogram()`. But do not forget that in this case the height of bins do no longer correspond to the relative frequencies. Because the area of the box of the bin are chosen such that they are proportional to the relative frequency.

3.3.4 Density Plot

A kind of expansion of the histogram or the scatterplot we build in the introduction of this section are density plots. They are another way to work with many datapoints but try to show more information by using transparency and overlays. Each data point is plotted transparent and the more points are on the same place the more colorful the data points become. This kind of "histograms" are termed density plots, because they represent the position and the density of your data points.

To illustrate the density plots, I suggest that we plot now the distribution of the online ratings in the different whisky shops (`REVIEWS`). The rating goes from `0` to `10` points, which stands for a five-star scale you know from social media websites. One point in `REVIEWS` equals ½ star, e.g., `4.5` stars would be equal to `9` points. For you and the Junglivet marketing team it would be interesting to see, if the users use the full range of ratings or if they gather around an average rating. A density plot is a very useful tool to illustrate this.

In base R we first have to make a trick and compute the density of our dataset, then we can build the density plot with:

```
foundation_density = density(whisky_collection$REVIEWS)
plot(foundation_density,
     xlab="Online reviews (0-10 points)",
     main = "User rating distribution in whisky shops")
```

In `ggplot2` this is straight forward possible with:

```
ggplot(data = whisky_collection) +
  aes(x=REVIEWS) +
  geom_density() +
  labs(x = "Online reviews (0-10 points)") +
  ggtitle("User rating distribution in whisky shops")
```

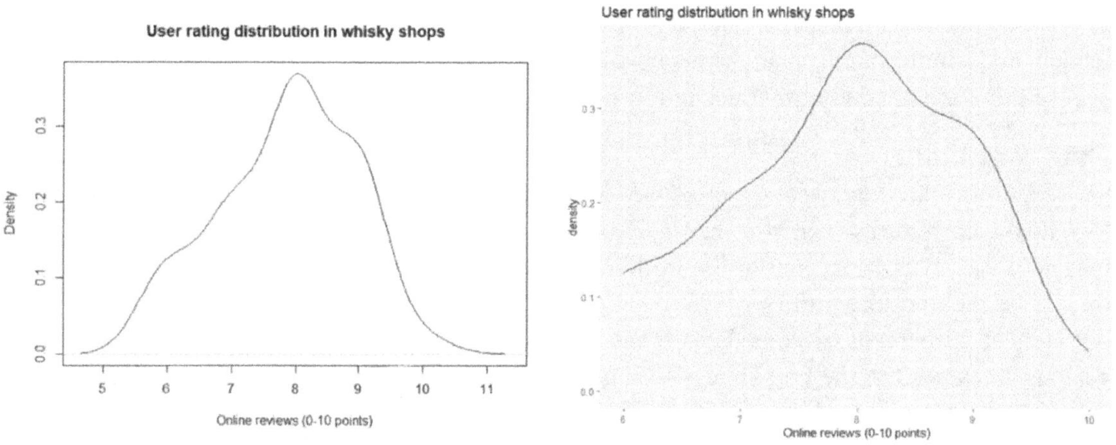

Figure 3-13 Density plot of distillery `FOUNDATION`s with base R and `ggplot2`.

A very useful expansion of these plots would be to add the whisky (`TYPE`) as a further variable.

```
ggplot(data = whisky_collection) +
  aes(x=REVIEWS, fill=TYPE) +
  geom_density(alpha=0.4) +
  labs(x = "Online reviews (0-10 points)") +
  ggtitle("User ratings by whisky type")
```

The result is the plot in Figure 3-14 and shows that blended and single malt whiskies are rated differently.

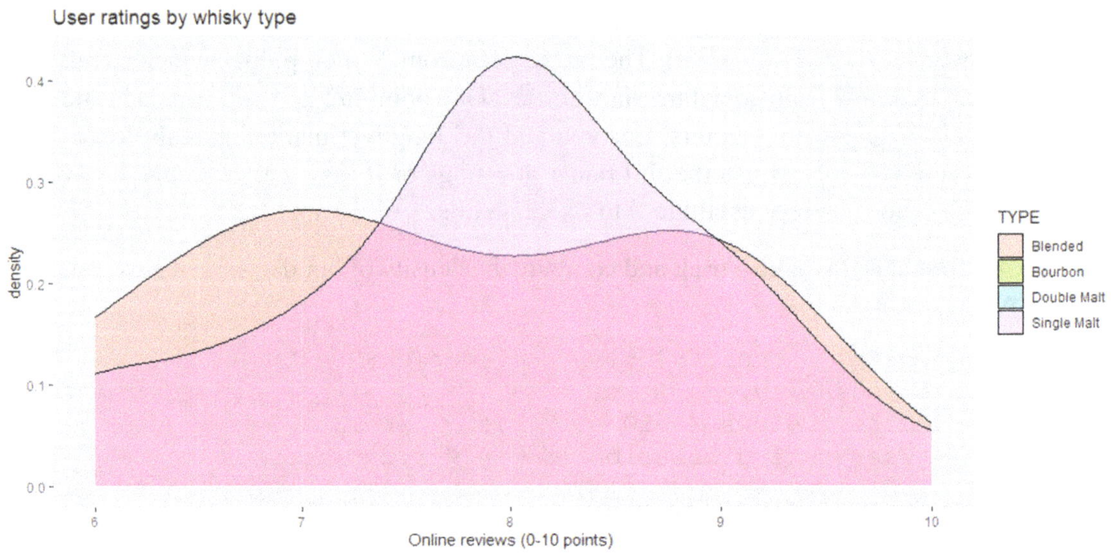

Figure 3-14 A density plot of user `RATING`*s by whisky type with* `ggplot2`.

For single malt whiskies, you can see a cluster of many ratings around `8` points or `4` stars. However, this then drops again as quickly as it has risen. With blended whiskies, on the other hand, there is a more or less constant plateau from `7` points or `3.5` stars to `9` or `4.5` stars. Thus, the blended whiskies tend to be rated worse by the users. While the best whiskies are equal blended and single malts. The whiskies of type `Bourbon` and `Double Malt` are dropped because there are not enough values in our dataset to compute a density plot for them.

3.3.5 Box Plot

One very informative way to visualize numerical data are boxplot. We already met this type of plots in the introduction of this book. Boxplots are very good to visualize and summarize numerical variables. The line in the middle of the boxplot is the median of your dataset. The box itself envelopes the middle 50 % of your data, marked by the top and bottom border of the box. Hence, 25 percent of your dataset has values below and 25 % has values above the box. The long lines besides the box are called whiskers. The maximum length of each whisker is 1.5 times the length of your box (interquartile range). If there is no data point at this value, the whisker gets shortened to the next available data point. We will call observations besides the two lines outliers. In the boxplot they are visualized as small circles.

A simple boxplot is generated with:

```
boxplot(PRICE ~ LOCATION, data = whisky_collection)
```

And in `ggplot2` with:

```
ggplot(data = whisky_collection) +
  aes(x=LOCATION, y=PRICE) +
  geom_boxplot()
```

The final plots are illustrated in Figure 3-15.

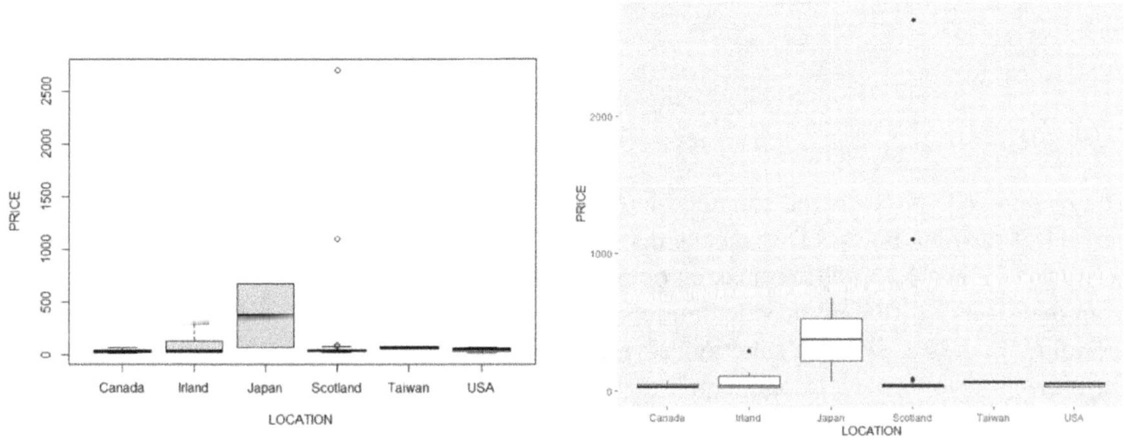

Figure 3-15 Boxplots illustrating the different `PRICE` levels of whisky distilleries by `LOCATION`.

The nice thing about our boxplots is that we can see directly in which countries we have outliers. In our case there are two outliers in Scotland above `1000` € and one in Ireland. A business data analyst would take a closer look at these three whiskies in a special analysis. Often there are measurement errors or the data points are very unique and require a separate analysis.

In addition, if we want to create a plot to generally learn about the distribution and not to look for outliers, the outliers in Scotland are a bit clumsy. They massively distort our y-axis and ensure that we can no longer accurately read the other boxplots of the other countries. We now remove them before we interpret the boxplots further!

In base R we rerun our boxplot again as follows:

```
boxplot(PRICE ~ LOCATION, data = whisky_collection, outline=FALSE)
```

And in `ggplot2` with:

```
ggplot(data = whisky_collection) +
  aes(x=LOCATION, y=PRICE) +
  geom_boxplot(outlier.shape = NA) +
  coord_cartesian(ylim =  c(0, 750))
```

We get new boxplots that allow us now to further analyze the prices of the whiskies in more detail.

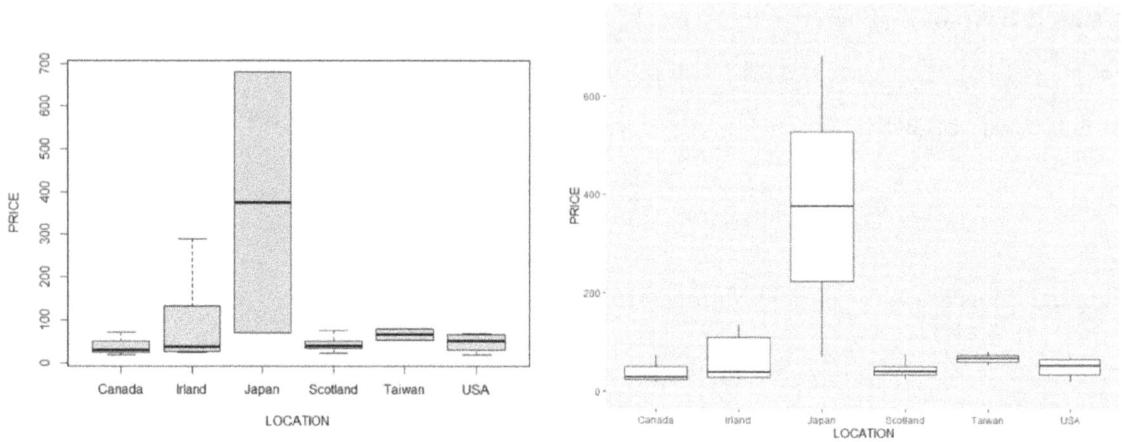

Figure 3-16 The previous boxplots without outliers in base R and `ggplot2`.

The range of the boxes is the so-called interquartile range and describes the value range of 50 percent of the data points. That means that half of our whiskies in Japan have roughly a value between `100` and `700`. Differences in boxplot appearance between `ggplot2` and base R stem from their use of different quartile definitions and outlier treatments. The plot in base R equals the values in the `summary()` function. If you want the same results in `ggplot2` you have to use the `coord_cartesian()` command.

Furthermore, we see the long lines beside the box (so called whiskers). They contain all values that are below 2.5 standard deviations away from the 25% and 75% quantiles (or box). Values besides this area, are termed outliers in analytics. However, there is a very vigorous debate between data scientist when or if a value is an outlier, and in particular how to measure it. But using boxplot and standard deviations to detect candidates that might be outliers is a simple but not very sophisticated way.

3.3.6 Line Chart and Stacked Line Chart

A line chart is a way of plotting data points on a line. Often, it is used to show trends or developments or the comparison of two datasets.

To illustrate line charts in base R and `ggplot2` let us compare the development of whisky foundations per century. It would be interesting to see, if there are epochs where many distilleries were founded (and which type of whisky they produce and where they come from).

For that purpose, we have to compute the number of whisky distillery foundations by century.

We can do that with:

```
> distillery_foundations = whisky_collection %>%
+   select(DISTILLERY, LOCATION, FOUNDATION) %>%
+   mutate(YEAR = as.numeric(substr(FOUNDATION, 1, 2))) %>%
+   group_by(YEAR) %>%
+   summarize(NUM = n())
```

And show the result with:

```
> distillery_foundations
# A tibble: 4 x 2
   YEAR   NUM
  <dbl> <int>
1    17     5
2    18    22
3    19     7
4    20     6
```

The new data frame `distillery_foundations` represents the number of distillery foundations per century. A simple line chart is obtained by setting the type parameter in the base R plot function as "l", which stands for line. Furthermore, we have to specify the x-axis to start at `16` and end at `20`. If we do not specify the new axis explicitly, we will get decimal points on our x-axis. Base R cannot handle our new numeric values and will add `16, 16.5, 17, 17.5,` etc. on the x-axis.

```
plot(distillery_foundations$YEAR,
     distillery_foundations$NUM,
     type="l",
     xaxp = c(16, 20, 4),
     xlab="Century",
     ylab="Number of distilleries",
     main = "Distillery foundations across the last centuries")
```

The line chart can be done easily in `ggplot2` in the following manner:

```
ggplot(data=distillery_foundations) +
  aes(x=YEAR, y=NUM) +
  geom_line() +
  geom_point(size=5) +
  labs(x="Century", y="Number of distilleries") +
  ggtitle("Distillery foundations across the last centuries")
```

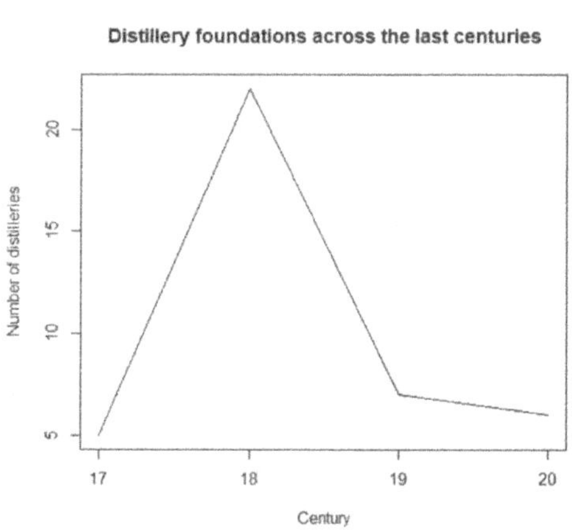

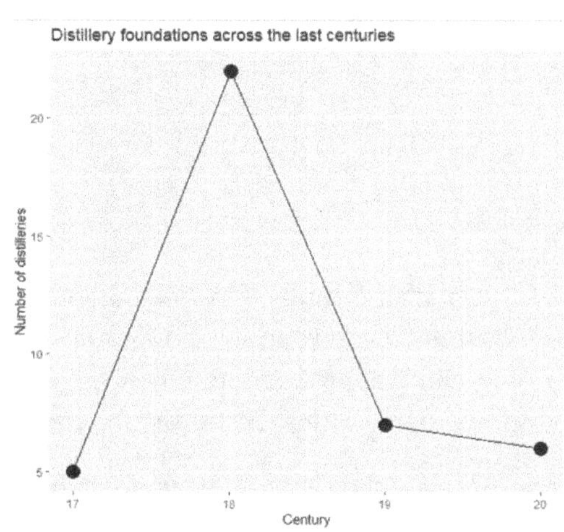

Figure 3-17 Line charts to illustrate the distillery `FOUNDATIONs` *across the centuries.*

As you can see I added a second geometric mapping `geom_point()`. This is not necessary in a classical line chart, but I prefer to see the different values of each year. If you want a line chart without the points just remove the second geometric mapping.

In some scenarios you might be interested to visualize multiple line charts in one chart. For instance, to represent the different foundations by whisky type. The following code snipped illustrates how to make a stacked line chart representing the distillery foundations per type.

```
distillery_foundations = whisky_collection %>%
  select(DISTILLERY, TYPE, FOUNDATION) %>%
  mutate(YEAR = as.numeric(substr(FOUNDATION, 1, 2))) %>%
  group_by(YEAR, TYPE) %>%
  summarize(NUM = n())

ggplot(data=distillery_foundations) +
  aes(x=YEAR, y=NUM, color=TYPE) +
  geom_line() +
  labs(x="Century", y="Number of distilleries") +
  ggtitle("Distillery foundations by type across the last centuries")
```

The resulting Figure 3-18 illustrates the foundations of distilleries focusing on `Blended` or `Single Malt` whisky. The other types `Double Malt` and `Bourbon` are not shown, because they have not enough data points per century (as before in Figure 3-14).

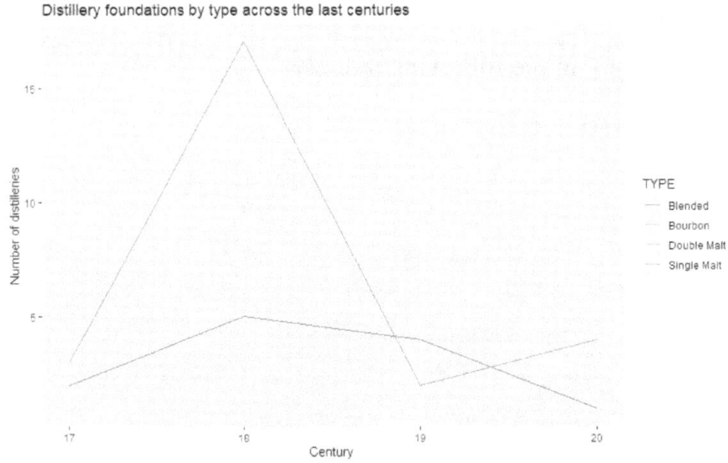

Figure 3-18 Stacked line chart to represent the type of whisky by each distillery that has been founded.

3.3.7 Area Chart and Stacked Area Chart

An area chart represents the evolution of a numeric variable. It is very close to a line chart. As a rule of thumb, you can use area charts when the magnitude of a trend is to be communicated rather than individual data values like in line charts.

In our case we want to use the same scenario from before and plot the number of whisky distillery foundations by year.

Base R allows to build area charts thanks to the `polygon()` function. The function requires the same two inputs as the plot function: x and y.

```
plot(distillery_foundations$YEAR,
     distillery_foundations$NUM,
     type="o",
     xaxp = c(16, 20, 4),
     xlab="Century",
     ylab="Number of distilleries",
     main = "Distillery foundations across the last centuries")

vertice_x=c(min(distillery_foundations$YEAR),
            distillery_foundations$YEAR,
            max(distillery_foundations$YEAR))
vertice_y=c(min(distillery_foundations$NUM),
             distillery_foundations$NUM,
             min(distillery_foundations$NUM))
polygon(vertice_x,
        vertice_y,
        col=rgb(0.3,0.3,0.3,0.3),
        border=F)
```

As you can see the computation is a bit complex. In a first step we have to define the polygon by setting the different borders. We use the min and max values for that. Then we combine them in the `polygon()` function together. In `ggplot2`, the procedure is straight forward:

```
ggplot(data=distillery_foundations) +
  aes(x=YEAR, y=NUM) +
  geom_area() +
  labs(x="Century", y="Number of distilleries") +
  ggtitle("Distillery foundations by type across the last centuries")
```

The results are illustrated in the following Figure 3-19.

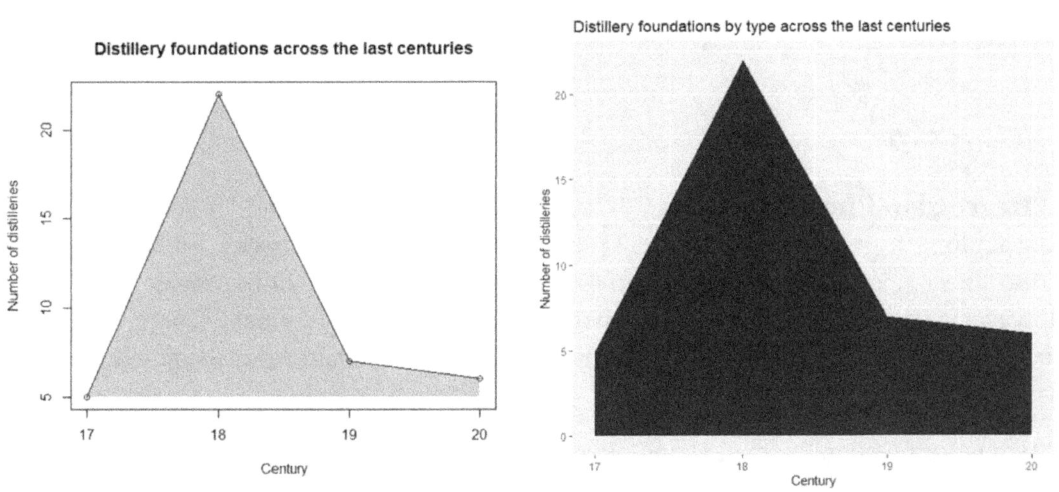

Figure 3-19 Area charts with base R and `ggplot2` to illustrate the distillery `FOUNDATION`s across the centuries.

In a next step it might be interesting to compute the whisky distillery foundations across the world. It would help to compare the different countries and developments. For that purpose, we need a stacked area chart. We can do that in `ggplot2` in the following manner:

```
distillery_foundations = whisky_collection %>%
  select(DISTILLERY, TYPE, FOUNDATION) %>%
  mutate(YEAR = as.numeric(substr(FOUNDATION, 1, 2))) %>%
  group_by(YEAR, TYPE) %>%
  summarize(NUM = n())

ggplot(data=distillery_foundations) +
  aes(x=YEAR, y=NUM, fill=TYPE) +
  geom_area(alpha=0.4) +
  labs(x="Century", y="Number of distilleries") +
  ggtitle("Distillery foundations by type across the last centuries")
```

The result is the plot in Figure 3-20 that shows the different type of whisky and the related distillery and their foundation century.

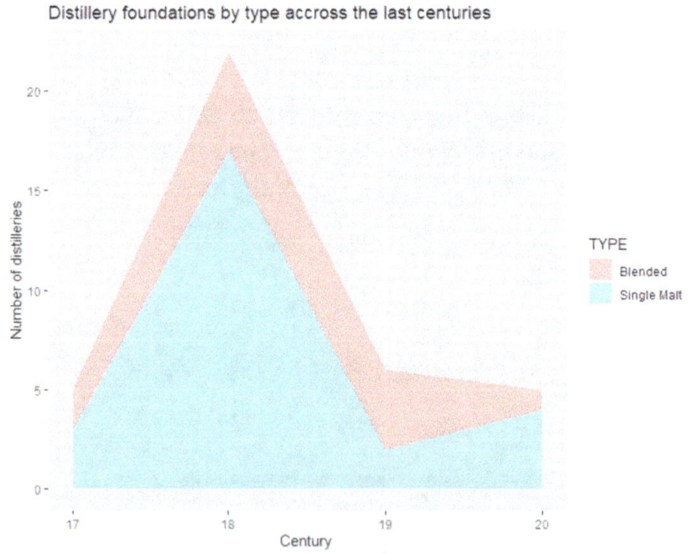

Figure 3-20 Area chart to represent the type of whisky by each `DISTILLERY` that has been founded.

3.3.8 Regression Plot

Regression plots, as the name suggests, plot a regression line between `two` parameters, visualizing linear relationships. In most cases, the term regression will not mean anything to you. However, this is not a big deal, as we will look at this topic in the next chapter. For now, you can think of a regression as a kind of model that predicts a certain value using other values.

As an example, we now want to use the expert ratings (`CRITIQUES`) to predict how well a whisky will be received in the market and rated by the users in the online shops (`REVIEWS`). We can do this using a regression, which we can then display directly in our plot.

In base R, we first need to create a simple plot. We then build in the prediction of the regression in a second step by adding a line using `abline()`.

```
plot(whisky_collection$CRITIQUES,
     whisky_collection$REVIEWS,
     pch=19,
     frame=FALSE,
     xlab="Expert rating",
     ylab="User rating",
     main = "Prediction of the consumer rating using expert ratings")
abline(lm(REVIEWS ~ CRITIQUES, data=whisky_collection), col = "grey")
```

In `ggplot2` it can be done in the following manner:

```
library(ggplot2)
 ggplot(data,aes(x=CPI,y=HDI))+
   geom_point()+
   geom_smooth(method = "lm")+
   geom_line(method="loess")
```

The resulting plots are illustrated in Figure 3-21.

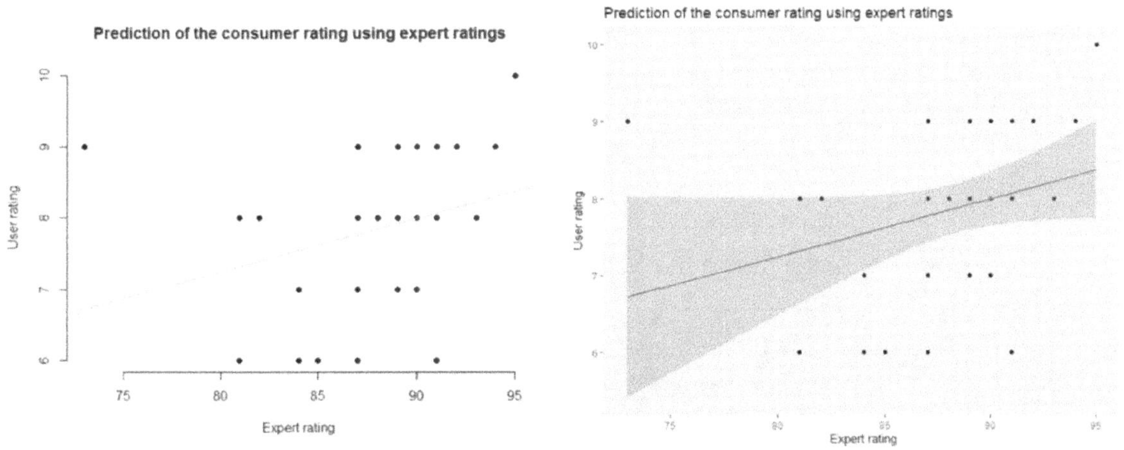

Figure 3-21 Regression plots to illustrate the model to predict user ratings with expert ratings.

3.3.9 Sankey Diagram

A Sankey diagram is like a flowchart that shows how things move from one place to another. It's often used to visualize the flow of energy, money, or materials in a system. The width of the arrows in the diagram represents the amount being transferred, making it easy to see where most of the flow is going.

At the moment there is no direct support of Sankey diagrams in base R. But together with `ggplot2` there is a package `ggsankey` that allows us to build Sankey diagrams. This package can be installed from GitHub with:

```
devtools::install_github("davidsjoberg/ggsankey")
```

To plot a Sankey diagram with `ggsankey` each observation has a stage (called a discrete x-value in ggplot) and can be part of a node. Furthermore, each observation needs to have instructions of which node it will belong to in the next stage. Hence, to use `geom_sankey` the `aestethics` `x`, `next_x`, `node` and `next_node` are required. The last stage should point to `NA`. This is all very complicated and difficult to build such a dataset by your own. But hey, good news! You can transform a tidy dataset easily with the `make_long()` function from the `ggsankey` package that only needs your variable sample as parameters.

```
whisky_collection_sankey = whisky_collection %>%
  make_long(..)
```

As an example, let us plot the relationship between the taste of a whisky like `SMOKENESS` and `RICHNESS` and their influence on the final `RATING` in a tasting. We can do that in the following manner:

```
df = whisky_collection %>%
  make_long(SMOKNESS, RICHNESS, RATING)

ggplot(df) +
  aes(x=x,
      next_x=next_x,
      node=node,
      next_node=next_node,
      fill=factor(node)) +
  geom_sankey()
```

Figure 3-22 Simple Sankey diagram illustrating the different tastes of whisky.

This Sankey is beautiful, but for the moment not quite useful. Let us now use all our `ggplot2` knowledge and build a real Sankey with labels and multiple variables to illustrate the flow.

```
df = whisky_collection %>%
  make_long(SMOKNESS, RICHNESS, RATING)

ggplot(df) +
  aes(x=x,
      next_x=next_x,
      node=node,
      next_node=next_node,
      fill=factor(node),
      label=node) +
  geom_sankey(flow.alpha=.6,
              node.color="gray") +
  geom_sankey_label(size=3, color="white", fill="black") +
  scale_fill_viridis_d() +
  theme_sankey(base_size=18) +
  labs(x=NULL) +
  theme_bw() +
  theme(legend.position="none",
        plot.title=element_text(hjust=.5)) +
  ggtitle("Whisky Quality Sankey")
```

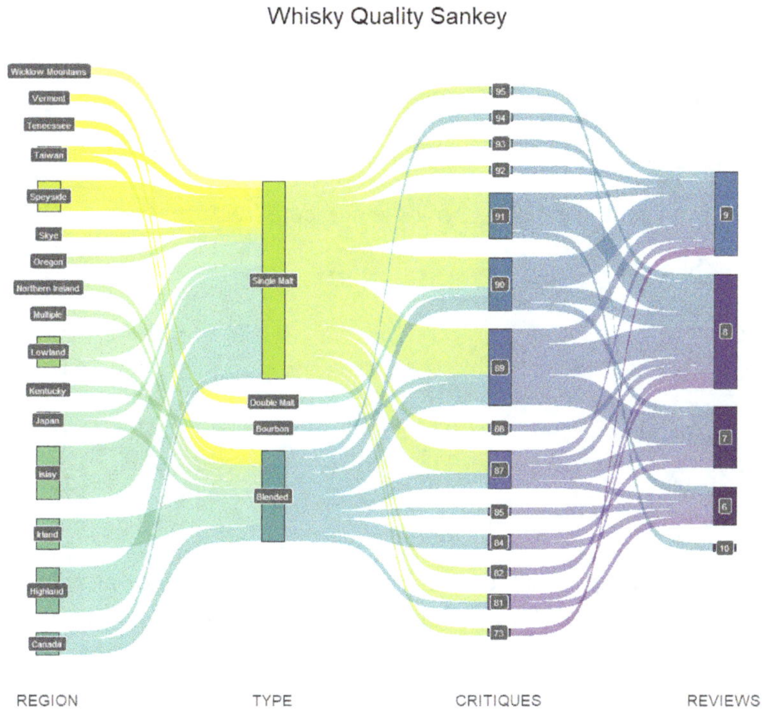

Figure 3-23 Whisky quality Sankey diagram with four features to illustrate relationships across.

3.3.10 Venn Diagram

A Venn diagram uses overlapping circles or other shapes to represent the logical relationship between two or more sets. They are mainly used in statistics but also in management theory. The applications for business data analytics are therefore diverse.

To illustrate the characteristics of the whiskies from our whisky collection we could use a Venn diagram. So far, there is no way to do so in base R, hence we will use `ggplot2` for that. To generate a Venn diagram, we will use `ggvenn`. As input, it needs a named list of integer vectors representing dummies of the values of our variable we want to plot, or a data frame consisting of logical dummy values.

Dummy (or binary) variables are commonly used in analysis and statistics. A dummy column is a column that has a value of `TRUE` or `1`, if a categorical event or value occurs and a value of `FALSE` or `0` if it does not. In most cases, a dummy variable is a characteristic of the event or object being described. For example, if the dummy variable represents a Scottish whisky, the dummy variable gets a value of `TRUE` if it is a Scottish whisky, if it is not, it gets a value of `FALSE`.

In a first step, we have to decide in which intersections of variable values we are interested in. As a possible example we can investigate the number of single malt whiskies and if they have won a critic's award (`RATING >= 90`). We can compute those dummy variables easily with `dplyr` from our `whisky_collection`.

```
whisky_characteristics = whisky_collection %>%
  filter(LOCATION=="Scotland") %>%
  select(NAME, TYPE, CRITIQUES) %>%
  mutate(SINGLE_MALT=ifelse(TYPE=="Single Malt", TRUE, FALSE)) %>%
  mutate(BLENDED=ifelse(TYPE=="Blended", TRUE, FALSE)) %>%
  mutate(AWARD=ifelse(CRITIQUES >= 90, TRUE, FALSE)) %>%
  select(AWARD, SINGLE_MALT)
```

Let us now make a check if our data frame has now the correct structure with:

```
> head(whisky_characteristics)
# A tibble: 23 x 6
   NAME         TYPE         CRITIQUES SINGLE_MALT BLENDED AWARD
   <chr>        <chr>            <dbl> <lgl>       <lgl>   <lgl>
 1 An Cnoc      Single Malt         90 TRUE        FALSE   TRUE
 2 Ardbeg       Single Malt         91 TRUE        FALSE   TRUE
 3 Auchentoshan Single Malt         91 TRUE        FALSE   TRUE
 4 Ballantine's Blended             81 FALSE       TRUE    FALSE
 5 Bowmore      Single Malt         88 TRUE        FALSE   FALSE
```

As you can see, we have now multiple new dummy columns. Each categorial value of our old variable `TYPE` has now become a new column with its value in the name and binary values.

To answer our question with a plot, let us now build a Venn diagram. This can then be done easily with:

```
whisky_characteristics %>%
  select(AWARD, SINGLE_MALT) %>%
  ggvenn()
```

Resulting in the plot in Figure 3-24:

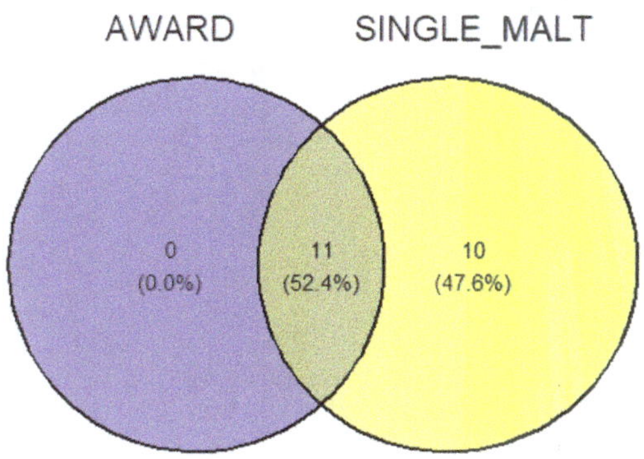

Figure 3-24 Venn diagram illustrating the proportion of awarded whiskies that are single malt.

At this point I would also like to point out the really very helpful package `fastdummies`. If you have the need to transform your categorial variables in a tidy dataset to dummy variables you can generate dummies also very easily with the `dummy_cols()` function from `fastDummies`, which is illustrated in the following code example:

```
whisky_characteristics = whisky_collection %>%
  filter(LOCATION=="Scotland") %>%
  select(NAME, TYPE) %>%
  dummy_cols(select_columns=c("TYPE"))
```

3.3.11 Heatmap

A heat map is a two-dimensional representation of data in which values are represented by colors. It therefore displays values for a main variable of interest across two axis variables as a grid of colored squares. The axis variables are divided into ranges, as in a bar chart or histogram, and the color of each cell indicates the value of the main variable in the corresponding cell range.

A very popular example of a head map is the contributions overview in GitHub. It represents how many contributions have been submitted in the last months.

Figure 3-25 The contribution overview on my private GitHub profile https://github.com/dominikjung42.

To generate basic heat maps you can use `heatmap()` in base R. This function needs a matrix as data input and the different features have to be scaled to have the same range. Otherwise, the heat map will not work correctly. Furthermore, we have to set explicit row names to have the names of our whisky on the y-axis instead of id numbers. The related code is:

```
whisky_collection_matrix = whisky_collection %>%
  select(RATING, REVIEWS, CRITIQUES, PRICE) %>%
  as.matrix() %>%
  scale()

row.names(whisky_collection_matrix) = whisky_collection$NAME

heatmap(whisky_collection_matrix)
```

The result is illustrated in Figure 3-26.

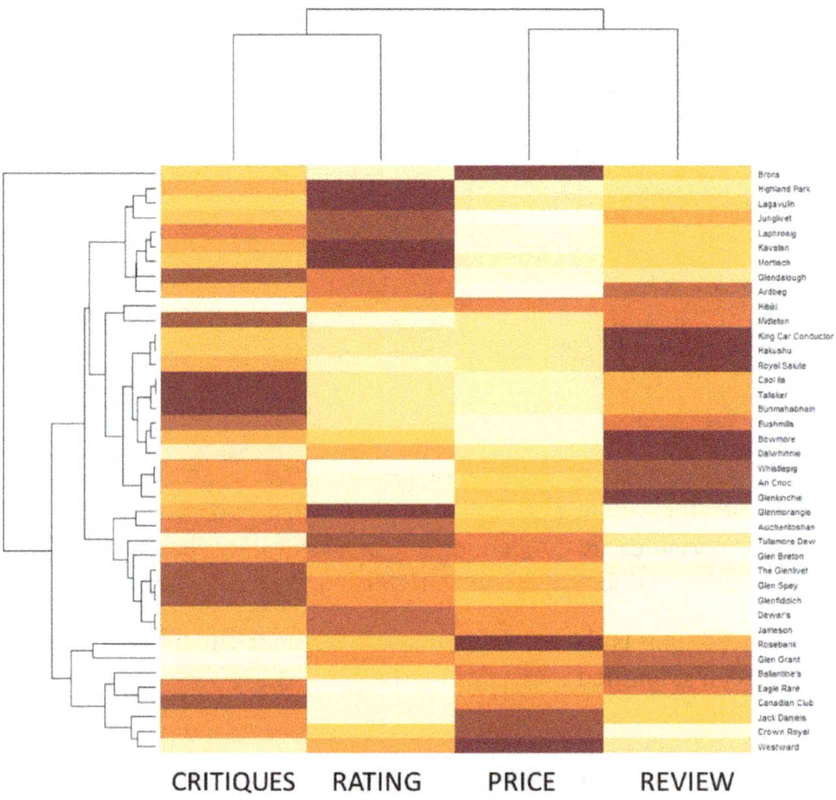

Figure 3-26 Heat map with base R illustrating key characteristics of our `whisky_collection`.

However, the base R heatmap function is a bit tricky and has some graphical issues. You can have a look at it, but I strongly recommend to use the `ggplot2` heat map in this case. In the next step we implement our own heat map in `ggplot2` and discuss this in detail.

The most basic heatmap you can build with R and `ggplot2` is using the `geom_tile()` function. For instance, we could be interested in the number of distillery foundations and the type of whisky they produce across the centuries. A heat map would help us to highlight relevant epochs in a simple plot.

```
distillery_foundations = whisky_collection %>%
  filter(TYPE %in% c("Single Malt", "Blended")) %>%
  select(DISTILLERY, TYPE, FOUNDATION) %>%
  mutate(YEAR = as.numeric(substr(FOUNDATION, 1, 2))) %>%
  group_by(YEAR, TYPE) %>%
  summarize(NUM = n())
distillery_foundations

ggplot(distillery_foundations) +
  aes(x=YEAR, y=TYPE, fill=NUM) +
  geom_tile() +
  labs(x="Expert rating", y="User rating") +
  ggtitle("Prediction of the consumer rating using expert ratings")
```

The resulting plot is illustrated in Figure 3-27:

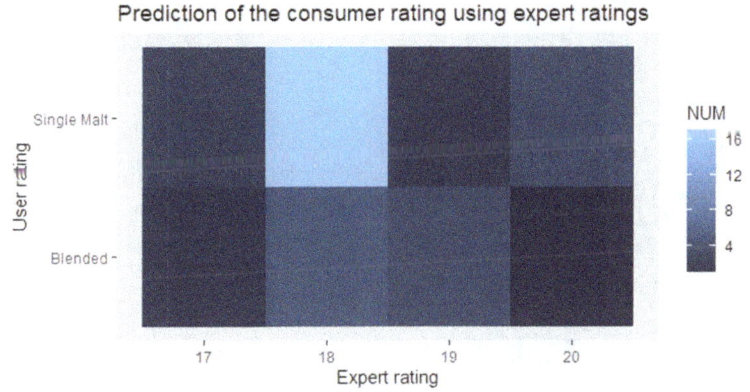

Figure 3-27 A heat map illustrating the centuries with the most distillery foundations.

3.3.12 Treemap

A treemap provides a hierarchical view of data, making it easier to see patterns, like which items sell best in a store. Branches are represented by rectangles and each sub-branch is represented by a smaller rectangle. The treemap shows categories by color and proximity and can easily display a variety of data that would be difficult with other chart types.

Let us make a short treemap to illustrate the consumer ratings across countries. We will do that by making a treemap with each tile representing a country in our whisky collection dataset. The area of the tile will be mapped to the number of whiskies and the tile's fill color to its review score. `geom_treemap()` is the basic geometric mapping for this purpose.

The `treemapify` package allows creating treemaps together with `ggplot2`. If you have not installed it, run the following commands before you continue:

```
install.packages("treemapify")
library(treemapify)
```

Then we run:

```
distilleries = whisky_collection %>%
  group_by(LOCATION) %>%
  summarize(PRICE=mean(PRICE), REVIEWS=mean(REVIEWS), NUM=n())

ggplot(data=distilleries) +
  aes(area=NUM, fill=REVIEWS , label=LOCATION) +
  geom_treemap() +
  geom_treemap_text(colour = "white", place = "centre", grow = TRUE)
```

The final result is illustrated in Figure 3-28.

Figure 3-28 Treemap of the number of whiskies by country and their averaged online shop rating.

As you can see in Figure 3-28, Scotland is the biggest whisky producer. It's even bigger than all the other markets in our dataset. You can see that directly by comparing the size of the different areas. Furthermore, we see that the worst whiskies seem to come from the US and Canada, while the best producer is surprisingly Japan! Please note that we used our `whisky_collection` dataset as data sample and an investigation with a bigger dataset could look differently.

3.3.13 Bar Plot and Stacked Bar Plot

A bar graph (or bar chart) is one of the most common types of visualization. It shows the relationship between a numeric variable and a categorical variable. Each unit of the categorical variable is represented as a bar. The size of the bar represents its numerical value.

Bar plots are very popular but very bad ways to visualize frequencies in your dataset. They are popular because they are simple and every analytic tool in the world supports bar plots. But they are a very bad way to visualize data because they have a very bad ink to information ratio (Inbar et al., 2007). That means that you will have very big graphical elements (many bars) in your visualization but very less information.

You can make a simple bar plot in base R illustrating the different whiskies by their origin (`LOCATION`) in our collection with:

```
distilleries = whisky_collection %>%
  group_by(LOCATION) %>%
  summarize(NUM=n())

barplot(height=distilleries$NUM, names=distilleries$LOCATION)
```

In `ggplot2` the computation of a distilleries dataset is not necessary, the `geom_bar()` mapping will do that for us, so the code is reduced to:

```
ggplot(data=whisky_collection) +
  aes(x=LOCATION) +
  geom_bar()
```

Both results are illustrated in Figure 3-29.

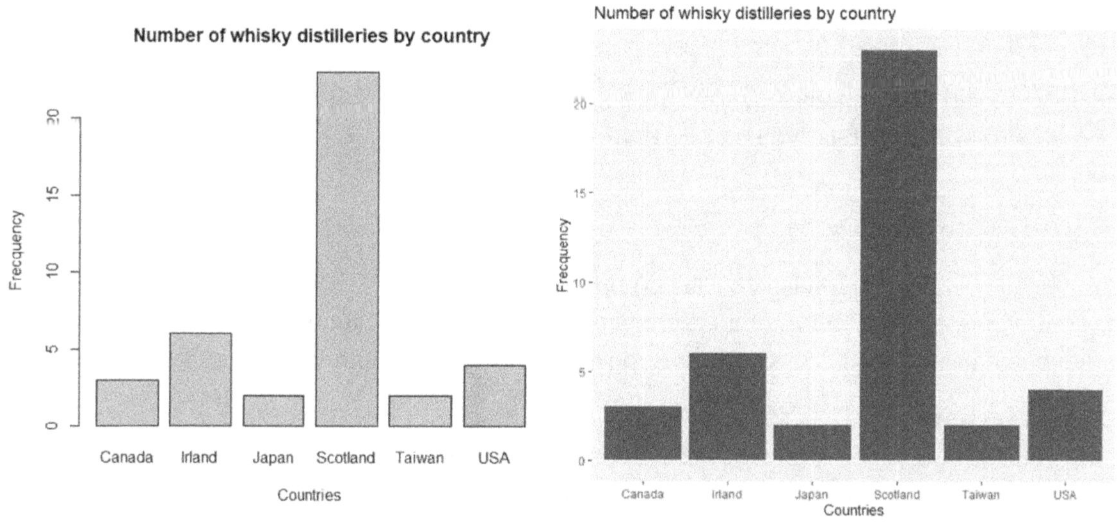

Figure 3-29 Barplots to illustrate the different countries with whisky distilleries in base R and `ggplot2`.

A bar graph can also show values for multiple levels of grouping. In the following plot, the number of whiskies by distilleries world-wide is shown by country (level 1) and per type of whisky (level 2). With this type of information, it is possible to create a stacked bar plot to visualize effects over time or in different groups.

```
ggplot(data=whisky_collection) +
  aes(x=LOCATION, fill=TYPE) +
  geom_bar() +
  labs(x="Countries", y="Frecquency") +
  ggtitle("Number of whisky distilleries by country")
```

Resulting in the visualization in Figure 3-30.

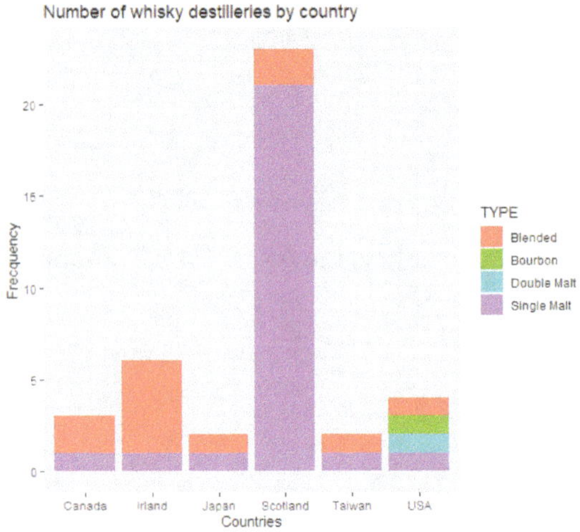

Figure 3-30 A stacked bar plot to visualize groups or developments over time with `ggplot2`.

Finally, I would like to point you towards a very useful feature of `ggplot2`. If you want to add error bars to your barplots, you can do that by adding the `geom_errorbar` to your `ggplot2` plot:

```
geom_errorbar(aes(ymin = lower, ymax = upper), width = .2)
```

Error bars in bar plots help visualize uncertainty around each bar's value, aiding in understanding the consistency and reliability of measurements. They communicate the level of uncertainty and assist in comparing groups or conditions, providing valuable insights for your stakeholder.

3.3.14 Piechart

A pie chart shows how a total amount is divided between the levels of a categorical variable as a circle that is divided into radial slices. Each categorical value corresponds to a single slice of the circle. The size of each slice (both in area and arc length) indicates the proportion of the total that each category level represents.

Let us now make a pie chart to illustrate the different types of whiskies in our collection.

Run the following code to make a pie chart in base R:

```
distilleries = whisky_collection %>%
  group_by(LOCATION) %>%
  summarize(NUM=n())

pie(distilleries$NUM,
    labels = distilleries$LOCATION,
    main="Pie Chart of Countries")
```

Or in `ggplot2`:

```
ggplot(data=distilleries) +
  aes(x="", y=NUM, fill=LOCATION, label=LOCATION) +
  geom_bar(stat="identity") +
  geom_text(position=position_stack(vjust = 0.5)) +
  coord_polar("y", start=0) +
  labs(x="", y="") +
  ggtitle("Number of whisky distilleries by country")
```

Resulting in the following visualizations in Figure 3-31.

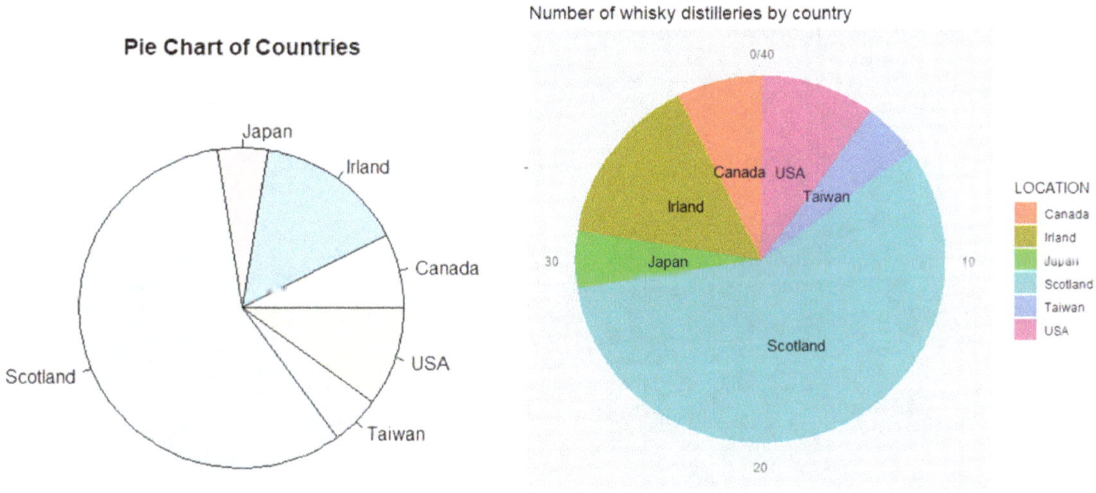

Figure 3-31 Pie charts to illustrate the different countries in base R and `ggplot2`.

3.3.15 Parallel Coordinates

One very popular visualization of multidimensional data are parallel coordinate plots. Parallel coordinates are gaining popularity as a data visualization technique in business data analytics. They allow for the visualization of multidimensional data in an intuitive manner. They enable the simultaneous representation of multiple features, helping us as business data analysts in the identification of potential patterns or relationships among our features. By observing how lines intersect or cluster along the axes, business data analysts can gain insights into how different features interact and contribute to the overall pattern.

From a technical point of view, they are based on scatter plots but with abstract and parallel coordinate axes. That means we draw the coordinate axes without labels on the same place, the y-axis.

So far there is no possibility to build parallel coordinates with base R, but `ggplot2` together with the `GGally` package can do the job. So, install the package before you continue with:

```
install.packages("GGally")
library(GGally)
```

We can build a parallel coordinates plot for to illustrate the different whisky characteristics across all countries with:

```
ggparcoord(whisky_collection,
           columns = c(9:14),
           groupColumn = "LOCATION",
           showPoints = TRUE,
           title = "Parallel Coordinate Plot for the Iris Data",
           alphaLines = 0.3) +
  labs(x="Countries", y="Frecquency") +
  ggtitle("Whisky characteristics by country")
```

Resulting in the plot in Figure 3-32 with parallel coordinates.

Figure 3-32 Parallel coordinate plot to visualize the different whisky characteristics across all countries.

3.3.16 Radar Plots

A radar plot or star plot is a way of showing a number of points of data and the variation between those points. Radar plots are comparable to the parallel coordinates plot, but with the difference that the axes are in star-like manner instead of the same place in parallel. As parallel coordinates plots, they are often useful for comparing two or more different variables.

Best way to build them is using the `ggradar` package together with `ggplot2`. So let us make sure that the package is installed before we continue with:

```
if (!require(devtools)) install.packages("devtools")
devtools::install_github("ricardo-bion/ggradar")
```

For the numerical variables to be displayed in the radar chart, the mean value must be calculated. These averages are then rescaled to the interval [0, 1] using the `normalize()` function from the `dstools` package (see the chapter 4.3 for more information). As soon as our dataset has this format, the function `radarchart()` will do the whole job for us:

```
radarchart(data)
```

Resulting in the following Figure 3-33:

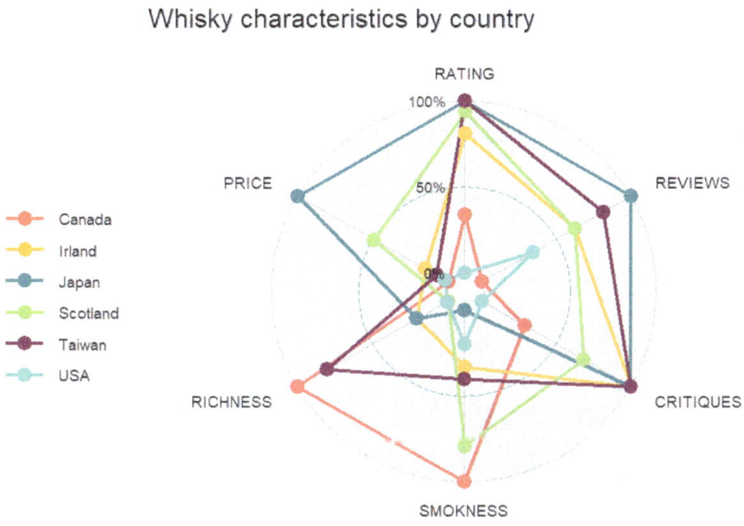

Figure 3-33 A radar chart to visualize the different whisky characteristics.

3.3.17 Maps

Maps are a very special type of visualization used to represent spatial information. In R, there are multiple approaches to plot maps, and I often need to revisit previous examples to refresh my memory on the steps involved. Let me show you, how you can setup a map fast without discussing all the different approaches.

Best way to go is using the `geom_map()` mapping from `ggplot2` to make world maps and plot your spatial information on it. Going this way you'll need two datasets: one that includes the map data (borders of the area you want to plot), and another one with your data you want to plot on this map.

To illustrate this procedure let us first make a simple world map and then add the location of whisky distillers from our dataset on it. We will use the world map from the `maps` package for that purpose:

```
install.packages("maps", dependencies = TRUE)
library(maps)
```

Let us now start to illustrate with:

```
# load map from ggplot2
world_map = map_data("world")
ggplot() +
  # geom_map() function takes world map as input
  geom_map(
    data = world_map, map = world_map,
    aes(long, lat, map_id = region))
```

As you can see, we made something unusual. We put the `aes()` function inside the `geom_map()` function. This is necessary because we will use multiple datasets in this plot. You remember? One for the map and one for the information on it. But let's first have a look at our result. The final map is illustrated in Figure 3-34.

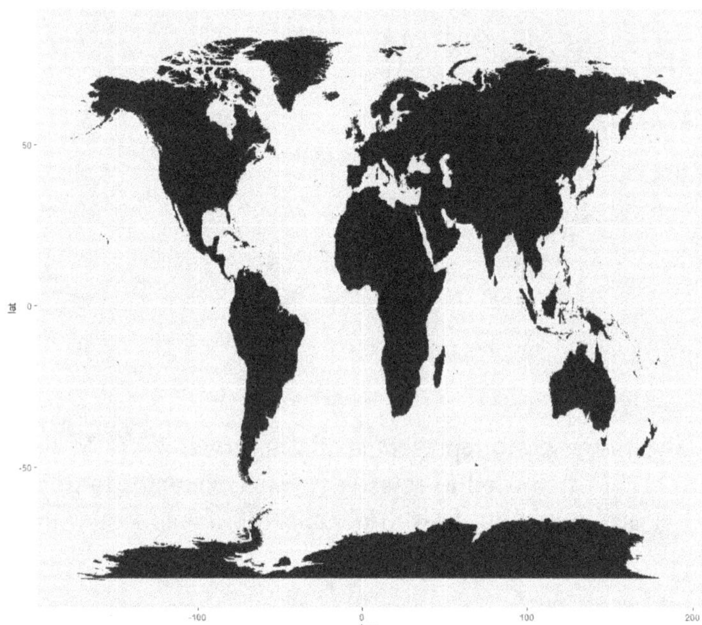

Figure 3-34 A world map with ggplot2's geom_map() mapping.

Now we want to add some spatial information on it. For instance, we can add some points with `geom_point()` for each distillery. We can use the `COORDINATES` features from the `whisky_collection` dataset for that purpose. Unfortunately, the data is not separated into two columns for longitude and latitude values it is in one column and separated by a delimiter:

```
> head(whisky_collection$COORDINATES)
[1] "57.78497402327632, -1.6207674406702595"
[2] "55.6406114241932, -6.107878725864291"
[3] "55.922066708629245, -4.437533995096446"
[4] "55.9659724419173, -4.568164585896366"
[5] "55.756718827510745, -6.289152434196375"
[6] "58.02538030697375, -3.863758501818989"
```

Therefore, we have to preprocess it a bit with `tidyr`, which we used also in other visualizations to prepare the data. The idea is that we separate it based on the comma, and then make two columns of it. The `separate_wider_delim()` function will do it for us:

```
mapdata = separate_wider_delim(whisky_collection,
                               COORDINATES,
                               names=c("LATITUDE", "LONGITUDE"),
                               delim = ",")
mapdata$LONGITUDE = as.numeric(mapdata$LONGITUDE)
mapdata$LATITUDE = as.numeric(mapdata$LATITUDE)
```

Then we can continue and plot our distillery locations on our world map with:

```
ggplot() +
  geom_map(
    data = world_map, map = world_map,
    aes(long, lat, map_id = region)) +
  geom_point(
  data = whisky_collection_mapdata,
  aes(LONGITUDE, LATITUDE, color = LOCATION, size=2)
)
```

Finally, we have our world map illustrated in Figure 3-35.

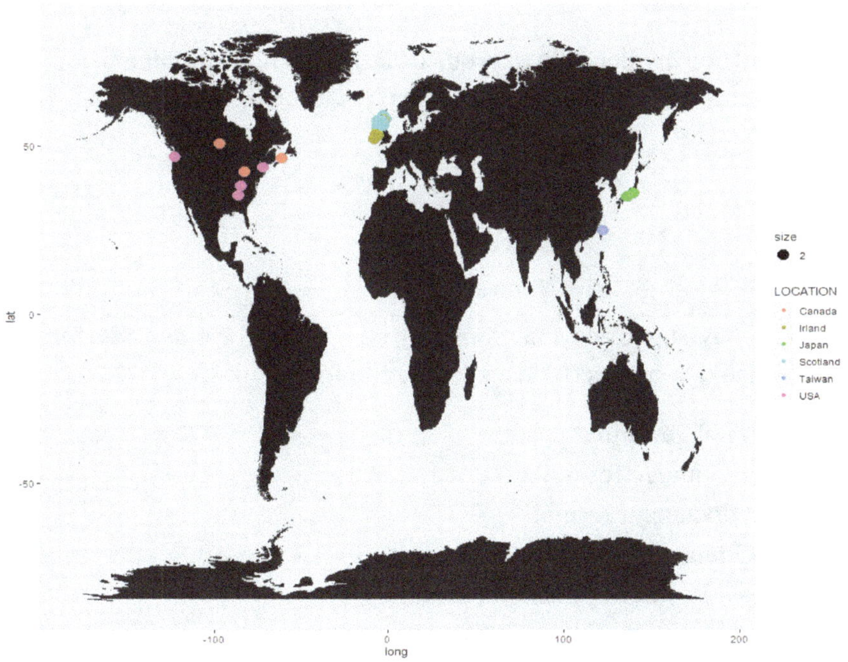

Figure 3-35 Worldmap with spatial information (whisky distillers).

Last but not least, I want to point out that it is also possible to plot only a specific part of the world map by sizing the plot to a specific area with:

```
...
+ coord_map(xlim=c(lon_min, lon_max), ylim=c(lat_min, lat_max))
```

Where the maxima/minima represent a rectangle of the world map where you are interested in.

3.3.18 Export your Visualizations

Once you have created your plot, you can save it to a file in your preferred format. The Export tab in the Plot window in RStudio will save your plots at a low resolution, which is not accepted by many journals and does not scale well for posters. The `ggplot2` extensions website provides a list of packages that extend the capabilities of `ggplot2`, including additional themes.

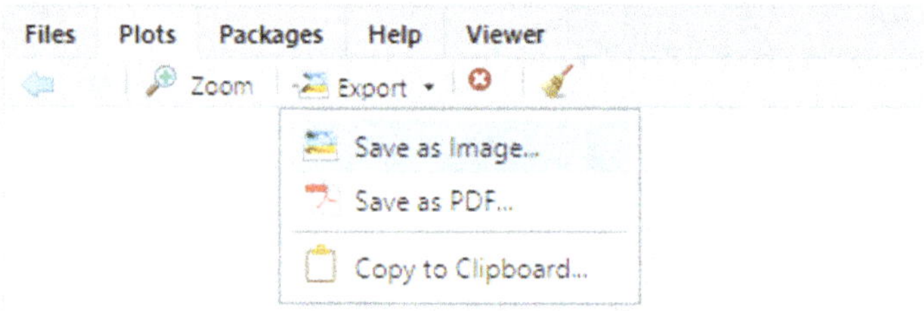

Figure 3-36 The export functionalities in RStudio.

Instead, use the `ggsave()` function, which allows you to easily change the dimension and resolution of your plot by adjusting the appropriate arguments (`width`, `height` or `dpi`):

```
plot_final = ggplot(data=whisky_collection) +
  aes(x=LOCATION) +
  geom_bar()

ggsave("whisky_distilleries.png", plot_final, width = 15, height = 10)
```

Alternatively, there exist also many supported functions in `grDevices` for many file formats that allow you to save your visualization directly in a specific file format like:

- `bmp()`: creates a bitmap image format
- `jpeg()`: generates a JPEG image format
- `pdf()`: produces a PDF document format
- `png()`: generates a PNG image format
- `postscript()`: creates a PostScript document format

Which you can use like this

```
ggplot(data=whisky_collection) +
  aes(x=LOCATION) +
  geom_bar()
png("whisky_distilleries.png")
dev.off()
```

3.4 Business Data Description

In the last section we have seen that visualizing business data helps us better to understand some of the main characteristics of our dataset. But it also helps us to start developing questions, assumptions and ideas for our analysis. Computing key values and characteristics allows you to identify any anomalies, patterns, or potential issues that might you not have seen in a plot or visualization. By examining the data's distribution, structure, and basic attributes, you set the stage for effective preprocessing, modeling, and understanding.

For instance, the visualizations of the `whisky_collection` dataset might rise the following questions:

- What is the highest price of whisky in our collection?
- What is the minimal rating of Scottish whiskies?
- Where do the most whiskies come from?

In a first step, to answer these questions, we reload the `whisky_collection` dataset:

```
library(dstools)
data("whisky_collection")
```

Then we compute the highest price of whisky with:

```
> max(whisky_collection$PRICE)
[1] 2700
```

As we can see the maximal price of a whisky in our collection is `2700` EUR. If we want to compute the minimal value, we can do that with the `min()` function. We can use that directly to find the minimal rating by country by making a subset and using the `min()` function on the `RATING`.

```
> scottish_whiskies = subset(whisky_collection, LOCATION=="Scotland")
> min(scottish_whiskies$RATING)
[1] 2
```

Our code shows, that the minimal rating of a Scottish whisky is `2`.

Finally, we want to taggle the questions, where the most whiskies come from. Therefore, we compute the number of whiskies for each `LOCATION`. We can use the `table()` function for that:

```
> table(whisky_collection$LOCATION)
  Canada    Irland     Japan Scotland    Taiwan       USA
       3         6         2       23         2         4
```

The `table()` function, returns the number of distinct values for an input vector. In our case it counts the distinct values in the column `LOCATION`. The result is the number of whiskies per country. And we can see that the most whiskies come from Scotland.

This short example should illustrate you that describing your data in business data analytics is about gathering facts and computing values about our business data to answer the questions that have risen. It is crucial because it provides a foundational understanding of the dataset's characteristics or helps to answer first business questions. Doing so you will have to prepare three types of descriptions as a business data analyst for that purpose (generally speaking). I put them together in an overview in Figure 3-37.

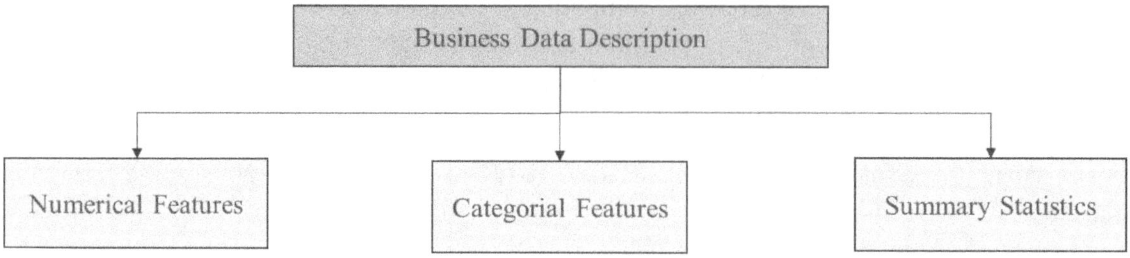

Figure 3-37 A simple overview of business data description techniques.

We have to investigate the numerical characteristics of our numerical business data and investigate our numerical features like the prices of our whiskies. Secondly, we have to take a look at categorial features like the location of a distillery. And last but not least compute generic summary statistics that give an overview of the whole dataset.

In conclusion, business data description is another and subsequent step besides business data visualization to understand your business data and match it with your requirements from the business understanding.

3.4.1 Describe Numerical Features

In R exits many functions that help you to compute descriptive statistics in addition to your business data visualization. As you have seen, they are very helpful to understand your data. The most relevant functions to start exploring the values and characteristics of a dataset are the minimum, maximum, quartiles, mean, median, standard deviation and range values.

We can compute them in R with:

- `min()`: Compute minimum
- `max()`: Compute maximum
- `quantile()`: Compute the quantiles
- `mean()`: Compute the arithmetic mean
- `median()`: Compute the median
- `sd()`: Compute standard deviation
- `range()`: Compute value range from min to max

For instance, if we want to compute the average price of a whisky in our collection, we use the descriptive function `mean()` together with the price values from our dataset:

```
> mean(whisky_collection$PRICE)
[1] 161.3
```

3.4.2 Describe Categorial Features

Besides the generic descriptive statistics, there exist many functions to investigate the values of categorial features. As we have seen in the example of this section, we can use `table()` to display the frequencies of the categories (or factor levels) of a feature:

```
> table(whisky_collection$LOCATION)
  Canada    Irland     Japan Scotland    Taiwan       USA
       3         6         2       23         2         4
```

This is particularly useful if we are interested in the combined frequencies of several factors as in a cross-tabulation:

```
> table(whisky_collection$LOCATION, whisky_collection$RATING)

             1  2  3  4  5
  Canada     0  2  1  0  0
  Irland     0  0  5  1  0
  Japan      0  0  1  1  0
  Scotland   0  3 11  6  3
  Taiwan     0  0  1  1  0
  USA        1  3  0  0  0
```

As we can see, these combined frequencies help us to compare the value distributions of two categorial features easily. In our case we see that there are `11` whiskies in Scotland that have a `RATING` of `3`, which is more whiskies than in each other `LOCATION`.

We can also transform this to proportions (relative frequencies) and get a proportion table:

```
whisky_ratings          =          table(whisky_collection$LOCATION,
whisky_collection$RATING)
prop.table(whisky_ratings)
```

which will look like this:

```
> prop.table(whisky_ratings)

               1     2     3     4     5
  Canada   0.000 0.050 0.025 0.000 0.000
  Irland   0.000 0.000 0.125 0.025 0.000
  Japan    0.000 0.000 0.025 0.025 0.000
  Scotland 0.000 0.075 0.275 0.150 0.075
  Taiwan   0.000 0.000 0.025 0.025 0.000
  USA      0.025 0.075 0.000 0.000 0.000
```

As we can see that the `11` whiskies with `RATING` of `3` represent `0.275` or `27.5%` of all whiskies in the collection. Generally, our focus lies not on the overall proportions, which sum up to 1 or 100% across the entire table, but rather on the relative frequencies based on specific rows or columns. To calculate the overall proportions, we will have to use the `prop.table()` function with `margin = 1`, while for relative frequencies based on rows or columns, we have to set `margin = 2`.

Another very popular method to explore the frequencies of categorial values is the mode. The statistical mode represents the most frequent value of a variable or feature. Unfortunately, there is no functionality to calculate the mode of a value in base R. But we can use the `statmode()` function from the `dstools` package for that:

```
> statmode(whisky_collection$LOCATION)
[1] "Scotland"
```

Which shows us that the most whiskies come from Scotland. But good news, we can also use this mode on numeric features like the `RATING`s of the whiskies:

```
> statmode(whisky_collection$RATING)
[1] 3
```

If you browse through the R documentation, you will almost certainly come across a `mode()` function. Now you may be tempted to simply use the `mode()` function in R to calculate the mode of a variable. However, R's `mode()` function is a false friend, as it calculates the storage mode of an R object.

Finally, I want you to point to the `unique()` function, which is very useful. You can use it to retrieve unique elements from a vector, or data frame. Which is very helpful, if you want the unique categories of a dataset. If we want a list of all countries in our whisky collection, we can run:

```
> unique(whisky_collection$LOCATION)
[1] "Scotland" "Irland"   "Canada"   "USA"      "Japan"    "Taiwan"
```

3.4.3 Summary Statistics

However, computing these descriptive values seems to be a lot of work to do for each feature of your dataset. Especially, if we will have to do that in every analytics project. Good for us that there exist many functions that can help us to compute the most relevant descriptive statistics and answer these questions directly. They are:

- `summary()`: This function is shipped with base R and computes the most relevant descriptive statistics for your whole dataset. Because this function is so popular other functions that compute descriptive statistics are also called "summary functions".
- `describe()`: Alternative from `psych` package to compute more detailed descriptive information about your dataset.

So instead of computing all the relevant descriptive statistics manually, we can just run `summary(whisky_dataset)` to compute means or number of observations per column.

```
> summary(whisky_collection)
     NAME               ...    FOUNDATION     ...       PRICE
 Length:40              ...   Min.    :1743   ...   Min.    :   19.0
 Class :character ...   1st Qu.:1823   ...   1st Qu.:   32.0
 Mode  :character ...   Median :1860   ...   Median :   41.5
                        ...   Mean    :1884   ...   Mean    :  161.3
                        ...   3rd Qu.:1961   ...   3rd Qu.:   70.5
                        ...   Max.    :2017   ...   Max.    :2700.0
```

The result is a huge list with the most relevant descriptive statistics for the numeric features, and some generic information about the non-numeric features like categorial features.

If you are interested in more detailed information, I recommend that you install the `psych` package:

```
install.packages("psych", dependencies = TRUE)
library("psych")
```

Then you can run the `describe()` function, which returns a more detailed report:

```
             vars  n    mean     sd median trimmed   mad  min  max range  skew kurtosis    se
NAME*           1 40   20.50  11.69   20.5   20.50 14.83    1   40    39  0.00    -1.29  1.85
DISTILLERY*     2 40   19.80  10.92   20.5   19.88 13.34    1   38    37 -0.07    -1.25  1.73
LOCATION*       3 40    3.67   1.31    4.0    3.69  0.00    1    6     5 -0.34    -0.23  0.21
TYPE*           4 40    3.02   1.39    4.0    3.16  0.00    1    4     3 -0.72    -1.47  0.22
REGION*         5 40    6.40   4.46    4.5    6.03  3.71    1   16    15  0.61    -1.02  0.71
FOUNDATION      6 40 1883.83  80.11 1860.5 1881.94 66.72 1743 2017   274  0.34    -1.29 12.67
COORDINATES*    7 40   19.52  11.65   19.5   19.50 14.83    1   39    38  0.01    -1.30  1.84
WIKIPEDIA*      8 40   20.18  11.31   20.5   20.22 14.08    1   39    38 -0.04    -1.28  1.79
RATING          9 40    3.12   0.91    3.0    3.09  1.48    1    5     4  0.16    -0.26  0.14
REVIEWS        10 40    7.88   1.04    8.0    7.94  1.48    6   10     4 -0.29    -0.79  0.16
CRITIQUES      11 40   88.33   3.98   89.0   88.81  2.97   73   95    22 -1.61     3.69  0.63
SMOKNESS       12 40    0.05   2.69   -1.0    0.03  1.48   -5    5    10  0.23    -0.88  0.43
RICHNESS       13 40    0.88   2.76    1.0    1.09  2.97   -5    5    10 -0.65    -0.50  0.44
PRICE          14 40  161.30 455.85   41.5   49.91 19.27   19 2700  2681  4.56    21.54 72.08
```

Besides the mean and standard deviation, the report computes the median, range and more advanced statistical moments like the kurtosis.

3.5 Business Data Quality and Validation

If it comes to business data understanding, you will recognize that the reliability and accuracy of business data are paramount. I can tell you that I had to manage projects which had well-define structured data from a database, but contained missing values or features where nobody knows what they stand for. In another case I had corrupt data from defect sensors or old pipelines. I even know an analytics project where a student was hired to manually type long rows of data into a excel sheet from another production machine because there was no interface. He stayed in the company for his whole study and it was his only job. How reliable this data is, however, could never be verified - apart from the fact that it is a completely abstruse idea to hire a human being to manually type out data for years.

In this last section we delve into the crucial process of ensuring that the business data you're working with is trustworthy, consistent, and aligned with your business objectives. From my experience, the validation of the business data, can and has to be done at three different points of your business understanding as illustrated in the following overview in Figure 3-38.

Before diving into data analysis, it's essential to comprehend how the business data was generated or created. This step helps uncover potential biases, errors, or irregularities that might impact the analysis. If the process is copying data manually or there is no documentation about the source or generation, then there might be serious concerns about the quality. By gaining insight into the business data's origin, you set the stage for a more reliable and meaningful analysis.

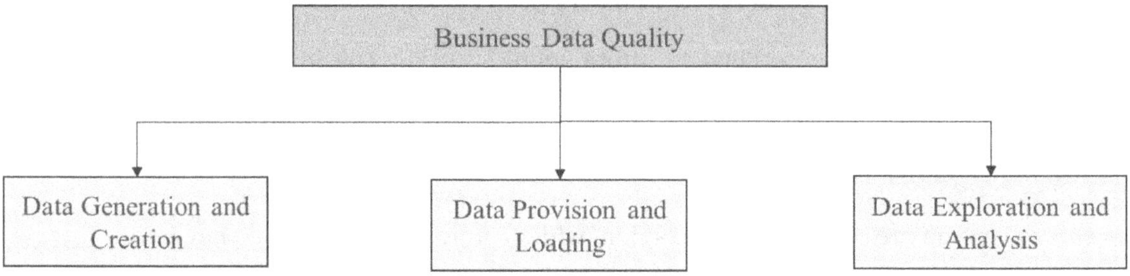

Figure 3-38 Validating business data quality throughout your business data analytics project.

To make this more visual for you, I list some of the most common problems you might face and identify in the context of data quality:

- Contains duplicate rows and values
- Is noisy and has noisy values
- Wrong data types for its values
- Contains missing values
- Contains inconsistent of contradictory values
- Contains outliers
- Not representative data
- Imbalanced (target) features
- Inconsistent naming scheme
- Data structure is not tidy
- Not ready for application (timeliness, relevance, etc.)
- No knowledge about the whole data
- …

Do not worry at the moment about these problems – you do not have to solve them now. Make sure that you document them carefully, and prepare a kind of data quality report (a short list or PowerPoint presentation with the issues and examples is enough). In the next chapter we will have a look at different strategies and how to handle these issues.

After you decided for the relevant data sources to answer the questions you defined in the business understanding, you have to check the quality of the data provision. When business data is provisioned for analysis, close collaboration with data engineers or data owners is key. Ensuring alignment on data formats, joins, and integrity safeguards is crucial. Asking questions and seeking clarification at this stage ensures that the data you're working with is clean, consistent, and well-prepared for analysis. Once data is loaded into your R environment, take a moment to conduct a preliminary assessment. Checking the head and tail of the data provides an initial impression of its structure, values, and possible anomalies. This quick examination sets the stage for further analysis and alerts you to any glaring issues.

Finally, as you delve deeper into data analysis, the importance of validation remains until the end of the business data analytics project. Generating graphs, statistics, and other analytical outputs enables you to gain insights into data trends and relationships. Use the tools from this chapter (visualizations and descriptive statistics) not only to understand the data. Use them to check the quality during the whole project. To this end validation involves ensuring that the results make logical sense and align with your expectation.

Let us now have a look at these steps in more detail!

3.5.1 Validate Business Data Generation

In business data analysis, the flying word "garbage in, garbage out" (GIGO) underscores the crucial role that data quality plays. Before starting your project, it's necessary to lay a solid foundation by understanding the process of data generation or creation. This step is fundamental cornerstone that can shape the entire trajectory of your analysis.

Figure 3-39 Why data validation is important: the garbage-in-garbage-out principle.

GIGO serves as a reminder of the critical role data quality plays in the outcomes of business data analytics right from the start. Imagine that the *Junglivet Whisky Company* plans to optimize its marketing strategy based on customer demographics. If the initial data used for analysis contains inaccuracies, duplicates, or missing information, the subsequent insights and conclusions drawn from this corrupted dataset will also be inaccurate and unreliable.

Furthermore, Alice Gleck from the marketing team informs you that the database contains outdated customer records, incorrect age entries, and even entries with missing income data. When our project is conducted using this faulty data, the results might suggest targeting strategies that are ineffective or misaligned with the actual customer base. This could lead to wasteful marketing expenditures, poor customer engagement, and missed opportunities.

Or as I always say to my students, just a chef cannot create a gourmet meal using spoiled ingredients, business data analysts cannot generate accurate insights from flawed or inadequate data. The insights and recommendations that stem from this flawed data are akin to the "garbage" output. This mismatch between the input quality and the expected output quality underscores the importance of data collection, cleaning, and validation in business data analytics.

As a result of illustrating GIGO in lecture, my students often ask why data quality is still such a big problem for companies today. You should know that the process of data generation is a complex interplay of various factors, ranging from data collection methodologies to the systems and technologies employed. Each step within this process can introduce its own set of challenges, including potential biases, errors, or irregularities. Without a comprehensive understanding of how the data was generated or created, you risk overlooking these subtleties that can significantly impact your analysis outcomes.

Thus, during business data generation you have to identify and address potential pitfalls that might otherwise go unnoticed. For instance, you might discover that data collection methods favored a particular demographic, leading to unintended biases in your dataset. Or you might stumble upon discrepancies arising from data migration processes that corrupted values. By uncovering such issues early on, you mitigate the risk of propagating errors throughout your analysis.

Validating business data quality during data generation is a strategic move that supports the credibility of your business data analytics project. It's a proactive approach to ensure that your insights are built upon a solid foundation of accurate and representative data. And your whole project will benefit from it.

3.5.2 Validate Business Data Provision and Loading

I promise you that the role of data provisioning cannot be overstated. It serves as the crucial bridge connecting the data generation stage with the eventual analysis, serving up data that is ready for interpretation. However, this bridge is not a solitary crossing; rather, it's a collaborative journey that demands close collaboration and alignment with data engineers, data architects, data stewards and data owners.

When data reaches the provisioning phase, the synergy between data analysts and the data stakeholders becomes the key success factor of data quality assurance. Close collaboration is essential to ensure that the data you want to use for analysis is not only available but also well-prepared and ready for the next steps of your project.

In these discussions, you have to consider that data formats need to harmonize across different sources, ensuring a seamless integration that doesn't trigger compatibility issues downstream. Furthermore, joins between datasets must be orchestrated, preventing the different sources from clashing and resulting in fractured insights. Keep those questions in mind, if we come to the next chapter *business data preparation*!

If the data provision has been defined, make some tests if everything works fine. Use a sample to conduct a comprehensive evaluation of the quality. These include assessing data completeness of the data provision. Use descriptive statistics to ensure that it encompasses all the necessary cases. Additionally, identify the presence of missing values within the dataset and report them to the data engineer. Write a short report, which means that you describe these missing values, their occurrence patterns, and their frequency within the dataset. Let me say that every question asked, every detail clarified, contributes to the success of your project. Are the data formats consistent across the data provision? Are there any idiosyncrasies to consider in the data integration? Are there integrity checks that warrant special attention?

In R exists many functions which help you to validate the correct data provision and loading. The most relevant are:

- `glimpse()`: Provides a concise overview of a data frame's structure
- `head()`, `tail()`: Displays the first and last rows of a dataset
- `View()`: Opens your data in RStudio
- `nrow()`, `ncol()`: Counts the number of rows or columns of your dataset
- `length()`: Counts the number of elements in a vector
- `nchar()`: Counts the number of characters in a string
- `arrange()` and `sort()`: Sorts your dataset to check the values

`glimpse()` is a function from the `dplyr` package. This function provides a concise overview of a data frame's structure, displaying a few rows and columns to give a sense of the data's content and layout.

The `glimpse()` function is a valuable tool for gaining a quick overview of a data frame's structure. By displaying a concise summary of the data's columns, data types, and the first few rows, it provides an immediate sense of the dataset's contents.

This can be especially helpful during the data validation process, as it allows you to spot potential anomalies or inconsistencies that may require further investigation. Utilizing `glimpse()` can aid in identifying data entry errors, incorrect data types, or missing values early in the analysis workflow.

If we want to get a generic overview of the `whisky_collection` dataset we have loaded in our RStudio environment, we can use `glimpse()` for that:

```
> glimpse(whisky_collection)
Rows: 40
Columns: 14
$ NAME        <chr> "An Cnoc", "Ardbeg", "Auchentoshan", …
$ DISTILLERY  <chr> "Knockdhu", "Ardbeg distillery", …
$ LOCATION    <chr> "Scotland", "Scotland", "Scotland", …
$ TYPE        <chr> "Single Malt", "Single Malt", "Single Malt",…
$ REGION      <chr> "Highland", "Islay", "Lowland", …
$ FOUNDATION  <dbl> 1894, 1815, 1823, …
$ COORDINATES <chr> "57.78497402327632, -1.620767440670", …
$ WIKIPEDIA   <chr> "https://en.wikipedia.org/wiki/Knockdhu", …
$ RATING      <dbl> 2, 4, 4, 2, 3, 4, 3, 3, 3, 2, 2, 3, 3, 2, …
$ REVIEWS     <dbl> 9, 9, 6, 8, 8, 10, 8, 8, 8, 7, 6, 8, 7, …
$ CRITIQUES   <dbl> 90, 91, 91, 81, 88, 95, 90, 89, 91, 89, …
$ SMOKNESS    <dbl> -1, 5, -2, 0, 3, 3, -4, 0, 4, 3, 3, 2, …
$ RICHNESS    <dbl> -3, -5, 4, 2, 1, 0, 1, 0, -4, 3, 3, 0, …
$ PRICE       <dbl> 32, 46, 25, 23, 35, 2700, 75, 36, 45, …
```

As a result, we can see for all columns of the dataset which represent the features, the data type (`character` or `double`) and some example values.

An alternative for this are the `head()` and `tail()` functions. They offer a straightforward way to inspect the initial and final records of a data frame. These functions allow you to directly view a portion of the data, enabling you to visually assess its structure and values. This visual inspection can be helpful to make a quick check and is sometimes preferred because it is more compact than `glimpse()`.

So, if we want to look at the first 6 rows of our dataset, we run:

```
> head(whisky_collection[,1:4])
# A tibble: 6 x 14
  NAME         DISTILLERY                LOCATION TYPE
  <chr>        <chr>                     <chr>    <chr>
1 An Cnoc      Knockdhu                  Scotland Single Malt
2 Ardbeg       Ardbeg distillery         Scotland Single Malt
3 Auchentoshan Auchentoshan distillery Scotland Single Malt
4 Ballantine's Pernod Ricard             Scotland Blended
5 Bowmore      Bowmore distillery        Scotland Single Malt
6 Brora        Brora distillery          Scotland Single Malt
```

When it comes to interactive exploration of data, the `View()` function is a powerful asset. By opening a new window in RStudio displaying the complete data frame, it facilitates hands-on inspection and analysis.

The view in RStudio allows you to navigate through the dataset, spot outliers, and identify irregularities that might not be evident through summary statistics alone. By actively exploring the data using `View()`, you can pinpoint potential data quality issues and assess their significance. For me it's one of the most used functions in the validation process.

The `nrow()` and `ncol()` functions provide straightforward solutions for counting the number of rows and columns within a data frame, respectively. I prefer to use them in my scripts that check if the data loading has been completed successfully. For instance, you can write some functions like this at the beginning of your data loading scripts:

```
check = (nrow(whisky_collection) == 40)
ifelse(check, "Data complete", "Data incomplete")
```

If you setup an automated data validation, you can use these functions to confirm that the number of records and columns align with your expectations. Any deviations from expected counts could indicate issues during data import, transformation, or data collection processes.

The `length()` function, while typically applied to vectors or lists, can also play a role in data validation. When used to calculate the number of objects in an input vector. Like `ncol()` or `nrow()` does it aid in verifying the consistency of data across columns. In cases where variables should have equal lengths, any differences might suggest data entry errors or inconsistencies.

nchar() calculates the number of characters in a given string. You can use it for individual strings, but also apply it to an entire vector of strings, making it useful for validating text-based data within a data frame. The nchar() function can be particularly valuable when working with textual data, such as customer names, addresses, or product descriptions. By applying nchar() to a column of text, you can quickly determine the length of each string. This is crucial for ensuring consistency and identifying data quality issues.

For instance, if a column contains webdata like email addresses or URLs you expect names to be within a certain range of characters, using nchar() can help you identify unusually long or short names that might require further investigation. In our example, we can use them to count the number of characters of the URL of the Junglivet whisky in our dataset:

```
> junglivet = whisky_collection[27,]
> nchar(junglivet$WIKIPEDIA)
[1] 24
```

Last but not least, there are also very useful sorting functions arrange() and sort(). You can use them to sort your data frame in a dplyr pipe:

```
whisky_collection_sorted = whisky_collection %>%
  select(NAME, PRICE) %>%
  arrange(PRICE)
```

Which will sort your data based on the PRICE in ascending order:

```
> head(whisky_collection_sorted)
           NAME PRICE
1 Canadian Club    19
2  Jack Daniels    19
3  Ballantine's    23
4       Jameson    24
5  Auchentoshan    25
6 Tullamore Dew    25
```

If you want to sort them descending, you have to use the desc() function inside the select() command:

```
whisky_collection_sorted = whisky_collection %>%
  select(NAME, desc(PRICE)) %>%
  arrange(PRICE)
```

Unlike other dplyr commands, arrange() largely ignores grouping, which means that you need to explicitly mention grouping variables (or use .by_group = TRUE) in order to group by and sort them.

The base R version of arrange() is sort(). You can sort a vector or factor (partially) into ascending or descending order like this:

```
sort(whisky_collection$PRICE, decreasing = FALSE)
```

3.5.3 Validation During Business Data Analysis

Last but not least, I want to strongly encourage you to think of business data understanding and validation as a continuous process and not just as a single activity in the pre-processing of business data. Validation also means making sure that the results of your analysis are logical and that they fit seamlessly with your original expectations and business goals.

To do this, use tools we introduced in this chapter, such as visualizations and descriptive statistics to check the coherence of your findings. Don't skip this step. This process prevents potential errors, inconsistencies or misleading results that lead to unnecessary work and reduce the credibility of your analysis.

As you move further through the different stages of the CRISP-DM framework, e.g. from data preparation to modelling and interpretation, you can think of continuous validation as a kind of safety check. It is a mechanism that ensures that the insights you gain are trustworthy, reliable and aligned with the goals of your business data analysis.

Whether you have used `ggplot2` to create graphs or charts to identify patterns, or whether you are using descriptive statistics to analyze business data, by continuously applying the techniques and methods discussed in this chapter, you will not only learn to analyze business data, but also develop an eye for critical questions and get a feeling for whether the data is valid or not.

3.6 Useful R Functions for Everyday Business Data Understanding

Here we go, you have finished the relevant content of this chapter. In this last section, I collected functions and best practices that you might find useful in your future projects but are not as relevant as the one we already discussed. If you need some time to learn and recapitulate the chapter than it's nothing wrong about to skip this subchapter and come back in some days to it.

The visualizations we created in this chapter are quite plain. Sometimes you might want to change some of the graphical elements. Here are some things you might want to do and I will show you how to do them in R.

First, you want to change the background of your visualization. With `ggplot2` background grids come as a default. However, we can alter them with the `theme()` function.

```
library(ggplot2)
ggplot(data = whisky_collection) +
  aes(x = REVIEWS, y = CRITIQUES) +
  geom_point() +
  theme(panel.background = element_rect(fill = "white", colour =
"black"))
```

By setting new parameters in the `theme()` function, you can change the graphical characteristics of your visualization as illustrated in Figure 3-41.

If you want to change the design and layout on a more general level, `ggplot2` provides a huge sample of themes. For instance `theme_bw()` changes the design into a black and white design so that your plot is more printer friendly, while `theme_minimal()` reduces the plot to the minimal number of elements.

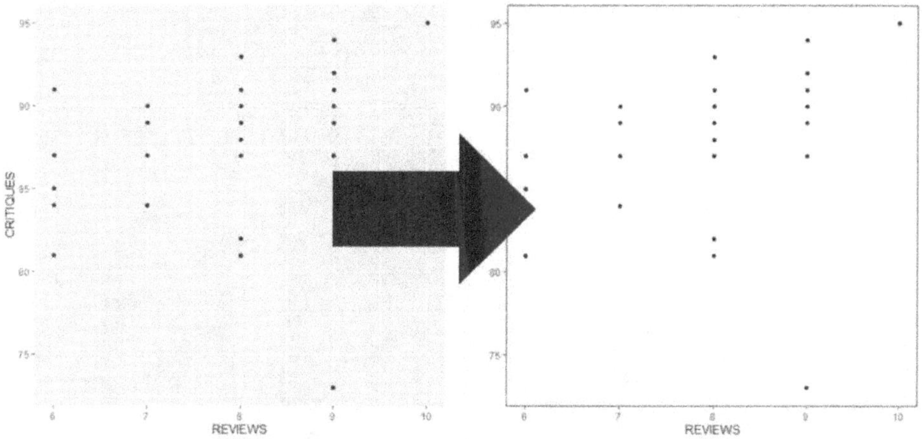

Figure 3-40 Changed theme of a `ggplot2` visualization.

Here is a full list of themes you can play around to find your preferred one:

- `theme_bw()`
- `theme_dark()`
- `theme_classic()`
- `theme_gray()`
- `theme_linedraw()`
- `theme_light()`
- `theme_minimal()`
- `theme_test()`
- `theme_void()`

Instead of the theme of a plot, you can also specify the color pallets of your visualization. In `ggplot2` we can use the `fill` and `color` parameter in `aes()` to select the feature that is used for the colors of each data point. If we select a numeric field, we will get a continuous gradient of colors in our plot. And if we select a categorial variable (like a factor or character), we will get contrasting colors for each group in our plot.

Here is an example, where we pass the `LOCATION` of each whisky to the color parameter:

```
ggplot(whisky_collection) +
  aes(LOCATION, fill=LOCATION) +
  geom_bar()
```

We can now change the colors that are used, the color palette, on two ways. We can add a manual color palette with:

```
+  scale_fill_manual(values = c("red", "blue", "green",…))
```

But you can also use an existing color palette like the color palettes from the `dstools` or `RColorBrewer` package:

```
+  scale_fill_manual(values = use_pal(name="bootstrap"))
```

Here is a comparison of the two resulting plots to compare the `bootstrap` palette with the default palette from `ggplot2`:

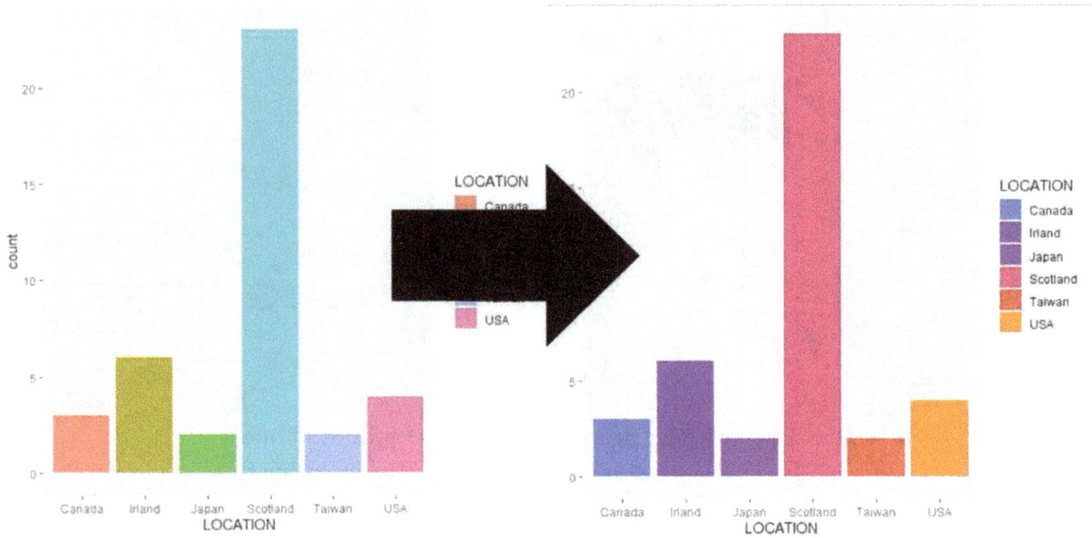

Figure 3-41 Two different color palettes, `ggplot2` default on the left and `dstool`'s bootstrap palette on the right.

If you are interested in the full list of color pallets from the `dstool` package, you can run:

```
> list_pals()
 [1] "acer" "adidas" "adobe" "aifrance" "airbus" "aldi" "amazon" ...
```

Sometimes, you also want that your visualization includes a legend (that is the little box that explains your plot for the viewer). `ggplot2` will automatically include legends, if you specify groups within the `aes()` function of your visualization.

If you do not have groups, `ggplot2` won't generate a default legend. If your intent is to compel `ggplot2` to display a legend, you have to add the `shape` or `line` type for your visualization. This can be used by `ggplot2` to present a legend featuring a single group.

Here is a working example for you:

```
ggplot(data=whisky_collection) +
  aes(x=REVIEWS, y=CRITIQUES, shape=LOCATION) +
  geom_point() +
  guides(shape=guide_legend(title="Awesome Plot"))
```

If you have a legend and want to remove it, add the parameter `legend.position = "none"` in your `theme()` function:

```
theme(legend.position = "none")
```

Sometimes your dataset contains many numeric variables. And you want to see scatterplots or boxplots for all pairs of your variables to see how they are related to each other. For that purpose, you might write a for loop and generate these plots. However, they are then splitted across many distinct plots. I recommend you instead to use the `ggpairs()` function from the `ggally` package.

```
library(GGally)
ggpairs(whisky_collection[, c("LOCATION", "TYPE", "RATING", "REVIEWS",
"CRITIQUES")])
```

Which will return a correlation matrix to investigate the plots of all your numeric variables.

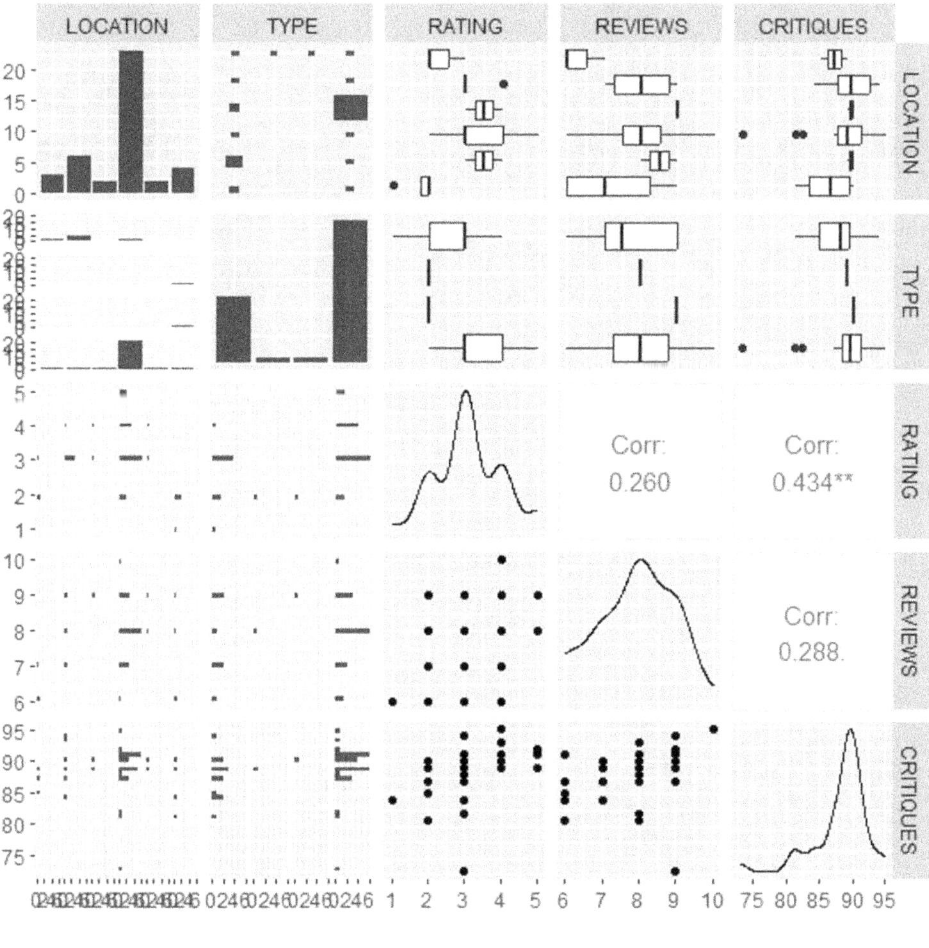

Figure 3-42 Interrelationships of the different variables in our `whisky_collection` dataset.

This kind of visualization is called a conditioning plot or facet. Facets are subplots where each small plot represents a subgroup of the data. In the example from above, each of the barplots, scatterplots, distribution plots and boxplots are an own plot. The faceting function computes them and combines them to one big visualization.

In `ggplot2` you can create such grids on your own. `ggplot2` has two facet functions for that. It is `facet_wrap()` and `facet_grid()`. The difference between these two is that the `facet_grid()` function will produce a grid of plots for each combination of variables that you specify, even if some plots are empty. The other function `facet_wrap()` will only produce plots for the combinations of variables that have values. If there aren't any values it won't produce any empty plots.

If you want to make a conditioning plot, you can do that easily by adding them to our `ggplot2` configuration:

```
ggplot(data=whisky_collection) +
  aes(x=REVIEWS, y=CRITIQUES, shape=LOCATION) +
  geom_point() +
  facet_wrap( ~ LOCATION)
```

Resulting in the plot in Figure 3-43.

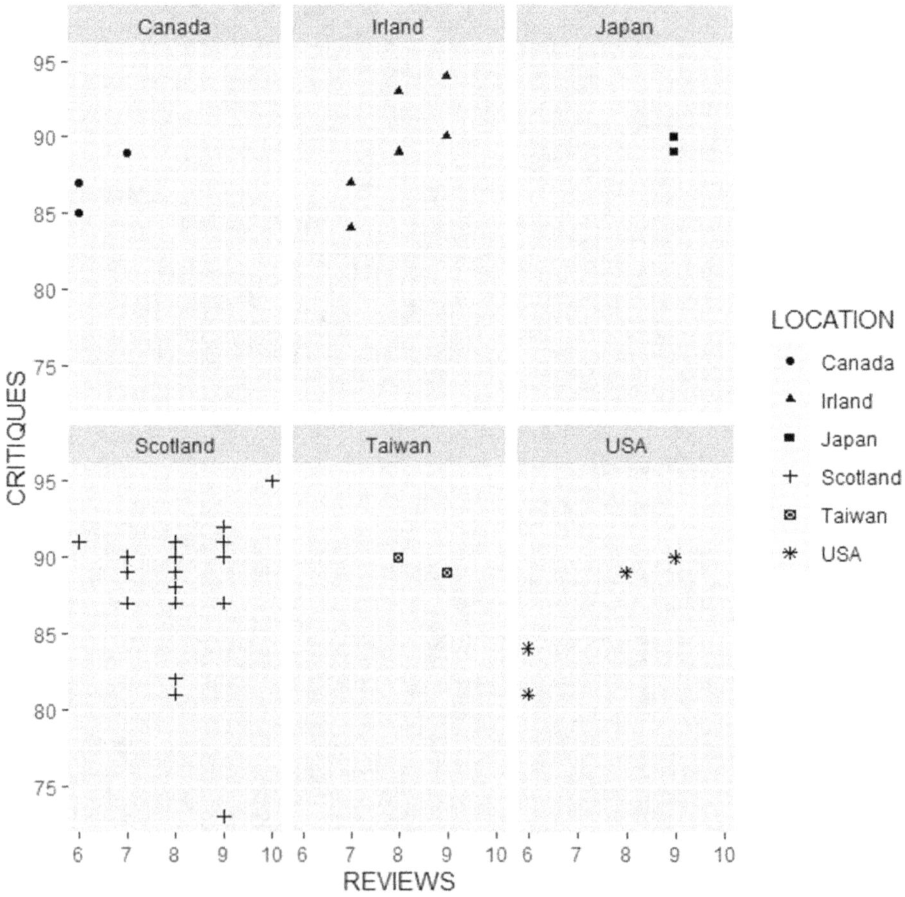

Figure 3-43 Conditioning plot or facet of different whisky ratings by country.

3.7 Checklist

1. Phase: Business Understanding			
1.1	Determine business objectives	❯	• Background • Business objectives • Business success criteria
1.2	Assess situation	❯	• Inventory of resources and capabilities • Requirements, assumptions and constraints • Risks and contingencies • Terminology • Costs and benefits
1.3	Determine analytics goals	❯	• Business data analytics goals • Business data analytics success criteria
1.4	Produce project plan	❯	• Project plan • Initial assessment of tools and techniques
2. Phase: Business Data Understanding			
2.1	Collect initial data	❯	• Initial data collection report
2.2	Describe data	❯	• Data description report
2.3	Explore data	❯	• Data exploration report
2.4	Verify data quality	❯	• Data quality report
3. Phase: Business Data Preparation			
4. Phase: Modeling			
5. Phase: Evaluation			
6. Phase: Deployment			

3.8 Business Case Exercise: Visualizing Whisky Data

At the end of each following chapter, you will find a short case study with a business problem you can solve to recapitulate the content of the previous chapter. These business cases are shorter than the full business cases at the end of the book. I did this because, my intention was to help you to reflect quickly what you have learnt in this specific chapter and not to conduct a whole business data analytics project. If you finish every business case exercise until the last chapter, you are well prepared for the more challenging full business cases in the last chapter.

3.8.1 Scenario

It is lunchtime as you sit with Sherry Young in the production hall. Tired, you dangle your legs from one of the whisky barrels and are discussing whether battered mars bars are more of a main course or a dessert when Alice Gleck strides enthusiastically into the hall. In her hand she waves a loose-leaf binder at you!

The hall buzzes with excitement as she shares her proposal to create a one-of-a-kind experience: the Junglivet Whisky Museum. This innovative endeavor aims to not only showcase the exquisite Junglivet whisky collection but to also immerse visitors in the art, heritage, and flavors that define the brand. The proposed museum will be nestled in the village adjacent to the Junglivet factory, creating a seamless journey from production to appreciation.

With the support of a dedicated team of data scientists, designers, and artists, Alice aims to craft an immersive experience that marries data-driven insights with visual storytelling. As the plan takes shape, Alice envisions a collection of visualizations that guides visitors through the diverse flavor profiles of whiskies world-wide.

3.8.2 Task

Your task is consisting of the following aspects:

- **Whisky Flavor Profiles**: Transform flavor profiles of the collection into vivid visual representations. Utilize color gradients and shapes to depict tasting notes, guiding visitors on a visual journey through the different aromas and tastes.
- **Investigate Ratings**: Leverage line graphs and bar charts to illustrate the rating of each whisky. Highlight the correlation between the different ratings and flavor and origin, giving visitors a deeper understanding of the craft.
- **Regional Mapping**: Employ geographical data to map the regional origins of each whisky. Use maps to display the distillery's location and the locations of various whiskies, offering a sense of terroir and history.

3.8.3 Remarks

Use the dataset `whisky_collection` for this exercise.

3.8.4 Solutions

You find the solutions of this exercise on the course website.

4 Business Data Preparation

4.1 Introduction

There is one very popular academic dataset which I struggle to use in my lectures. It's the `Iris flower dataset` from the British statistician Ronald Fisher (Fisher, 1936). Fisher collected `50` samples from each of three species of Iris flowers (`setosa`, `virginica`, and `versicolor`). The data set consists of different features of the flowers and they can be used easily to separate the data into the three flower types. And thus, his dataset became a typical example to explain analytics techniques and is often used in many analytics lectures. Most likely you have seen it before:

Figure 4-1 Iris flower dataset from Fisher (1936) with the three types of species (blue, red, green).

In my opinion Fisher's dataset illustrated in Figure 4-1, is not that useful to teach real-world business data analytics, because it doesn't represent one fundamental aspect of real-world business data: it is clean. You can use just two features and the data separates clearly into the categories of our target feature. While it may help me as a lecturer to explain the basics of clustering or classification (we will come back to that in the next chapter) without annoying questions from students, it may also lead to many students getting a completely wrong impression of clustering or even analytics in general.

Real world business data analytics faces dirty business data. I mean the kind of ugly, unformatted, erroneous, nasty business data. The kind of business data that will not let you sleep

D. Jung, *The Modern Business Data Analyst*, https://doi.org/10.1007/978-3-031-59907-1_4

at night and will hunt you the most time of your analytics projects. You will spend hours, days, or sometimes weeks to clean and prepare your data before you can even think at analyzing it. Real-world business data is dirty.

Thus, this chapter is about this exciting process to make beautiful well-formatted, understandable data out of huge pile of manure that you normally get in real-world.

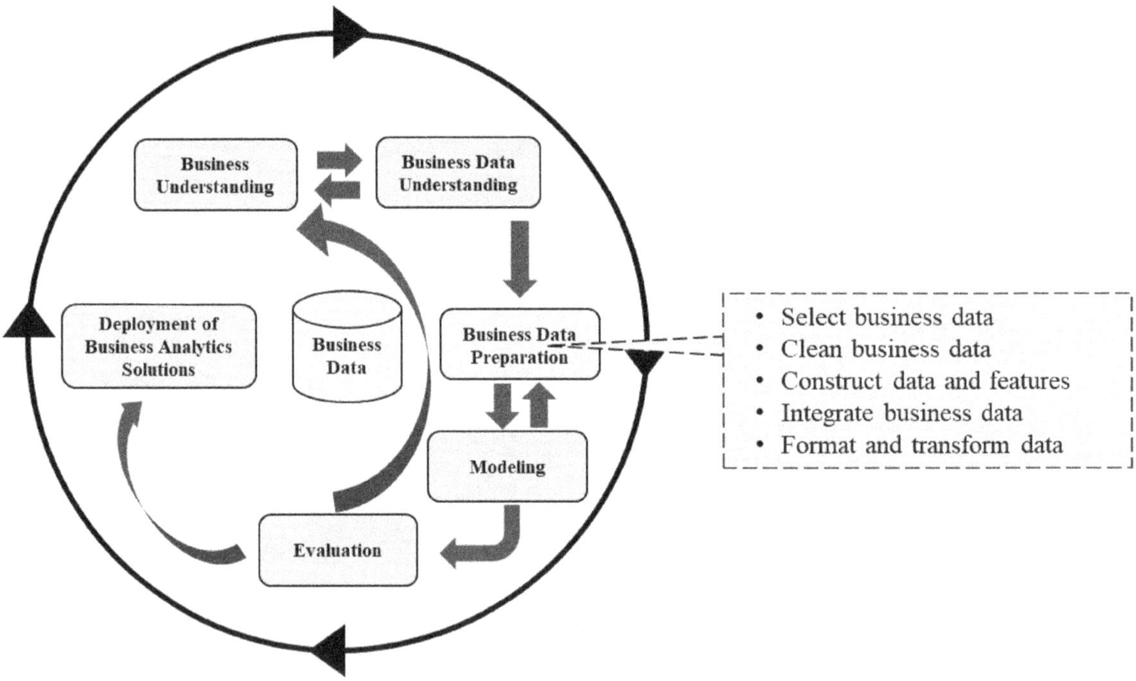

Figure 4-2 The generic tasks of the business understanding phase.

As in the previous chapters, we will base this chapter also on the CRISP-DM framework (see Figure 4-2). And consequently, we will now taggle the issue that business data comes in various facets and quality. Besides, as a business data analyst, you will also have to combine the various business data sources and prepare them for analysis. And that's exactly what the following chapter is about:

- **Select business data**: Decide which data to use in this project based on a clear and documented rationale for inclusion or exclusion (e.g. selection algorithm)
- **Clean business data**: Clean your data based on transparent rules and report them
- **Construct data and features**: Derive features and generate records if necessary
- **Integrate data**: Merge data from different sources or include external datasets
- **Format and transform**: Format and transform the data for the modeling phase, make one or multiple final datasets and document them

The central step of this phase is to select the business data that will be used for the subsequent phases. Consider your identified objectives from the business understanding, the data quality, and any technical limitations like data volume or types (big data!). It's important to note that data selection entails both picking features (columns of your data frame) and determining which observations (rows of your data frame) you should include.

Then, check the business data quality to meet the requirements by the chosen analytics method in the next phase. This process might encompass opting for clean subgroups of the data, introducing appropriate default values, or employing advanced strategies like specific models to estimate missing data. Discuss your choices and steps taken to resolve the data quality concerns. Also, analyze the alterations applied to the data for cleansing purposes and their potential repercussions on the analytics outcomes.

Furthermore, it might be necessary to construct new features and sometimes even new data. This includes generating new features based on existing ones, crafting entire fresh records, or adjusting values of existing features to your needs. For instance, instead of keeping multiple values like `customer_birthday`, `customer_age`, `customer_age_years` you should decide to keep one of these values. Creation of entirely new data is not that often necessary but sometimes useful if you do wait for future data and do not want to delay the project or if it its helpful for the modeling type you choose. For instance, generating data for customers who have not made any purchases in the last year can be a useful action. While such data might not be present in the original dataset, it could be beneficial for modeling to explicitly indicate the scenario where specific customers have made zero purchases.

After you have prepared your different datasets, the challenge is to integrate them into one data source that is used in the modeling phase. This process is called data integration. Data integration involves gathering information from various data frames or databases to build new features or values. For that purpose, you might want to merge or join your data frames and remove values containing distinct details about the same entities. For instance, the marketing team from the *Junglivet Whisky Company* wants to learn more about their customers and clusters them. Therefore, they might combine insights from the shopping behavior in the online shop but also use data from a demographic database. With each data frame containing information about each customer, they can be merged to create a new more informative data frame with information about each customer, combining relevant fields from both data frames.

If you work with different data sources and merge them into data frames for modeling it is best practice to transform and format them into tidy data. Tidy data is a structured format for organizing data sets that facilitates ease of analysis and visualization (Wickham, 2014). In tidy data, each variable is a column, each observation is a row, and each type of observational unit is a table. This standardization simplifies data manipulation tasks and promotes clarity, enabling analysts to quickly and efficiently explore and understand the data's underlying patterns and relationships. We will look at that in detail, but it is obvious that it requires structural adjustments applied to your business data without altering its significance. Additionally, specific modeling techniques might require purely structural alterations to meet their specifications. Examples include truncating all values to a maximum of `32` characters or removing commas from text fields within comma-delimited data files (we have done such stuff already in the business understanding, e.g., to handle our `COORDINATES` feature to make a map). The `dstools` package contains also many functions for that and will also be our friend in this last step.

4.2 Business Data

In the last few chapters, we have created graphs and explored data as a matter of course without realizing a simple fact: In order to do business data analytics, we need to prepare our business data! Business data is all data that is related to your business, and in our case every data that has to do with the *Junglivet Whisky Company*. Production logs, sales, barley prices, staff data and whisky bottles, all these information is business data. But this data is in different formats, spread across different data types and sources and probably corrupted, missing and incomplete.

If you decided to become a business analyst you will probably spend a large fraction of your workdays to handle these issues to integrate different data sources, format the data, gather the data from these sources, clean the data and check the data quality, and transform the data for your analytics models. And it's not surprising that most business data analysts and data scientists find this to be the most boring aspect of their work (Press, 2016).

To speak from my experience, yes, this process is quite embarrassing, but in big companies there are data engineers who do this exciting work the whole day (let's say thank you to our fallen brothers). And let me say that data engineering will always be a relevant and huge part of analytics projects. Different sources state that up to 80% of your business data analytics project time will go down for tasks related to business data preparation (Press, 2016). Sounds exciting, yay!

So, before we continue our journey and visit the highlight of a business data analysts work, we have to talk about business data in general. How does it look like in practice? What are common data sources in business data analytics projects? How do I handle them?

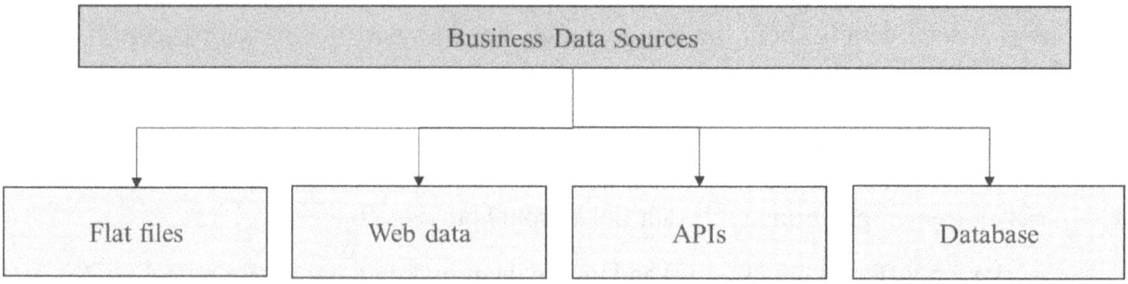

Figure 4-3. Most common business data sources.

In the following section, we will now look at the most popular business data sources in business data analytics projects, as illustrated in Figure 4-3. We will learn different methods and techniques how to get these data sources into your R-environment. And how to integrate different data sources, clean and format it, build new features, reduce irrelevant observations and features, and finally transform it for your analysis!

4.2.1 Flat Files

Most flat files are text files that are an extract of a database or a local excel file. CSV files, which we already know from the previous chapters are also flat files. In your business data analytics projects these kinds of files will be the most common data source.

You can explicitly read flat files from your local machine in your R environment. The most common flat file data type are comma-separated values (CSV files). I prepared such a file which you can use to test and read like that:

```
whisky_collection_csv = read.csv("whiskycollection.csv")
```

However, CSV files are used less and less in modern projects. The reason is that the comma (",") separator is not very usable because the comma itself has become a very common character often used in the business data value itself. Therefore, other separators such as ";" or "|" are increasingly used. All these files that use some forms of such separators or delimiters are called DSV or delimiter-separated values in the technical language. In R these can easily be read in with the wizard (see chapter 2) or directly in the following steps.

First, we install the `readr` package:

```
install.packages("readr", dependencies = TRUE)
library (readr)
```

And then load the dataset into our R environment:

```
whisky_collection_dsv = read_delim("whiskycollection.dsv", delim=";")
```

The last parameter of the function `read_delim()` of the `readr` package is used to specify the delimiter. If your flat file contains values that are separated by "|" you just change the parameter to `delim="|"`.

Furthermore, there exists a specific R data format, that can be used to save your data.

You can load such R data files with:

```
load(file="whiskycollection.RData")
```

You can also write your results to your local machine, which generates new flat files for other analytic projects. Its best-practice to save your data in the R-specific format, which will decrease loading time a lot. But it also has some cons like that it can be only used in R projects.

To save your objects, plots, variables etc. from your project locally, you can do that with `save()`:

```
save(name_of_the_object, file="name_of_the_object.RData")
```

4.2.2 Web data

In most cases, as a business analyst, you will be working with files from your companies. However, there may be scenarios where you will need to access data from the web. Either because you want to use current information in your analyses (such as stock market prices or weather data) or because you are analyzing certain websites.

So, in most cases you will either

- load files from the web into your R project or
- read in a website in the form of an HTML file and extract the information relevant for you.

Entire R books have been written about these scenarios, and I certainly don't claim to cover them completely in this book. However, I think it can't hurt to take a short look at the procedure for the two scenarios.

In the first scenario, we will now look at how one or more files located on a website can be downloaded and used in your project. For this we want to download the files from the git repository *https://github.com/dominikjung42/BusinessAnalyticsBook* which you can also find on the course website *junglivet.com*.

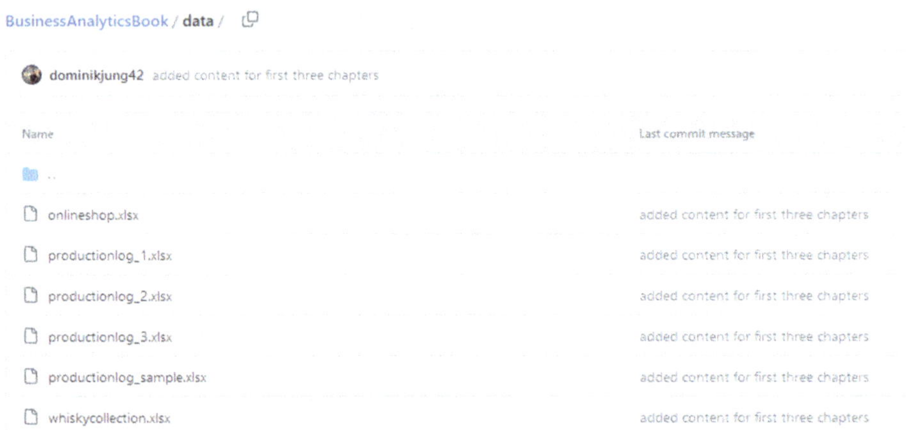

Figure 4-4 The git repository contains all the file for this book.

As you can see there are different files on the git repository (`whiskycollection.xlsx`, `onlineshop.xlsx`, etc.). To make a live connection to the repository and download them directly in our R environment, we need to bring in some packages. The R packages that we'll be using are `readr` and `rvest`. The last one is new and you have to install it, when it comes to web scraping.

But for now, we can start with downloading the file and loading it into R with:

```
# Read from the web
urlfile="https://raw.githubusercontent.com/dominikjung42/BusinessAnal
yticsBook/main/data/whiskycollection.csv"
dataset = read_csv(url(urlfile))
```

As you can see the procedure is straight forward. You specify the URL with the variable `urlfile` and then you can access the file in the web with the `url()` function.

Alternatively, you could download the file and save it in your R project with base R and then import it locally. The disadvantage is that you have to download the file and it might be outdated if you do not so. On the other side this procedure makes sure that you always have a working copy of the dataset on your local machine:

```
download.file(urlfile, destfile="./downloaded_whiskycollection.csv")
dataset = read_csv("downloaded_whiskycollection.csv")
```

In other cases, the data is not saved as a local file, it is provided on a website e.g., as a table or text. In such cases you have to scrape the website for the information you need.

Let us assume you want to download a table with all whisky distilleries of Scotland. The source is available at: *https://en.wikipedia.org/wiki/List_of_whisky_distilleries_in_Scotland.*

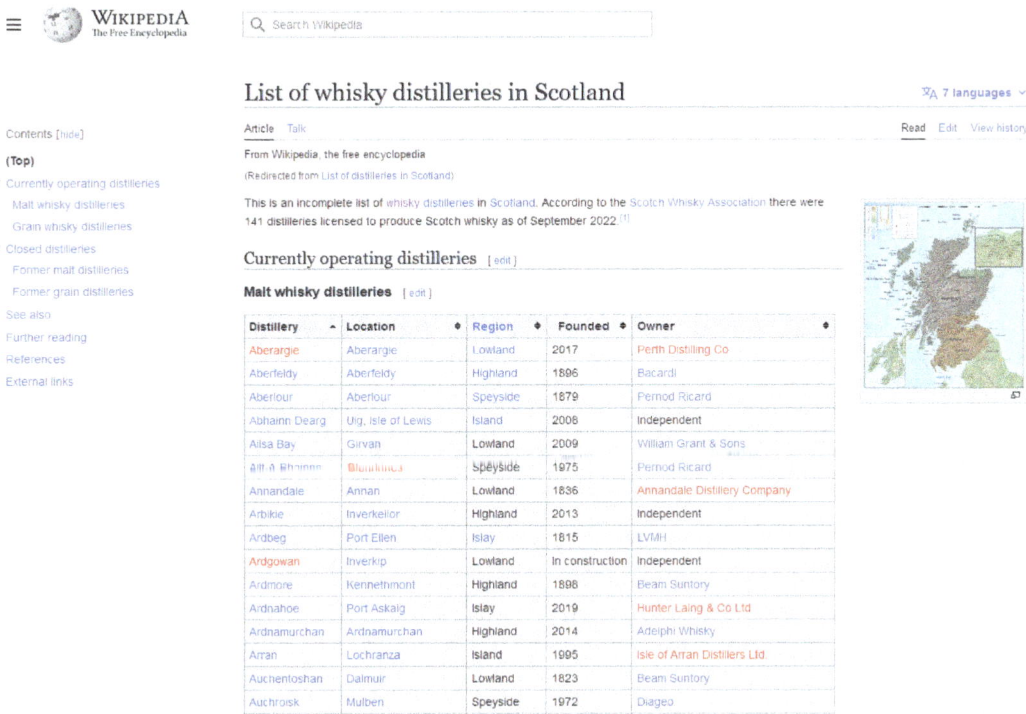

Figure 4-5 The whisky distilleries in Scotland are available at Wikipedia on a long list we want to download.

A first good practice is to first inspect the structure of the website you want to scrape. This is necessary to understand if you have to download a table or a paragraph or something else. If you open the Wikipedia page it is quite obvious that the whisky producers are stored in a table for which the associated html-tag is `<table>`.

We can read the entire website using `read_html(url)` and filter all tables using `read_html(html_nodes(...,"table"))`.

If you did not install `rvest` so far, you have to do it now:

```
install.packages("readr", dependencies = TRUE)
library(readr)
```

And then we can start scraping the table from the Wikipedia page with:

```
url="https://en.wikipedia.org/wiki/List_of_whisky_distilleries_in_Sco
tland"
website = read_html(url)
tables = html_nodes(website, "table.wikitable")
tables = html_table(table, header = TRUE)
```

In this example we choose the Wikipedia site instead of the git repository as URL. Then we parse all tables in an object called tables.

You can then access them by indexing the number in the `tables` object. For instance, if you want to have the "Former grain distilleries", the fourth table on the website, you can do that with:

```
> tables[4]
  Distillery         Location             Region  `Year closed`
  <chr>              <chr>                <chr>   <chr>
1 Caledonian         Haymarket            Lowland 1988
2 Cambus             Tullibody            Lowland 1993
3 Carsebridge        Alloa                Lowland 1983
4 Dumbarton          West Dunbartonshire  Lowland 2002
5 Garnheath          Airdrie              Lowland 1986, demolished 1986
6 North of Scotland  Tullibody            Lowland 1980
7 Port Dundas        Glasgow              Lowland 2010
```

Next step would be to store the scraped table into a data frame and save it locally or use it directly for further analyses.

4.2.3 APIs

A very convenient way to download and access data in the web are APIs. API stands for application programming interface and means that the data you want to access is provided directly for usage. So, you can think of an "API" as a kind of procedure for how one of our R scripts interacts with another computer.

APIs thus provide business data analysts with a very sophisticated way to request clean data from a website. When a website like Reddit, Twitter, or Facebook provides an API, there is an explicit server that is essentially just there to wait for data requests and provide them to interested users.

The procedure for doing this is relatively simple. Once the server receives a data request, it processes the data and sends it to the computer that requested it. So, from our perspective, we need to write code in R that creates the request and tells the computer running the API what we need. The server then reads our request, processes it, and returns nicely formatted data that can be easily read by existing R packages.

Now where is the advantage in this approach over the web scraping, we just learned about? When we as analysts read a web page, we often have to write very elaborate code that looks in the web page for the data we need. And as a result, we get the data in the form of a messy piece of HTML. While there are packages that simplify the parsing of HTML text, these are all cleaning steps that must be performed before we can even get our hands on the data we want! Often, we can use the data we get from an API right away, saving us time and frustration. Most of the websites offer an API, to reduce unnecessary traffic to their websites through web scraping.

If you work as a business data analyst you will probably have access to the APIs of your company or your company will buy you access to APIs you are interested in. Because I do not want that you have to pay money for a random company to get access to their API, I decided to use here the free API from omdb – the open movie database.

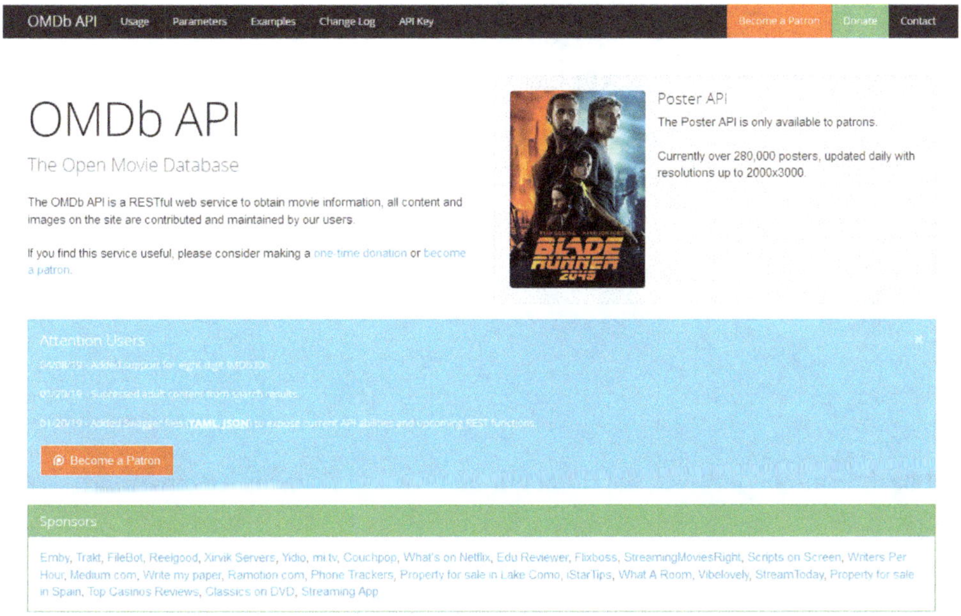

Figure 4-6 The open movie database is a free to use web service to obtain movie information.

If you visit the website, you can generate your own API key (just go to the menu option "API Key"). If you generated your own API key you can continue this code example, otherwise the code won't work for you. The final key is an URL that should look somehow like this:

```
http://www.omdbapi.com/?i=123123123&apikey=8888ABCD
```

If you do not want to setup an API key, you can now continue quickly and just read over the code example to understand the general concept behind APIs.

Whatever you have decided, as before, we need some packages. In this case to have access to functions to use APIs from the web. The R packages that will help us in this section are `httr` and `jsonlite`. `httr` is to access APIs and the other package `jsonlite` is necessary to read and handle the JSON data format which is very common in web APIs.

If you don't have these packages installed, you will have to download them first:

```
install.packages("httr", dependencies = TRUE)
install.packages("jsonlite", dependencies = TRUE)
```

After downloading the packages, we will now load them in our R-environment and make a first API request. There exist many types of requests you can send to an API. In our case we are interested in receiving data, so we will send a "`GET`" request, which stands for "get some data".

We will send this request to the server that has the API, and assuming everything goes smoothly, the server will send back a response with our data.

To create such a GET request, we need to use the `GET()` function from the `httr` package. Then we need to specify the `GET()` function. It needs a URL that specifies the address of the server to send the request to.

Here is a working example, if you have an API key:

```
api_key = "xxxxxxx"
movies_raw = GET(url = api_key)

# get status
status_code(movies_raw)

# download
str(content(movies_raw))
movies_text = content(movies_raw, "text", encoding = "UTF-8")
str(movies_text)
```

The result we got back from the API is a list of length `25` movies. Of these, two parts are important:

- `status_code`: That tells us, if the call worked network-wise. For a list of possible status codes, see *https://en.wikipedia.org/wiki/List_of_HTTP_status_codes*.
- `content`: The API's answer in raw binary code, not text. Alas, the answer could also be an image or a sound file.

If we examine the status code:

```
movies_raw$status_code
## [1] 200
```

We see that the response is `200`, which means, all worked out fine. Note that this status code only tells us, that the server received our request, not if it was valid for the API or found any data.

The requested API data is displayed using the `content()` function. This is in JSON format and if you would run just `get_movie_text` this will return a lot of non-readable text:


```
[1] "{\"Title\":\"Guardians of the Galaxy Vol. 2\",\"Year\":\"2017\",\"Rated\":\"PG-13\",\"Released\":\"05 May 2017\",\"Runtime\":\"136 min\",\"Genre\":\"Action, Adventure, Comedy\",\"Director\":\"James Gunn\"...
```


Therefore, we need the other package `jsonlite` to reformat the data from JSON to a data frame, which we can use in R:

```
# parsing json
movies_json = fromJSON(movies_text, flatten = TRUE)
movies_json

# convert to data frame
movies_dataframe = as.data.frame(movies_json)
movies_dataframe
```

JSON stands for the term "JavaScript Object Notation". This means that the data is stored as JavaScript objects. The JSON format is used on the one hand because it is easily readable by a computer, on the other hand Javascript is the relevant language to build web applications. And for this reason, the Javascript proprietary data format JSON has become the main method for transporting data via APIs and has now replaced the previous XML standard. In short, most APIs send their responses in JSON format.

4.2.4 Databases

So far, we have dealt with manageable data sets like tables or CSVs. These are at most a few GB in size and therefore fit easily into your computer's memory. In practice, however, you will encounter data sets that are too large for your computer to process as a whole. Companies store this type of data in a database. In some cases, a copy of the database is provided to you, the analyst.

However, it is much more convenient if you access the data in the database directly. The advantage is obvious. When you connect to a database you can always retrieve the parts you need for the current analysis. In addition, many large data sets are already available in public or private databases. You can easily query these and integrate the current data set directly into your R script.

The R programming language is compatible with almost every existing database type. By now, there are R packages for most common database types, with which you can easily set up a database connection. In addition, there are packages that extend existing packages to work natively with databases.

A very popular combination is the `dplyr` package in conjunction with `dbplyr`. `dbplyr` allows `dplyr` to connect to widely used open-source databases such as sqlite, mysql and postgresql as well as many other databases. Also, the provider of RStudio has collected all the important information especially on this topic and provided detailed documentation and best practices for working on database interfaces on its website: *https://rviews.RStudio .com/2017/05/17/databases-using-r/*.

In this section, we will now look together at how to load records from a database by generating “select SQL statements”. To do this, we'll look at how to interact with a database using `dplyr`, using both `dplyr`'s verb syntax and SQL syntax.

As before it might be necessary that you install the required package:

```
install.packages("dbplyr", dependencies = TRUE)
```

The SQLite database is contained in a single file `whisky_collection.sqlite` that contains the `whisky_collection` dataset in the form of a database. If you don't have it, you can download it manually from the GitHub repository or run the following code to download it in your current R-project:

```
urlfile="https://raw.githubusercontent.com/dominikjung42/BusinessAnal
yticsBook/main/data/whiskycollection.csv"
download.file(urlfile,
destfile="./downloaded_whisky_collection.sqlite")
```

After downloading and installing the database tools you can connect to the database with:

```
con = dbConnect(SQLite(), "downloaded_whisky_collection.db")
```

Probably the `dbConnect()` command will catch your eye right away. This function means that we connected to the database using the `dbConnect()` method in the `DBI` package. `DBI` is not a package you will use directly as a beginner in business data analytics. The `DBI` package helps connecting R to database management systems (DBMS).

With this command the advantage of a database comes into play. It does not load the data into the R session (as with the `read_csv()` function from the previous section). Instead, we simply tell R to connect to the SQLite database contained in the `downloaded_whisky_collection.db` file. When we need the data, it is then provided. Practical, not?

Now let's take a closer look at the different tables in the database we just connected to in a next step:

```
> as.data.frame(dbListTables(con))
  dbListTables(con)
1 whisky_collection
```

As we can see there is one table called `whisky_collection`. In this database there is only one table, but in your business databases there will be probably hundreds of tables. Now that you know how to connect to the database, let's look at how we can use the data in R.

To connect to the individual tables within a database, you can use the `tbl()` function of `dplyr`. This function can be used to send SQL queries to the database. To demonstrate this functionality, we select the columns `NAME` and `RATING` from the table `whisky_collection`:

```
tbl(con, sql("SELECT NAME, RATING FROM whisky_collection"))
```

If you are familiar with SQL you can use any of the common SQL queries you know with this approach. If you are not familiar with SQL you can alternatively use the `dplyr` syntax.

As mentioned earlier, one of the strengths of `dbplyr` is that it extends the `dplyr` package to work directly with databases and their tables.

To do this, we select the table on which to perform the operations and then use the standard R syntax as if it is a data frame:

```
db = tbl(con, "whisky_collection")
head(db)
```

However, if you run:

```
> nrow(db)
[1] NA
```

You see that R does not know how many rows the table has. The reason is that compared to data that is loaded locally like the CSV with `read_csv()` the database data is not loaded on the memory of your computer. This addresses a common problem with R, which is that all operations are performed in memory, and thus the amount of data you can work with is limited by the available memory. The `dplyr` package remove this limitation, because you can connect to very huge databases, run your queries directly, and use on your computer only the data needed for analysis.

In the next chapters we will learn more about data preprocessing and the related steps: data integration, data cleaning, and data engineering.

4.3 Business Data Cleaning

As illustrated in the introduction, business data unfortunately comes in all shapes and sizes, with all sorts of issues. Building on the business data quality report (see chapter 3) from the business data understanding, we have to clean it and fix these issues.

For that purpose, we start with the raw data, which is in most cases the data we have identified in the business data understanding. Then, we work on preprocessing and transforming it into tidy and clean data that's ready for the specific needs. This clean data will give us accurate results when we use it for statistical analysis (GIGO-principle).

The steps we have to take can differ depending on the type of data cleaning problem we're dealing with. It's important to be mindful of the specific data you are using. Different datasets might require different steps to clean them up.

In general, there are four types of data cleaning problems in business data analytics:

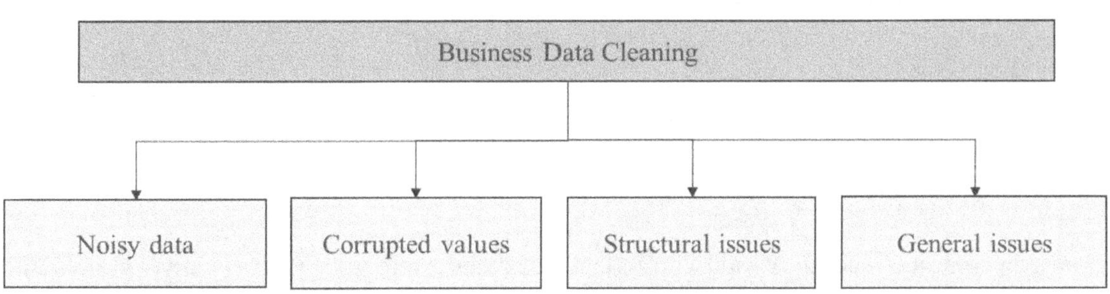

Figure 4-7 The different problems that have to be handled in business data cleaning.

Unclean data can have many shapes like noise, corrupted values, structural issues or general quality issues (see Figure 4-7). These problems can change depending on the specific dataset we're working with. So, keeping an eye out for these issues is crucial to getting reliable results in our analysis.

Noisy data is one of the most relevant issues. Outliers and unusual values can introduce noise into your dataset, while duplicate entries can distort analysis and lead to erroneous insights. Thus, you have to identify and remove duplicated rows and values, ensuring that your data remains accurate and representative. And you have to develop strategies or use techniques to identify and handle outliers, including data smoothing and transformation methods.

Second, corrupted data like missing values or wrong data types can result in low performance of your analytics model or throw errors so that your modeling algorithms won't work. Therefore, ensuring the correct data type for each variable is crucial for accurate analysis. You have to apply methods to impute missing values and understand the implications of different imputation techniques.

Structural issues might mostly rise when you combine different data sources into one single data frame for your analysis. Inconsistent naming conventions will complicate merging your data and use it directly for analysis. Furthermore, tidy data is a central requirement in business data analytics and if the data is from other sources your data might be untidy. We have to look at different techniques to reshape and reformat your data to tidy data, making it easier to explore and visualize it.

Last but not least, we have to address general data issues like timeliness and relevance. And to ensure that our data reflects the current context and is aligned with our objectives from the previous phases. It is also a good point to check if you have a holistic understanding of the business data you are using. Including its characteristics, like potential biases, and trends.

To illustrate the problem of corrupted data I prepared a corrupted version of the `whisky_collection` dataset we used in the previous chapter. You can download the corrupted dataset as "whisky_collection_corrupted.xlsx" from the GitHub repository and load it into your R environment:

```
download.file(urlfile, destfile="./whisky_collection_corrupted.xlsx")
whisky_collection_corrupted                                          =
read_excel("whisky_collection_corrupted.xlsx")
whisky_collection_corrupted
```

Or just from the `dstools` package by running:

```
data("whisky_collection_corrupted")
whisky_collection_corrupted
```

Before you continue reading, I want you to load the dataset into your RStudio environment and investigate it with the techniques you learnt in chapter *3 - Business Data Understanding*. You can make plots, investigate it manually by running the `View()` command or compute descriptive overviews with `summary()` or `glimpse()`. Write down the issues you identified. And then continue reading – we will have a look at them now!

4.3.1 Fix Corrupted Values

Fixing corrupted values is one of the most challenging parts of data cleaning. But what values are corrupted? And which value do I use instead? That are questions that are not easy to answer. Let us have a look at this now step-by-step!

If we load the `whisky_collection_corrupted` and run a simple `summary()`, there are a lot of quality issues related to corrupted values that come to our mind.

First the data type of `RANKING` is `character` and not `numeric`. By checking the column with the `table()` function we see that it contains only numeric values:

```
> table(whisky_collection_corrupted$RATING)
 1  2  3  4  5
 1  8 19 11  4
```

We can now use this information to make a new dataset `whisky_collection_pre` which is the name of our prepared dataset. In the prepared dataset we want to fix the quality problems.

In this example, we can solve this issue easily by recasting it to a numeric column with:

```
whisky_collection_pre = whisky_collection_corrupted
whisky_collection_pre$RATING=as.numeric(whisky_collection_corrupted$R
ATING)
```

In the best case your dataset has no further issues, which means it is complete and has all values. But as we learnt real-world datasets are ugly. They will probably have errors and be incomplete. Which means they have values that are not plausible or have missing values.

For instance, if we look at the column `FOUNDATION`:

```
> summary(whisky_collection_corrupted$FOUNDATION)
   Min. 1st Qu.  Median    Mean 3rd Qu.    Max.    NA's
   1743    1823    1860    2316    1969   20111       1
```

We see that there is an issue with a whisky that has as foundation the year `20111`. This seems to be a typo and we have to handle this issue. Most straight forward procedure is to drop the row with the unusual values. For instance, if the column `FOUNDATION` should only contain values between `1743` and `2024`, we can remove outliers or corrupted values easily with:

```
whisky_collection_corrupted %>%
  filter(between(FOUNDATION, 1743, 2024))
```

This method may seem intuitively correct, but it has the problem that we may lose important data. Unusual data very often has a plausible reason and can represent important information beyond the feature. Moreover, just because a single value of a single row is wrong, it may be excessive to discard the whole row or column!

A better approach is to replace the unusual value or outliers with NA or even a meaningful one like 0 and come back to this row if you start handling missing values:

```
whisky_collection_corrupted %>%
mutate(FOUNDATION = ifelse((FOUNDATION <0 | FOUNDATION >2050), NA,
FOUNDATION))
```

In R missing values are indicated as NA in your data table. As we know from the previous chapters, NA stands and represent an unknown value and that most operations you compute on NA values will return a NA value:

```
> 100 + NA
NA
```

Furthermore, it has also implications for data preprocessing, because the `filter()` function only includes rows where the condition is TRUE, which excludes both FALSE and NA values in your columns.

As a consequence, it is very important to find the missing values because many algorithms in business data analytics have problems with it. For instance, if you run a computation of the arithmetic mean with `mean()` it will return no usable value:

```
> mean(c(1,NA, 3))
NA
```

To handle this issue, you have to set the parameter `rm.NA=TRUE` to get usable results.

```
> mean(c(1,NA, 3), rm.NA=TRUE)
2
```

If you want to determine if a value is a NA value, you can use `is.na()` to test for rows with NA values:

```
is.na(c(1,NA, 3))
output fehlt !!!
```

By investigating the dataset with `summary()`, you should have noticed there are also many NA or missing values in our dataset:

```
> summary(whisky_collection_corrupted[,c(6,14)])
   FOUNDATION           PRICE
 Min.    : 1743    Min.    :   19.00
 1st Qu.: 1823    1st Qu.:   34.25
 Median : 1860    Median :   43.50
 Mean    : 2316    Mean    :  219.21
 3rd Qu.: 1969    3rd Qu.:   71.50
 Max.    :20111    Max.    :2700.00
 NA's    :1        NA's    :1
```

There is also the column `origin` (sic!) in the dataset that contains mostly missing values, which are not detected because the column is marked as character column. But by plotting or inspecting with `View()` you will have probably noticed this issue.

Handling missing values is a more complicated issue. We have to identify the reason for that and decide if we remove the row, the whole column, the whole source dataset (if your dataset is a combination of multiple datasets).

These methods are straight-forward when you want quick results and reduce the effort of the business data preparation. If you want to do this you can use filter or remove `NA` values in specific columns like `FOUNDATION` of your dataset:

```
whisky_collection_corrupted %>%
  filter(!is.na(FOUNDATION))
```

Or you can remove rows with `NA` value in any column with:

```
df %>%
  na.omit()
```

However, this is not the best way to go. From my long-time experience as data scientist, I can say you that it seems like an easy solution, but it can have broader implications. It might require you to reconfigure your analysis, and it could potentially waste the effort that went into collecting or processing the data.

Deleting columns with missing values brings you faster towards building models but can lead to a significant loss of information. Each column in your dataset may provide valuable insights or context to your analysis, and removing them could result in a less comprehensive understanding of the data. In particular, the missing data could be relevant to your business problem or offer insights that contribute to your understanding of trends and patterns.

If you think at the production log from Mr. Gumble, it becomes clear that the reason for missing data could be the production machines and there could be an issue. As a result, you should talk to him and identify the reason for the missing data. If the missing values are due to data collection issues or data entry errors, these problems might persist in future data collection efforts. Deleting columns won't address the underlying issue, and you'll likely encounter the same problem in future datasets.

Another problem is that by removing specific data you will start to build analytics models with biased data. The reason for this bias is that removing columns with missing values does distort the representation of your data. If the missing values are not randomly distributed, the data left after column removal may no longer be representative of the population you are analyzing.

So, after reading this annoying plaidoyer to avoid easy solutions and invest more effort into the data preprocessing, I recommend that you instead make a so called "educated guess". This guess could be to fill the missing value or values manually, or based on our business knowledge. Some guidelines recommend to replace the missing value with a constant like `0`. If you have some more motivation, I recommend you to fill them by a computed value.

There are many approaches to compute values to make an educated guess. The most common ones are the measures of central tendency, like the arithmetic mean, the median or the mode.

Mean is best when your data is numeric and follows a relatively symmetric distribution. The mean is affected by all values in the dataset, making it a suitable choice for a balanced distribution. However, the mean is sensitive to outliers (we will come to that topic later). If your dataset contains outliers that could significantly skew the mean, you might consider using a more robust measure of central tendency, like the median.

Use the mode when dealing with categorical or nominal data. The mode represents the most frequently occurring category or value. It's suitable when the data has distinct categories with one or more being dominant. For example, in the column `TYPE`, if `"Single Malt"` is the most common type, using the mode would make sense.

In our case, we can fill the missing values of the columns `PRICE` and `FOUNDATION` in R with the median because both features are not symmetrically distributed.:

```
whisky_collection_pre[7,14] =
median(whisky_collection_corrupted$PRICE, na.rm = TRUE)
whisky_collection_pre[15,6] =
median(whisky_collection_corrupted$FOUNDATION, na.rm = TRUE)
```

In some cases, you might even consider using different methods and comparing their effects on your analysis. For example, if you have missing values in a numeric variable, you could fill them with the mean to maintain the average, and then check if the mode (or another category) is a better choice to fill missing values in a categorical variable.

Another issue is the column `LOCATION`. It contains missing values for the whiskies `Laphroaig` and `Caol Ila`. You could also say that there are missing values for the whiskies `Eagle Rare`, `Jack Daniels`, `Westward`, and `Whistlepig`. But there is a column origin, which contains the `LOCATION` for these whiskies, so this is a structural problem and no data corruption problem. Hence, we will fix the values for the last ones later. But it would be also fine to fix this issue in this step.

To compute the missing values of Laphroaig and Caol Ila for `LOCATION`, we can make an educated guess by looking in the other columns and notice that they all are from regions that are in Scotland. Hence, we can fix that easily with:

```
whisky_collection_pre[3,3] = "Scotland"
whisky_collection_pre[15,3] = "Scotland"
```

In our cases good values for the missing values could be found by computing means or investigating the data. Alternatively, you could also use the column `WIKIPEDIA` to look up the information in the internet. Which is fine, because it shows that you understood the data.

In other cases, it might be much more difficult to find suitable values. In this case you can use the most probably value for an attribute derived by some modeling algorithm from other attributes. We will have a look at different modeling techniques in the next chapter. But I want to point out to use them (e.g. clustering, regression, classification) not only to model business analytics problems, but that you can use them also to find suitable values for missing values.

Ultimately, the decision should be driven by your understanding of the data, the distribution, and the specific requirements of your analysis. It's always a good practice to explore the data, consider potential biases, and think about the implications of using either the mean or the mode to fill missing values.

If you work with real data, it might be very likely that you have unusual values in your dataset. These are rows with values that are too high or too low or contain values that are unrealistic. Sometimes the reason for them might be import errors or data entry errors.

4.3.2 Reduce Noise and Outliers in your Data

Another relevant problem besides corrupted data is noise. Reducing noise in your dataset can help you better understand the distribution of your data and increase the accuracy of your predictive models.

One of the most common origins of noise in your data are duplicates. Duplicate data can skew your results and compromise the integrity of your analysis. And the most straight forward approach to check for duplicates is by the base R function `duplicated()`. It returns a logical vector where `TRUE` specifies which elements of a vector or data frame are duplicates.

```
> duplicated(whisky_collection_corrupted)
 [1] FALSE FALSE FALSE FALSE FALSE FALSE FALSE FALSE FALSE FALSE FALSE
FALSE FALSE FALSE FALSE FALSE FALSE FALSE FALSE FALSE FALSE FALSE FALSE
FALSE FALSE
TRUE FALSE FALSE FALSE FALSE TRUE FALSE FALSE FALSE FALSE TRUE FALSE
FALSE FALSE FALSE FALSE FALSE FALSE
```

If you want to extract duplicate elements to understand where the duplicate rows are you can run:

```
> rows = duplicated(whisky_collection_corrupted)
> whisky_collection_corrupted[rows,1]
[1] "Glenmorangie" "Brora"        "Junglivet"
```

And there we go, there are three duplicate rows and we have the names of the duplicate whiskies. We can now use this information as before to make a new dataset `whisky_collection_pre`, which will contain our fixed values. Hence, let us now remove the duplicate rows with:

```
whisky_collection_pre = whisky_collection_corrupted[!rows,]
```

Another straightforward way would be to use the `unique()` method:

```
whisky_collection_pre = unique(whisky_collection_corrupted)
```

Or to run the `distinct()` command in a `dplyr` chain:

```
whisky_collection_pre = whisky_collection_corrupted %>%
  distinct()
```

Another relevant origin of noise is noisy values. You will mostly detect them by plotting your data in the business data understanding phase. Value noise can distort your analysis and lead to erroneous conclusions. Therefore, you have to identify and address noisy values using R.

In our case the features RATING, CRITIQUES and REVIEWS represent one similar thing – a general scoring of the whisky. However, they use different value ranges and scorings. They seem to be good candidates for the noisy column award. Furthermore, the feature PRICE also needs some further investigation. One could argue that it's noisy, because it has some values in the range of 0 - 100 and some very expensive whiskies that are far away from this range. This indicates that the whiskies with prices far above 100 EUR are potential outliers or at least there seems to be some noise in the feature PRICE, which we have to reduce.

We have now different options to handle this. One is simplification or binning. This means that we use techniques to group our values into different categories to reduce the complexity in the data. This approach builds on the idea that by categorizing our continuous values into discrete groups or intervals we make the feature "easier" to understand and to analyze. Furthermore, binned data tends to be less sensitive to noise or small variations in the original values, which can help smooth out fluctuations and reveal broader trends.

In our case we could compute a new column RATING_CAT, CRITIQUE_CAT and REVIEWS_CAT ("cat" stands for categorization) and transform the values in categories like "ok" and "excellent". This process is called manual simplification and can be done like this, e.g. for the feature REVIEWS:

```
whisky_collection_corrupted %>%
  mutate(RATING_CAT = ifelse(REVIEWS < 9, "normal", "excellent"))
```

However, such manual approaches, while not fundamentally wrong, require a great deal of understanding about the data and distribution. It is therefore more common to follow either an equal width (distance) binning or equal depth (frequency) binning approach.

For equal frequency binning, I suggest using the cut() function:

```
whisky_collection_corrupted %>% mutate(REVIEWS_CAT = cut(REVIEWS,
breaks=2))
```

If you want to do equal width binning, you have to do some calculations before using cut():

```
bins = 2
reviews_min = min(whisky_collection_corrupted$REVIEWS)
reviews_max = max(whisky_collection_corrupted$REVIEWS)
width = (reviews_max - reviews_min)/bins

whisky_collection_corrupted %>%
mutate(REVIEWS_CAT=cut(REVIEWS,breaks=seq(reviews_min, reviews_max,
width)))
```

Another way to reduce noise is building concept hierarchies. Concept hierarchies provide context to your data. Using them you align data values with more general concepts, enhancing your analysis. For example, you could map prices to ranges, providing a broader perspective on your data. However, as before in the manual binning approach, you need to have a profound knowledge about the data and the domain you are working.

Last but not least, later in the next chapter, we will also learn more about modeling and how using modeling techniques like regression or clustering can help you to reduce noise in your data.

Besides, this specific noise reduction techniques there exist some transformation techniques on a lower level that can help you to improve the quality of your dataset for the modeling. Many analytics techniques like clustering or decision trees are sensitive to the scale of your data or won't work if the columns in your data have different units.

For instance, if the values of one variable range from `0` to `1.000.000` and the values of another variable range from `0` to `100`, the variable with the larger range will be given a larger weight in the analysis and reduce our predictive power. For that purpose, business analysts scale or normalize the data before analyzing it.

Let us now have a look at the two most important techniques to handle different scales or units in your dataset. Two common ways to normalize (or "scale") variables include:

- Min-max normalization
- Z-score standardization

Normalization techniques allow us to reduce the scaling of the variables. In min-max scaling, we scale the data values in a range from `0` to `1`. This suppresses the influence of outliers on the data values to a certain extent. It also helps us to obtain a smaller value for the standard deviation of the data scale.

The `dstools` package has a normalization function to scale the values in a vector, matrix, or data frame to the range from `0` to `1`.

```
normalize(whisky_collection$Rating)
```

An very popular alternative is the `scale()` function in base R, it can be used to scale our values with the z-score standardization or transformation. In this type of scaling, we scale the data values such that the overall statistical summary of every variable has a mean value of `0` and a standard deviation of `1`. The `scale()` function enables us to apply standardization on the data values as it centers and scales the business data:

```
scale(whisky_collection$Rating)
```

For the moment you will probably see no direct benefits of scaling your data. But please keep this section in mind, when you start building your own models. If you scale your business data before modeling, you will normally increase the accuracy of your model.

4.3.3 Unify your Data Structure

A very common challenge in business data analytics is that we have to use data from different sources that we have identified in the business data understanding. The result is that the related data sources may have different naming conventions. Another very common problem is that the resulting dataset has other data in their headers or uses special characters in their names. Even if the datasets are formatted and named in the same manner, the names of the columns might be in capital and small letters mixed.

In base R there is the `make.name()` function to handle some of these problems. However, the function cannot handle spaces and uncommon characters correctly. It also adds a "." Between the different elements of a name, which makes it difficult to export your dataset, because many databases and programs have problems to handle tables with points in their names.

The `dstools` package provides an updated function of base R's `make.name()` function. The improved `make_names()` function generates R-compatible column names by removing spaces, points and other uncommon characters from data frame column names and unifies their typography. We can test that by using the `names()` function:

```
names(whisky_collection_corrupted)
whisky_collection_corrupted = make_names(whisky_collection_corrupted)
names(whisky_collection_corrupted)
```

If you run this code snippet, you will see that we could replace the `distillery` column name in our data frame with `DISTILLERY`.

Another challenge, we discussed in the introduction, is that datasets will come in different structures. This is bad because we want that our data sets are tidy (Wickham, 2014). This means in particular that each variable is a column and each observation is a row. Let me illustrate this with an example.

Let us take a look at the average rating of each whisky in the different regions of each country. We compute that with the following pipeline:

```
whisky_ratings = whisky_collection %>%
  group_by(LOCATION, REGION) %>%
  summarize(MEAN_RATING = mean(RATING))
```

The result looks like that:

```
> head(whisky_ratings)
  LOCATION REGION            MEAN_RATING
1 Canada   Canada                  2
2 Irland   Irland                  3
3 Irland   Northern Ireland        3
4 Irland   Wicklow Mountains       4
5 Japan    Japan                   3.5
6 Scotland Highland                3.67
```

As you can see the same country is multiple times in each row. In this case, we have to reshape your dataset to become tidy again. The `tidyr` package provides useful functions to reshape your datasets to make them tidy again. There is the the `pivot_wider()` and the `pivot_longer()` function to do so. Let us now start with the `pivot_wider()` function.

In the next step of our example, we want to further analyze the whiskies of each country by the most relevant types. So, we group by the columns `LOCATION` and `TYPE` before we compute the mean `RATING`. A possible pipeline doing so could look as follows:

```
whisky_ratings = whisky_collection %>%
  group_by(LOCATION, TYPE) %>%
  summarize(MEAN_RATING = mean(RATING))
```

As a result, we get a new data frame (or tibble) with the average rating per type and country:

```
> whisky_ratings
# A tibble: 12 x 3
# Groups:   LOCATION [6]
   LOCATION TYPE         MEAN_RATING
   <chr>    <chr>              <dbl>
 1 Canada   Blended             2
 2 Canada   Single Malt         3
 3 Irland   Blended             3
 4 Irland   Single Malt         4
 5 Japan    Blended             4
 6 Japan    Single Malt         3
 7 Scotland Blended             2.5
 8 Scotland Single Malt         3.48
 9 Taiwan   Blended             3
10 Taiwan   Single Malt         4
11 USA      Blended             1
12 USA      Bourbon             2
13 USA      Double Malt         2
14 USA      Single Malt         2
```

Our data frame with `40` whiskies has been summarized into a data set of only `14` rows, containing information of the `6` countries in our dataset. Each country has two related rows, one for blended whiskies and one for single malt whiskies except `USA`. The `USA` have four rows, because they have two further types of whiskies (`Double Malt`, `Bourbon`) that do not exist in the other countries.

The data format of our results is called "long". It has this name because it is longer as it should be (compared to tidy data). It contains multiple rows for each whisky but only three columns. The opposite is "wide" data. Wide data is called "wide" because it typically has a lot of columns which stretch the dataset. But now each row represents one single whisky. I illustrated you this relationship in the following Figure 4-8:

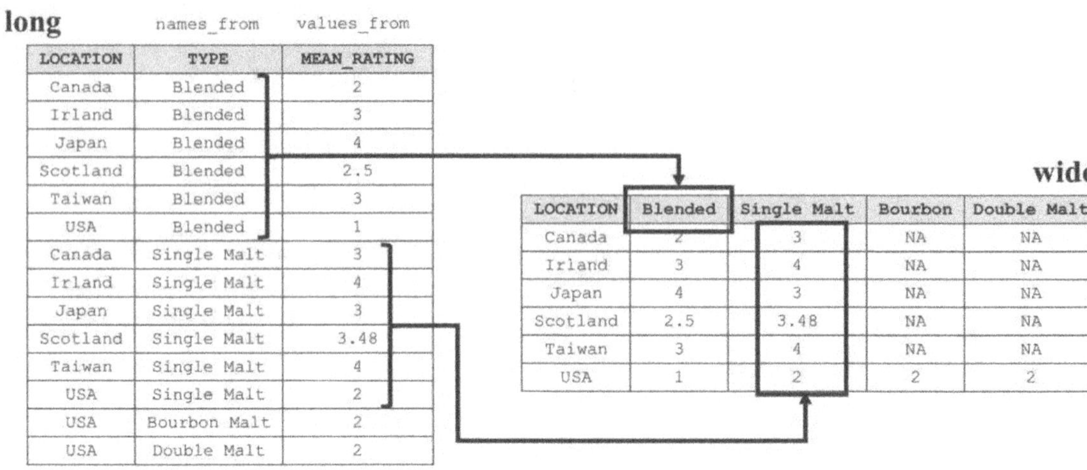

Figure 4-8 Pivoting from long to wide format with `pivot_wide()`.

Consequently, we can now transform our data frame in long with the `pivot_wide()` command pivoting from long to wide format. The function needs two parameters: `names_from` and `values_from`. As illustrated in Figure 4-8, you have to select the column you want to use as new column names and the columns where the related values should come from:

```
whisky_ratings_wide = whisky_ratings %>%
  pivot_wider(names_from = TYPE, values_from = MEAN_RATING)
```

The result is a new data frame which contains less rows, but more columns:

```
> whisky_ratings_wide
# A tibble: 6 x 5
# Groups:   LOCATION [6]
  LOCATION Blended `Single Malt` Bourbon `Double Malt`
  <chr>      <dbl>         <dbl>   <dbl>         <dbl>
1 Canada       2             3        NA            NA
2 Irland       3             4        NA            NA
3 Japan        4             3        NA            NA
4 Scotland     2.5           3.48     NA            NA
5 Taiwan       3             4        NA            NA
6 USA          1             2         2             2
```

If we want to do it the other way round we can use the `pivot_long()` function to retransform our wide dataset. Or in other words the `pivot_long()` function will transform our wide dataset back into a long format.

As before, I made a visualization of the process in Figure 4-9 to illustrate the function:

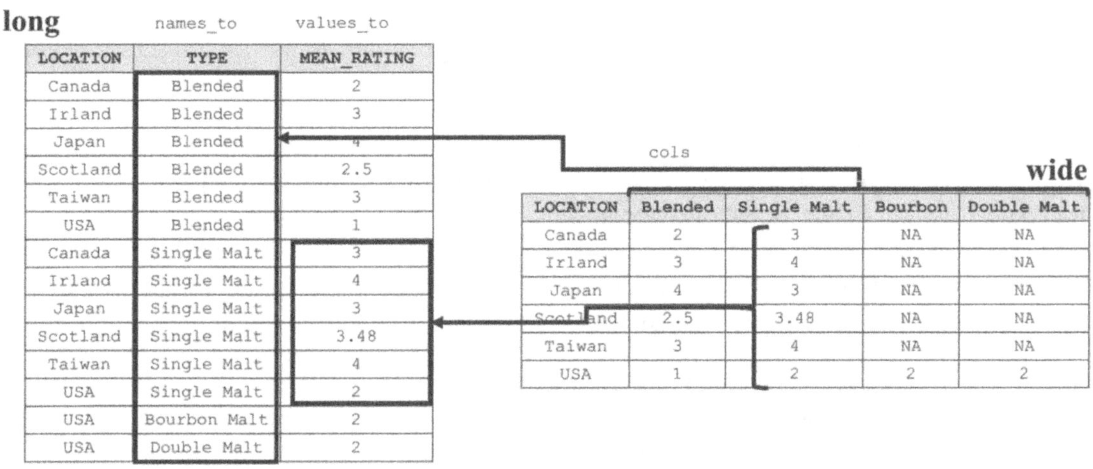

Figure 4-9 Pivoting from wide to long format with `pivot_long()`.

To make a long dataset we have to define which columns have to become levels of a new variable. We can do that by setting the columns with `cols` that have to be transformed (in our case all columns except the `LOCATION`), with `names_to` the name of the columns to store the column names as values and with `values_to` where to store our values from the wide format.

```
whisky_ratings_long = whisky_ratings_wide %>%
  pivot_longer(names_to="TYPE",     values_to="MEAN_RATING",     cols=-
LOCATION)
```

Finally, we reengineered our previous dataset:

```
> whisky_ratings_long
# A tibble: 24 x 3
# Groups:   LOCATION [6]
   LOCATION TYPE         MEAN_RATING
   <chr>    <chr>              <dbl>
 1 Canada   Blended                2
 2 Canada   Single Malt            3
 3 Canada   Bourbon               NA
 4 Canada   Double Malt           NA
 5 Irland   Blended                3
 6 Irland   Single Malt            4
 7 Irland   Bourbon               NA
 8 Irland   Double Malt           NA
 9 Japan    Blended                4
10 Japan    Single Malt            3
...
```

Our previous dataset has now more rows than before. The reason is that R generated also entries for types that do not exist in the related countries (Bourbon exists only in the US and is made of corn). The entries don't really matter for now, but if you want you can remove them with a simple `na.omit()`.

4.3.4 Overcome General Data Issues

In this last section, I want to give you some generic types and hints that might help you to overcome other issues in the data preparation.

First, when working with data, it's important to establish clear criteria for its validity. You have to define conditions that determine whether the data is accurate and relevant to your analysis. This could involve setting thresholds for acceptable values or identifying data points that fall outside expected ranges. Additionally, consider incorporating expiry dates for data, especially if you're dealing with time-sensitive information. This helps you maintain up-to-date and accurate datasets, reducing the risk of basing decisions on outdated or irrelevant data.

Furthermore, inconsistent data can wreak havoc on your analysis, leading to misleading insights and erroneous conclusions. To ensure data consistency, implement consistency checks during preprocessing. These checks can identify duplicate records, conflicting information, or irregularities within the data. By identifying and resolving inconsistencies early in the process, you can improve the overall quality of your analysis and boost confidence in your results.

From my experience, it also makes sense to define a general data format for integration. When combining datasets from different sources, each dataset may have its own structure and format. To ensure seamless integration, define a general data format that accommodates various data sources. This format could include standardized column names, data types, and units of measurement. By establishing a common structure, you streamline the integration process and make it easier to analyze data from diverse sources.

Last but not least, domain expertise is invaluable when preprocessing data. Reach out to experts who possess in-depth knowledge of the subject matter. They can provide insights into the significance of certain data points, help identify anomalies, and guide you in making informed decisions about handling missing or ambiguous data. Collaboration with domain experts enhances the accuracy and relevance of your analysis, making your results more actionable.

4.4 Feature Engineering

In the last sections, we rolled up our sleeves to tackle the data quality issues with data cleaning. We discussed techniques to remove duplicates, handle noisy values, and ensure the data was tidy and reliable. Now, with our clean data, we are ready to take the next step on our analytical project and prepare our features with feature engineering.

Imagine just as Ted Gumble crafts a unique whisky blend by selecting the finest ingredients, we, as business data analysts, will carefully engineer our features to extract the most valuable information for our model. Doing so, feature engineering is about transforming our raw data into meaningful attributes that can fuel our modeling engine. We can do that with four relevant techniques, as illustrated in the following Figure 4-10:

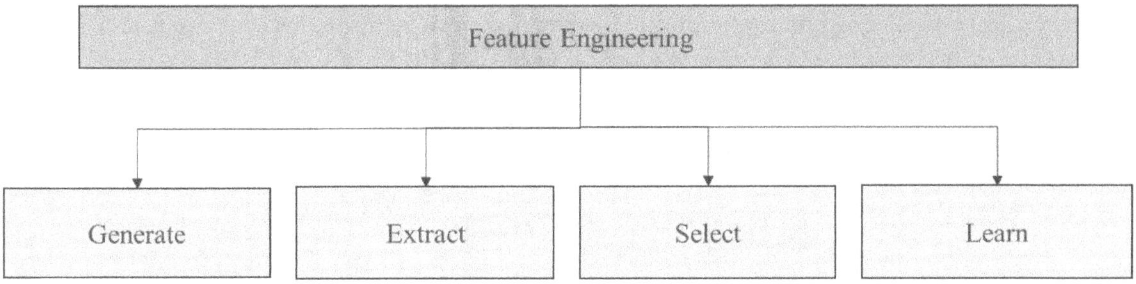

Figure 4-10 Different feature engineering techniques.

Generating new features involves creating entirely new columns for your dataset derived from existing data. This can provide fresh insights and dimensions to your analysis. For instance, you might multiply the columns age and income of your Junglivet customers to create a new feature called wealth.

Alternately, you can also generate new features by extracting them. This technique revolves around extracting valuable information from existing features. By distilling data down to its core, you can uncover hidden patterns and relationships. For example, you can extract "WEEKDAY", "MONTH", or "SEASON" from a "PURCHASE_DATE" column.

If you think that you have now (or already) enough features, you can use feature selection or reduction techniques. Feature selection is all about identifying the most relevant attributes for your analysis and excluding less significant ones. It simplifies your dataset, focusing only on essential factors. For instance, when predicting whisky taste, you might choose "age", "alcohol_content", and "barrel_type" while omitting less impactful attributes like "bottle_color".

Last but not least, there are relevant developments in machine learning research proposing automated feature engineering methods. These techniques are called feature learning and the related algorithms can sometimes autonomously discover important features within your business data. This means your models can identify valuable attributes without explicit feature engineering. For example, deep learning models like neural networks can learn intricate features from raw data like images or text.

4.4.1 Generate New Features

Create new features by usage of raw features is a very common task in business data preparation, because sometimes there is information missing or hidden in your business data. We have to generate new features which we can add to our dataset to help your modeling algorithm to find these patterns.

Think for instance at our whisky dataset `whisky_collection`. It contains many information about rating and taste of the whisky but you might be interested in adding your own rating to it (missing information). Or you might want to transform the price from Euro into US-Dollar (transform to show hidden information).

In dplyr and base R exist many functions for creating new features that you can use. The most relevant is mutate(), that you already know from chapter 3.

As a quick refresher, you can create a new column with mutate() in a dplyr pipe. For instance, if you want to add a new column AGE to represent the distillery's age you run:

```
whisky_collection_new = whisky_collection %>%
  mutate(AGE = 2023 - FOUNDATION)
```

This adds a new column at the end of the data frame with the name AGE. Do not forget to run the View() command or the head() function to check if everything worked correctly. If you want to compute more complex features you can also use control functions like ifelse() or even your own functions in the pipe. This can be helpful, if you want to write generic functions that automatically remove errors.

For instance, we have the Glendalough whisky which has a wrong date in the column FOUNDATION:

```
         NAME                    DISTILLERY LOCATION FOUNDATION
1 Glendalough Glendalough Irish Whiskey   Irland      20111
```

We can now use a custom function get_age() in mutate to check if the date has a correct value and otherwise replace it for the computation of the correct AGE:

```
get_age = function(foundation){
  actual_year = as.integer(format(Sys.time(), "%Y"))
  foundation = ifelse(foundation <= 1608, 1608,
         ifelse(foundation > actual_year, actual_year, foundation))
  age = actual_year - foundation
  return(age)
}

whisky_collection_new = whisky_collection_corrupted %>%
  mutate(AGE = get_age(FOUNDATION))
```

The oldest distillery is the Bushmills distillery in Northern Ireland with official records dating back to 1608. We have therefore added a check to see if the year of foundation is lower than that of Bushmills distillery. Furthermore, the year should also not be in the future. Therefore, we have added a check for this as well. The function format(Sys.time(), "%Y") takes the current date and extracts the year from it, with as.integer() we make sure that the results has the correct data type.

Another common use case is to calculate a difference across a column, row or even group. And again, we can use mutate() to compute features that represent an offset of another feature that stands in another column or row.

To illustrate this let us compute the price differences of a whisky compared to the priciest one by region (price offset) in the whisky_collection.

You can do that with `lag()` on the sorted data frame in the following manner:

```
whisky_collection %>%
  group_by(REGION) %>%
  arrange(REGION, desc(PRICE)) %>%
  mutate(DIF_PRICE = PRICE - lag(PRICE))
```

In this example we group our dataset by `REGION`, and then sort them by `PRICE`. Because `arrange()` sorts ascending by default we have to specify that it should be sorted descending by `PRICE`. Finally, we compute our new feature `DIF_PRICE`, which stands for difference price. Alternatively you can use the `lead()` function to refer to the other value.

Furthermore, there are cumulative or rolling functions to compute running sums (`cumsum()`), products (`cumprod()`), minima (`cumin()`) , maxima (`cumax()`) and even cumulative means (`cummean()`). Their application is very special, but I thought it would make sense to mention them here.

In other cases, you might want to rank your rows by a ranking function and generate new features based on that. The reason is that sometimes a feature that represents ranked values instead of absolute values is more suitable for the modelling algorithm. There exist many different variants like `min_rank()` to rank by their values starting with the smallest one with `1`. Or `row_number()` to rank them by row number, `percent_rank()` to compute percentual ranking `percent_rank()` or its generalization `ntile()`.

4.4.2 Select Relevant Features

Having dealt with the generation of new features in the previous section, let's now shift our focus to an opposing challenge. The more features a dataset has, the more difficult it becomes for both you and your statistical model to find the really important features. And if the number of features is higher than the number of rows, many algorithms will fail to produce reliable results at all. This problem has long troubled data scientists and is known as the "curse of dimensionalitmy" (Verleysen & François, 2005).

Let me discuss this over a good whisky. Imagine walking into the pub near the *Junglivet Whisky Company* and seeing a huge selection of whiskies. Which of them would you try and which would you not? There are a lot of whiskies in the collection, but not all of them are equally interesting and unless you're an alcoholic, not all are worth the taste. Some pieces are really expensive like Brora, others are relatively cheap and maybe even a real letdown in terms of taste like the Jim Beam in the tarnished bottle on the bottom shelf.

It's the same with business data. There are rather usable features and rather useless ones. The key to valuable insights lies in choosing the right whiskies, or in the language of business data analytics: the right features. Let us now explore why feature selection is so important, what techniques exist to use it effectively, and how it can greatly improve the quality of your models.

While visually inspecting the data can be an effective approach, at least to decide on the number of features, there also exist three algorithmic approaches to select relevant features in business data analytics: filter methods, wrapper methods and embedded methods.

Filter methods are a family of techniques that rely on statistical measures to evaluate the relevance of each feature independently. The goal is to rank or score features based on their individual significance. The most established filter methods in business data analytics are the entropy or information gain of a feature and the correlation with other features.

Entropy is a valuable metric for understanding the information contained in a set of values. The idea behind entropy is a fundamental concept in information theory, introduced by Claude Shannon (Shannon, 1948):

$$H(p_1, \dots, p_n) = \sum_{i=1}^{n} p_i \log_2 p_i$$

Where p_i is the probability of value i and n is the number of possible values.

In the context of features, entropy measures the level of disorder or randomness in the values of a feature. High entropy indicates greater disorder, while low entropy signifies more order and predictability. Or in other words high entropy in features is associated with high variance and thus we can assume that the feature contains a lot of information.

For instance, we want to build a model that predicts the rating of whisky. We will use our `whisky_collection` dataset for that. To keep our model as simple as possible, we have to decide on a single feature. Due to data availability, we only have the choice between `REGION` and `LOCATION`. Without assessing the whole technicalities we can calculate the entropy of the two features with the `entropy()` function from the `dstools` package:

```
> entropy(whisky_collection$REGION)
[1] 3.624024
> entropy(whisky_collection$LOCATION)
[1] 1.914263
```

As we can see, the `REGION` feature contains much higher entropy than the `LOCATION` feature. Thus, it would tend to be more suitable. The reason for this is that it contains many more different values and, on the other hand, they are distributed more differently.

Entropy serves as a foundation for calculating an even better indicator of information in features which is called information gain. Information gain measures how much a feature reduces the uncertainty (entropy) in the target variable if we would segment the dataset (parent) on this feature (child). Features with high information gain are considered valuable because they provide more information for the feature, we are interested in to predict. You can think of them as that they are very good in helping us to differentiate between different outcomes or classes within the target variable. So, how do we calculate information gain?

$$IG = H_p - \sum_{i=1}^{n} p_{ci} H_{ci}$$

Where H_p is the entropy of the complete dataset (parent), n is the number of values of our target variable, p_{ci} is the probability that an observation is in child i, and H_{ci} is the entropy of the child or segment i.

Thankfully, this is fairly simple to do using the `dstools` package:

```
> information_gain(whisky_collection, "RATING", "PRICE", bins=5)
[1] 1.503185
```

As an alternative to information gain, you can calculate Pearson's correlation coefficient (`cor()`) to measure the linear relationship between numeric features and the target variable. We will come back to the concept of correlation later in the next chapter; hence we go straight forward to the implementation in R:

```
cor(whisky_collection$RATING, whisky_collection$PRICE)
[1] 0.1773374
```

To identify the best features, you compute the correlation to the target feature and compare them. Features with high absolute correlation values are considered more relevant. But instead of computing the individual correlation of your feature-pairs manually, most business data analyst will compute a correlation matrix (which you know from the chapter 3) for the numeric features to investigate them:

```
numeric_features = whisky_collection[,c(6, 9:13)]
correlation_matrix = cor(numeric_features)
```

However, if you work with big datasets your correlation matrix will get very difficult to interpret and understand. Good for you, the `caret` package provides the `findCorrelation()` function which will analyze a correlation matrix and highlight the features that can be removed.

To test the correlation matrix and other feature selection concepts, we will use the `caret` package. If you did not install it, please do so and then load it into your R environment:

```
install.packages("caret", dependencies = TRUE)
library(caret)
```

Then you can run the code again and save the column number of the best features:

```
best = findCorrelation(correlation_matrix, cutoff=0.25)
```

With `cutoff` we specify the pair-wise absolute correlation cutoff. Best practice is a value `> 0.7` but in our case our data is not good enough and we have to reduce the cutoff. Then we can print the features that are the most relevant:

```
> print(names(numeric_features)[best])
[1] "RATING"    "CRITIQUES"
```

If the features you want to compare are categorical, chi-squared tests (`chisq.test()`) can assess the dependency between each feature and the target. Features with significant p-values (best practice `p <= 0.05`) are retained.

```
> chisq.test(whisky_collection$TYPE, whisky_collection$LOCATION)
      Pearson's Chi-squared test
data:  whisky_collection$TYPE and whisky_collection$LOCATION
X-squared = 35.27, df = 15, p-value = 0.00225
```

Another bundle of techniques is summarized under the term wrapper methods. They evaluate feature subsets by employing business data analytic algorithms to train and test models iteratively. These methods aim to find the optimal feature subset that leads to the best model performance by testing different feature combinations against each other. Unfortunately, we did not introduce business data analytics algorithms like regressions and decision tress so far. But nevertheless, I want shortly demonstrate how they are implemented. If you want, you can ignore this section or at least the part about the modeling algorithm and come back to this section later.

Common wrapper methods are forward selection, backward selection and recursive feature elimination. The forward selection approach begins with an empty feature set and iteratively adds features that improve model performance. The opposite is backward elimination. In this case you start with all features and progressively remove the least valuable ones until model performance peaks.

Best-practice in business data analytics is recursive feature elimination (RFE). In recursive feature elimination you train models with all features and recursively remove the least important feature until the desired subset size is achieved. To implement recursive feature elimination in R, we will rely on the `rfe()` function from the `caret` package:

```
control  = rfeControl(functions=rfFuncs, method="cv", number=10)
results  = rfe(whisky_collection[,1:13],
               whisky_collection[,14],
               sizes=c(1:13),
               rfeControl=control)
```

Before running the `rfe()` function, we first have to specify the control options with `rfeControl()`. For instance, the modeling algorithm we want to use to test the feature importance.

The `caret` package includes a number of algorithms, such as random forest or linear regression. In our case, we use random forest (called `rfFuncs`). In addition, we will use repeated 10-fold cross-validation with `10` repeats. The `rfe()` function gets our dataset without the target variable as first parameter, then the target variable to be predicted, with `sizes` number of features that should be retained in the feature selection process and with `control` a list of options for the feature selection algorithm.

After running the code, you can print the results with:

```
> print(results)

Recursive feature selection

Outer resampling method: Cross-Validated (10 fold)

Resampling performance over subset size:
```

```
 Variables  RMSE Rsquared   MAE RMSESD RsquaredSD MAESD Selected
         1 311.2   0.4059 192.6  443.5     0.3388 255.6        *
         2 318.8   0.3385 197.0  449.1     0.2493 255.8
         3 363.8   0.5976 266.1  420.9     0.3901 229.0
         4 378.0   0.3481 288.0  404.8     0.2787 228.6
         5 393.4   0.3071 309.7  391.7     0.3218 210.5
         6 400.8   0.4130 315.5  386.6     0.3948 214.8
         7 415.6   0.3361 333.0  374.4     0.3723 201.5
         8 399.9   0.2351 319.6  382.3     0.3120 206.4
         9 408.2   0.3323 321.6  379.0     0.3475 205.0
        10 410.6   0.3954 329.6  378.9     0.3623 203.5
        11 424.6   0.3251 349.2  369.5     0.3239 196.3
        12 406.1   0.3823 332.9  376.8     0.3243 218.7
        13 392.0   0.3912 322.8  374.0     0.3131 217.8

The top 1 variables (out of 1):
   CRITIQUES
```

The little asterisk sign under the `Selected` column in the output indicates that our recursive feature selection recommends only `1` feature for the model to predict the `PRICE`. Error measures like root mean squared error (`RMSE`) and mean absolute error (`MAE`) reach the minimum level when `CRITIQUES` is the only feature in the model.

The last set of feature selection methods are summarized under the term “embedded methods”. Embedded methods incorporate feature selection within the model building process. These methods use algorithms that inherently select features as they learn. Popular embedded methods in R include the random forest algorithm and XGboost. We will look at these methods in detail in the next chapter. But I want to point out that the `randomForest` package in R provides a versatile ensemble learning algorithm that automatically ranks feature importance during model training. The extreme gradient boosting, is accessible through the `xgboost` package, which employs feature selection by pruning trees based on their importance.

Finally, feature selection is an indispensable step to enhance model performance and gain deeper insights from your data. By mastering filter methods, wrapper methods, and embedded methods, you'll be equipped with a robust toolkit to handle diverse datasets and extract the most relevant information. Experiment with these techniques in your R projects to identify the feature selection approach that best suits your data and modeling goals.

The course of dimensionality is a popular problem in analytics. When you work with data that has many different dimensions or features, things can get complicated. The more features you have, the more data you need to describe the space adequately. This can lead to problems like sparsity, where you have too few data points in high-dimensional space to make reliable predictions or conclusions. In essence, it's a challenge that arises when you have lots of features in your data, making analysis more complex.

4.4.3 Extract New Features

In contrast to feature generation, feature extraction strives to diminish the number of features within a dataset through the creation of new features summarizing existing ones and then reducing the number of features by discarding the original features. These new reduced set of features should then contain a significant portion of the information present in the original set of features. If you want you can see it as a combination of feature generation and feature selection.

One popular method to reduce the number of features is by computing its principal components. Principal component analysis (PCA) is a powerful tool in business data analytics, often used to simplify complex datasets. It works by reducing the number of features while preserving essential information.

To illustrate the importance of the principal component analysis, let us investigate our `whisky_collection`. We have many features like `RATING`, `REVIEWS`, and `CRITIQUES` representing the results of different whisky tastings. To reduce the number of columns (the dimensionality) of our dataset we want to combine them into one new feature that contains the same information. Here's where principal component analysis comes to the rescue. Let's go through the process step by step using the `prcomp()` function from `stats` included in base R.

```
tasting_pca = whisky_collection %>%
  select(RATING, REVIEWS, CRITIQUES, ) %>%
  prcomp(., scale = TRUE)
```

We used three tasting related features and run the principal component algorithm on them. We can now look at the results with:

```
> summary(tasting_pca)
Importance of components:
                          PC1    PC2    PC3
Standard deviation     1.2887 0.8800 0.7516
Proportion of Variance 0.5536 0.2581 0.1883
Cumulative Proportion  0.5536 0.8117 1.0000
```

The first row gives the standard deviation of each component. The second row shows the percentage of explained variance. Accordingly, the first principal component explains around `55%` of the total variance, the second principal component explains about `25%` of the variance, and the last one only 18%. Finally, the last row, `Cumulative Proportion`, calculates the cumulative sum of the second row. By all, we are done with the computation of PCA in R. As a consequence, `55%` of the "information" of three features can be represented by just one feature the `PC1`.

Now, it is time to use this new feature `PC1` and delete the other ones. Principal components are linear combinations of the original features, and you can use them to represent the data in a lower-dimensional space. To create a new feature, we have to multiply the selected principal components by the original feature data. This results in a projection of the data onto the space defined by the selected principal components.

We can now look at the distinct feature by running:

```
> tasting_pca
Standard deviations (1, .., p=3):
[1] 1.2887246 0.8799555 0.7515765

Rotation (n x k) = (3 x 3):
                PC1        PC2         PC3
RATING     0.6034282 -0.4149902  0.68092404
REVIEWS    0.5058837  0.8593007  0.07539298
CRITIQUES  0.6164058 -0.2989741 -0.72846300
```

Using this information, you can compute our feature `TASTING_55` representing `55%` of the original information with the following formula:

$$TASTING_{55} = PC1 \cdot RATING + PC2 \cdot REVIEWS + PC3 \cdot CRITIQUES$$

$$= 0.6034282 \cdot RATING + 0.5058837 \cdot REVIEWS + 0.6164058 \cdot CRITIQUES$$

Here, the variables `RATING` etc., represent the values of your original features on each row, and the variables `PC1` etc., are the selected principal components. Congratulation you made your first principal component analysis!

However, in our case one could argue that reducing three features with about `40` rows to one feature with `45%` information loss might not be a big deal. But imagine you work for the product development team of the *Junglivet Whisky Company*, and you're interested in understanding what makes a whisky unique. You have a dataset with a lot of features representing different whisky characteristics, such as alcohol content, peatiness, sweetness, and more. This dataset is quite high-dimensional, making it challenging to discern meaningful patterns. If we run now a principal component analysis and are able to summarize hundreds of features to `1` or `2` important ones this becomes a big deal.

4.4.4 Learn New Features

Feature learning or representation learning is a set of techniques to automatically discover features that represent the relevant information in a dataset without manually generating, selecting or extracting them.

The key idea behind feature learning is to let the learning algorithm itself determine what features are important. However, the application of these methods is machine learning rather than business data analytics and hence there exists not many R packages that implement them.

Therefore, we will only briefly discuss the most important approaches in this section, as they would otherwise go beyond the scope of the book.

- **Autoencoders**: Neural networks used for unsupervised learning. They learn to encode the input data into a lower-dimensional representation and then decode it back to the original data. The bottleneck layer in an autoencoder is where the learned features are extracted.

- **Deep Learning**: Neural networks, such as convolutional neural networks (CNNs) for images and recurrent neural networks (RNNs) for sequences, automatically learn hierarchical representations of data. These models can uncover complex features in the data, which are useful for various tasks like image recognition and natural language processing.
- **Independent Component Analysis (ICA)**: ICA aims to find statistically independent components in the data. It is often used in signal processing and blind source separation to uncover hidden factors contributing to the observed data.
- **Feature Selection with Embedded Methods**: These methods are building on the fact that some machine learning algorithms, like decision trees and random forests, have embedded feature selection mechanisms. They evaluate the importance of each feature during model training and select the most relevant ones automatically.

4.5 Business Data Integration

In chapter 2, we learnt that data frames are data structures with the purpose to store tidy data. In general, we can assume that our data is in tidy form. That means our data comes in the form of a table or matrix. Rows contain observations or data objects, while columns represent features or variables. Tidy data makes us as analyst happy, because these kinds of data can be most easily used to develop analytics pipelines.

In the real-world scenarios, to work with the data, we often come across situations wherein we find the datasets that do not come in this nice form. The datasets will be distributed in different data sources. And we have to extract it from there and join it into a single table that we can use for further analysis. If we are interested in the likelihood if a customers will ship back its current online purchase, we have to build a statistical model that uses information from our sells, customer and storage data tables. If we want to improve it, we might use further information from other data sources like geographical or time-related information.

To use the information from these different data sources we have to integrate them in one single dataset. There are many practical business cases to combine relevant information from your business data into one single dataset and use it for further analysis or modeling. This process is termed data integration and a relevant part of the data preprocessing and preparation.

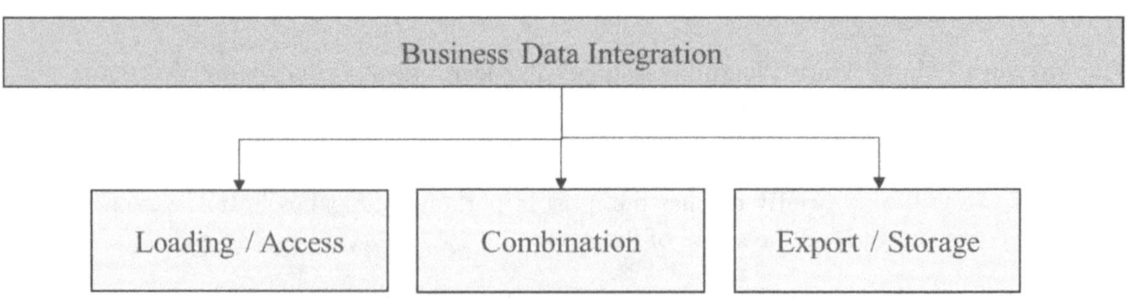

Figure 4-11 Different aspects of data integration we taggle in this section.

Considering the different steps of business data integration illustrated in Figure 4-11, there are different relevant challenges in data integration, we have to:

- clarify how to access and load the different datasets
- solve the issue that the datasets have different schemes and names
- and that the features in the datasets have different scales and formats
- combine different datasets that are possibly not tidy
- export or save them in a suitable manner for our project

In the following section we will use the `tidyverse` and the `dstools` package. It has many functions that help us to handle these problems. If you read this chapter, you should have these two packages already installed. Run the following code to check if the packages are installed and can be loaded.

```
library(dstools)
```

If you have `tidyverse` or `dstools` not installed, run the following code to install it. Afterwards you should be able to load the `dstools` package.

```
install.packages("tidyverse", dependencies = TRUE)
install_github("dominikjung42/dstools")
```

4.5.1 Loading Different Datasets

Best practice to load a dataset is to use the RStudio wizard for that. You can open him by clicking on a dataset in your project in the left bottom pane. Or in the menu by clicking "File" and then "Import Dataset".

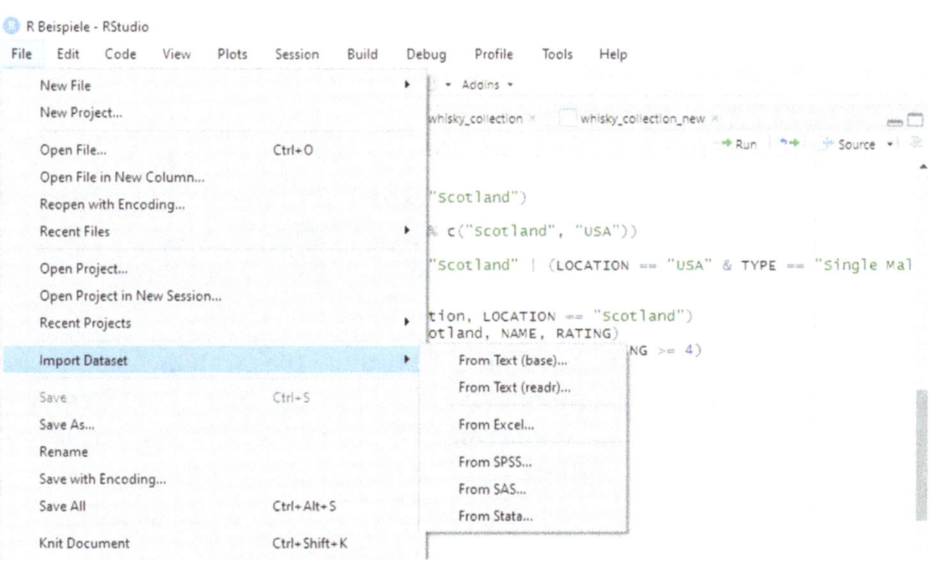

Figure 4-12 RStudio has many helper functions to import datasets from many different formats.

After you opened the wizard, you have many options to import datasets of many types. As you can see in the following Figure 4-13, RStudio gives you a preview of the current configuration and the resulting dataset. The preview of RStudio helps you to specify your configuration to load a dataset.

The wizard has always different options and configurations depending on the imported data format. For instance, if you select to import data from a CSV file the wizard has many CSV related options. As illustrated in Figure 4-13, you can select between different delimiters (if you read none-CSV, but delimited files), skip specific lines or quotes.

You can also explore a preview like a data frame in RStudio. And RStudio gives you some suggestions of the variable type of each variable in your dataset. You can see the suggestion in brackets below the name in the preview. If the type is not correct you can change the type by selecting another one.

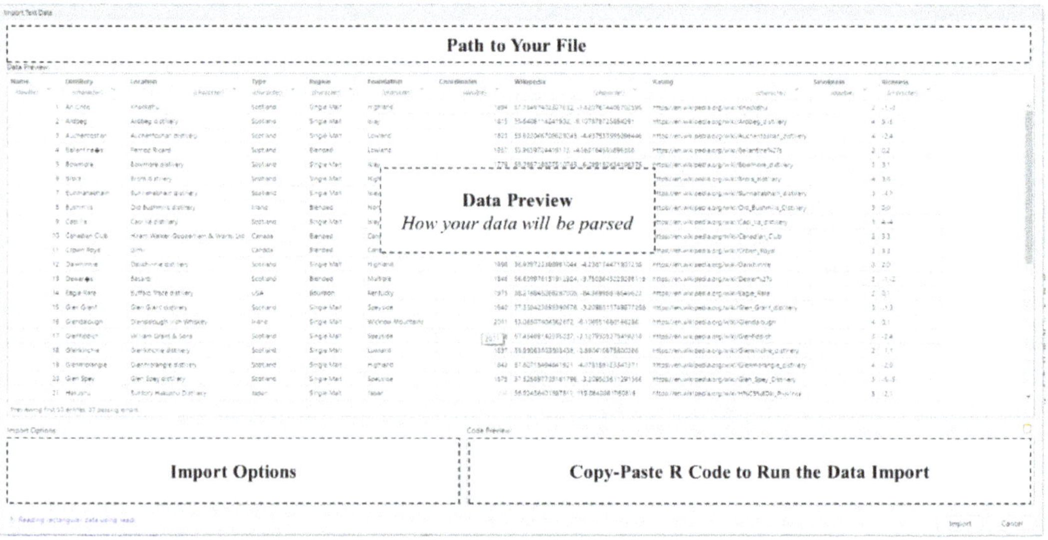

Figure 4-13 The RStudio data import wizard has many options to configure your data import.

After specifying your import, you can also copy the R-code in the right, bottom corner. The code represents your configuration and is executed when you click the "Import" button. If you save the code in your script, you can load your dataset again and again without opening the wizard each time.

However, if you want to load different datasets, it might be that the files with the datasets are spitted into many distinct files. For instance, if you want to analyze the online shop data from the *Junglivet whisky company* the current transactions might be stored monthly and send to you by a script. Then you have many distinct files like "`productionlog_1.xlsx`", "`productionlog_2.xlsx`" and "`productionlog_3.xlsx`".

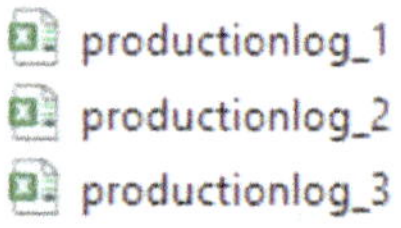

Figure 4-14 Example of bad business data: The production information from the last three months of the Junglivet whisky company is stored in distinct files.

That is not best practice, but unfortunately very common in most companies. As a consequence, you might face the problem to import many single files which have all the same format. But good news: The `dstools` package allows you to do that easily. You can read all files in a specific directory into your environment and bind them into a data frame with the `read_dir()` function. For that we have to give the path to the folder with the data as parameter. If you do not know the path of your working directory you can run `getwd()` to check it. Then run the following code to load all files that end with "`.xlsx`" from the folder "`productionlogs`":

```
getwd()
path = "C:/users/dominik/productionlogs/"
data_all = read_dir(dir = path, type = "xlsx")
```

The `dir` parameter specifies the path to the folder with the datasets, the `type` parameter the data type of the files. If the folder contains also other files that should not be loaded like a "`metadata.xlsx`" you can specify a pattern to exclude other files in the folder. If we are only interested in the `productionlog` data we change our code snippet in the following manner:

```
data_all = read_dir(dir = C:/logs, pattern = "productionlog", type =
"xlsx")
```

Congratulation, you are now ready to continue with the next section!

4.5.2 Combining Different Datasets

In business data analytics, rarely do we find all the information we need in a single dataset. Data is often scattered across multiple tables, files, or databases. This is where data joining comes into play. Joining is the process of combining data from different sources to create a unified dataset.

In business data preparation, data joining is a fundamental skill that allows us to merge, link, or integrate datasets based on common columns or variables. It's like piecing together a puzzle to reveal a more comprehensive picture of your data.

In this section, we will investigate the most relevant methods and functions from the `dplyr` package illustrated in Figure 4-15 to perform these data combinations efficiently.

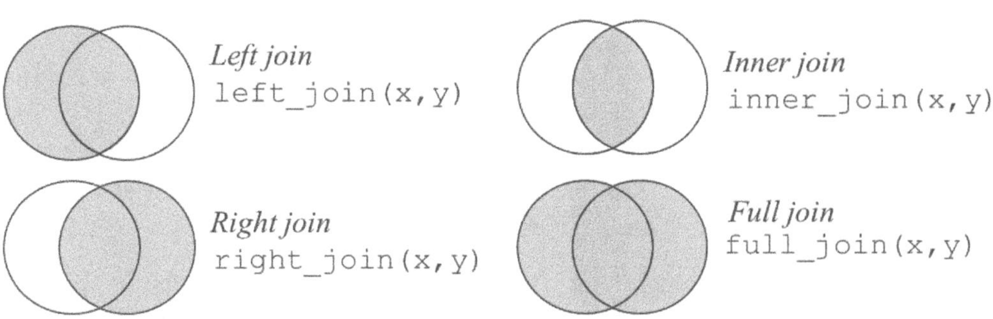

Figure 4-15 Most relevant joins to integrate different datasets.

To illustrate the different joins in R, let us extend our `productionlog_sample` from the introduction with another file we got from Sherry Young from the production team. The file productionlog_extension contains further information about the production process and might be helpful for us to generate deeper insights. Let us have a short look at the files by loading them directly from the dstools package:

```
> data("productionlog_sample")
> head(productionlog_sample)
  DAY MONTH MANUFACTURER             PRODUCT   SHIFT COLOR          MALTING TASTING
1   1     4      Leonard           Junglivet Morning  0.27          Inhouse     895
2   1     4      Carlson Junglivet Premium Evening  0.27 Burns Best Ltd.     879
3   2     4      Leonard           Junglivet Morning  0.28          Inhouse     938
4   2     4      Carlson           Junglivet Evening  0.32          Inhouse     900
5   3     4      Leonard           Junglivet Morning  0.32       Matro Ltd.     917
6   3     4      Carlson           Junglivet Evening  0.28          Inhouse     900
```

The productionlog_sample should be familiar to you. It contains some information about the whisky production and the production team and the quality of the outcome. The new dataset productionlog_extension from Sherry contains further information:

```
> head(productionlog_extension)
  DAY MONTH DAYTIME OFF-FLAVORS CLOUDINESS ALCOHOL
1   1     4 Morning       FALSE      FALSE      OK
2   2     4 Morning       FALSE      FALSE      OK
3   3     4 Morning       FALSE      FALSE      OK
4   3     4 Evening       FALSE      FALSE      OK
5   4     4 Evening       FALSE      FALSE      OK
6   5     4 Morning       FALSE       TRUE    HIGH
```

This other dataset contains further information that extends our original dataset. For instance, there is information about different whisky quality indicators like off-flavors. OFF-FLAVORS are unusual or unpleasant flavors and aromas and can be an indicator of issues in the fermentation or distillation process. These off-flavors might include notes of sulfur, rotten eggs, or excessive bitterness – nothing you want to have in your final whisky. The other indicator is CLOUDINESS. Whisky should be clear and not cloudy. Cloudiness may suggest issues with filtration or improper blending. And finally, ALCOHOL or alcohol content represents the expected alcohol content by volume. A significantly lower or higher content than expected could indicate problems during distillation or dilution.

Unfortunately, we also see that the new file is incomplete. There is no information about the evening shifts on day 1 and 2. And some minutes later you detect that the morning shift of day 4 is also missing. The reasons for this could be manifold, and we discussed them in this chapter in the section about data cleaning. For now, we want to ignore the missing entries and just integrate these two datasets into a new combined dataset.

A first approach could be to join our old dataset with the new one (left join). To do this we have to specify a single or multiple columns as key. The key is used to decide which rows should match with which rows in each dataset. In our case we see that both datasets contain a column DAY and hence the day seems to be a good pick as a key.

However, if we try to join, we get the following error:

```
Warning message:
In left_join(productionlog_sample, productionlog_extension, by = "DAY)
  Detected an unexpected many-to-many relationship between `x` and `y`.
i Row 5 of `x` matches multiple rows in `y`.
i Row 1 of `y` matches multiple rows in `x`.
i If a many-to-many relationship is expected, set `relationship = "many-
to-many"` to silence this warning.
```

The reason for this is that day is not unique. There are multiple entries for each day. One for the morning and one for the evening shift. Hence, we have to use the time as additional parameter for our key:

```
productionlog_all = left_join(productionlog_sample,
                              productionlog_extension,
                              by = c("DAY" = "DAY", "SHIFT" =
"DAYTIME"))
```

Finally, we have a complete dataset containing the information of both files:

```
> glimpse(productionlog_all)
Rows: 21
Columns: 12
$ DAY <dbl> 1, 1, 2, 2, 3, 3, 4, 4, 5, 5, NA, 6, 6, 7, 7, 8, 8,
$ MONTH.x <dbl> 4, 4, 4, 4, 4, 4, 4, 4, 4, 4, NA, 4, 4, 4, 4, 4, 4,
$ MANUFACTURER <chr> "Leonard", "Carlson", "Leonard", "Carlson",
"Leonard", "Carlson", "Leonard", "Gumble", "Leonard",
$ PRODUCT <chr> "Junglivet", "Junglivet Premium", "Junglivet",
"Junglivet", "Junglivet",
$ SHIFT <chr> "Morning", "Evening", "Morning", "Evening", "Morning",
"Evening", "Morning", "Evening", "Morning",
$ COLOR <chr> "0.27", "0.27", "0.28", "0.32", "0.32", "0.28", "0.29",
"0.29", "0.33", "0.27", NA, "0.3", "0.31", "0.26"
$ MALTING <chr> "Inhouse", "Burns Best Ltd.", "Inhouse", "Inhouse",
"Matro Ltd.", "Inhouse", "Inhouse", "Matro Ltd."
$ TASTING <dbl> 895, 879, 938, 900, 917, 900, 934, 951, 852, 850, NA,
991, 950, 863, 936, 996, 978
$ MONTH.y <dbl> 4, NA, 4, NA, 4, 4, NA, 4, 4, 4, NA, 4, 4, 4, 4, 4, 4
$ `OFF-FLAVORS` <chr> "FALSE", NA, "FALSE", NA, "FALSE", "FALSE", NA,
"FALSE", "FALSE", "FALSE", NA, "FALSE", "FALSE", "FALSE",
$ CLOUDINESS   <chr> "FALSE", NA, "FALSE", NA, "FALSE", "FALSE", NA,
"FALSE", "TRUE", "TRUE", NA, "FALSE", "FALSE", "FALSE"
$ ALCOHOL <chr> "OK", NA, "OK", NA, "OK", "OK", NA, "OK", "HIGH", "OK",
NA, "OK", "OK", "LOW", "OK", "OK", "OK"
```

Because we did a left join, and the "left" dataset of the `productionlog_sample` contains all morning and evening shifts, the missing data from the "right" dataset is replaced with `NA`.

Alternatively, you can join them the other way round with:

```
productionlog_all = right_join(productionlog_sample,
                               productionlog_extension,
                               by  =  c("DAY"  =  "DAY",  "SHIFT"  =
"DAYTIME"))
```

In this case, the additional rows of the "left" dataset will be omitted. Hence the new dataset will have less rows:

```
> nrow(productionlog_all)
[1] 17
```

The inner join selects records that have matching values in both tables. Because in our example the `productionlog_sample` has every shift that exists in the `productionlog_extension` it will result in the same dataset.

You can check that with:

```
inner_join(productionlog_sample,
           productionlog_extension,
           by = c("DAY" = "DAY", "SHIFT" = "DAYTIME"))
```

Finally, you can also join regardless of existing values with a full join. The full join will return all production records when there is a match in left (`productionlog_sample`) or right (`productionlog_extension`) dataset:

```
full_join(productionlog_sample,
          productionlog_extension,
          by = c("DAY" = "DAY", "SHIFT" = "DAYTIME"))
```

4.5.3 Saving and Exporting your Data

Now that we have learned how to use `dplyr` to prepare raw data and extract or summarize relevant information from it, let's look at how to export these new datasets. This way, the prepared dataset can be made available to other collaborators and data preparation does not have to be done every time. Also, the data can then be archived for later analysis or review.

There are several ways to do this in R. The two most relevant are:

- `save()`: This is a base R function that allows to store data in the `.RData` format
- `write_`: This group of writing functions from `readr` package (part of the `tidyverse` package) supports many different export formats like tab-separated or Excel CSV.

Now, before we turn to saving the data with these functions, we create a new folder called `data_export` in our working directory. In this folder we want to store the generated datasets. While this is not strictly necessary, I always recommend to my students that the generated datasets should not be in the same directory as the raw data. It is a good practice to keep them separate to keep the project clean and small.

Here is an example structure of a cleanly structured R project with appropriate folders, which you can download from the course website:

```
name_of_your_project
├── 1 Data Preprocessing.R
├── 2 Data Analysis.R
├── 3 Reporting.R
├── data_raw
│   ├── sells.csv
│   ├── customer.db
│   └── demografics.xlsx
├── data_export
│   └── dataset_all.RData
└── visualizations
    ├── prediction_ytm.jpg
    └── prediction_Q2.jpg
```

It is general convention in business data analytics that the `data_raw` folder contains only the unmodified raw data, and the data in it is not (!) modified. The reason is that by doing so, we ensure that we do not delete or modify it unintentionally. Our preprocessed data is used in the scripts and saved in the `data_export` directory to provide it to other team member or analytics solutions (e.g., reports or dashboards). If the files are deleted, we can always regenerate them by rerunning our analytics pipeline.

Sometimes it makes also sense to add further folders, for instance a folder for visualizations or general project information. The number of folders and subfolders depends on the project size and your personal style, but to distinguish between raw and export data is extremely useful, even in the smallest projects.

Let's finally turn to the topic of exporting data. The most practical and fastest solution is to save your data directly in R's own format "`.RData`". This way the data can later be reloaded very fast directly by R using `load()`.

```
save(whisky_ratings, file="whisky_ratings.RData")
```

As a result, a new file in your working directory should appear.

If you plan to make the data available to other users who do not have R installed, the `read_csv()` function can be used to export it to the CSV format. Similar to the `read_csv()` function we used in the introduction to read CSV files in R, we can use `write_csv()` to save our data as CSV files.

```
library(readr)
write_csv(whisky_ratings, file="data_export/whisky_ratings.csv")
```

But the `readr` package has also many other export functions like `write_delimiter()`, if you want to export it in another CSV like format like values that are separated with a specific delimiter like the delimiter "`|`".

You can to that with:

```
library(readr)
write_delim(whisky_ratings,file="data_export/whisky_ratings.csv",
delim="|")
```

Besides there exist `write_xlsx()`, `write_excel_csv()` for exports to Microsoft Excel and `write_tsv()` to export it to tab-separated values.

4.6 Useful R Functions for Everyday Business Data Preparation

In this section I would like to present some useful functions that might help you in preparing business data. They are very useful, but the situations in which you might want to use them are rather rare. So, you are free to just skip this section and come back to it at a later time.

If you are working with a lot of files, and your working directory has become very big, you can use `list.files()` to get a list of files in your project folder:

```
> list.files()
 [1] "whisky.csv" "whisky.db" "whisky.RData" …
```

This function is very useful for retrieving the names of all files within your working directory. However, its primary utility lies in two main applications: as a reference to recall your file names or, more commonly, as input for another task, such as importing data files. To utilize `list.files()` effectively, you provide it with a path and an optional pattern. This enables you to list files located in a particular directory and those that match a specified regular expression pattern

```
path = paste0(getwd(), "/productionlogs/")
list.files(path=path, pattern="\\.xlsx")
```

Another relevant issue is that if you are working with very long numbers, or if you convert a character column of numbers in scientific notation, the decimals aren't preserved. For instance, a number like 12345678987654321, becomes this in R: `1.23456e+16`.

If you want to avoid this you can turn off scientific notation for numbers using the option:

```
options(scipen = 999)
```

Or if you want to reverse it, you can run:

```
options(scipen = 0)
```

4.7 Checklist

1. Phase: Business Understanding			
1.1	Determine business objectives	❯	• Background • Business objectives • Business success criteria
1.2	Assess situation	❯	• Inventory of resources and capabilities • Requirements, assumptions and constraints • Risks and contingencies • Terminology • Costs and benefits
1.3	Determine analytics goals	❯	• Business data analytics goals • Business data analytics success criteria
1.4	Produce project plan	❯	• Project plan • Initial assessment of tools and techniques
2. Phase: Business Data Understanding			
2.1	Collect initial data	❯	• Initial data collection report
2.2	Describe data	❯	• Data description report
2.3	Explore data	❯	• Data exploration report
2.4	Verify data quality	❯	• Data quality report
3. Phase: Business Data Preparation			
3.1	Select data	❯	• Rationale for inclusion/exclusion
3.2	Clean data	❯	• Data cleaning report
3.3	Construct data	❯	• Derived features • Generated records
3.4	Integrate data	❯	• Merged data

3.5	Format data	❯	• Reformatted data • Dataset • Dataset description
4. Phase: Modeling			
5. Phase: Evaluation			
6. Phase: Deployment			

4.8 Business Case Exercise: Investigating Quality Production Problems

4.8.1 Scenario

No time has passed since you looked at Sherry's data set when Ted Gumble appears at the door: "Hey, are you all right?" he asks and enters your office. "Say, could you maybe take another look at your analysis from the first day? I think the data from Sherry could possibly identify further issues. Besides, the analysis wasn't quite finished yet. Or?". You know Ted very well and you know that was more of a request than a question. "No problem, Ted, I'll get to work!"

4.8.2 Task

Your task is to redo the analysis from the whisky production data from the introduction. For that purpose, try to solve the following tasks:

- Try to structure your work according to the process model from the chapter
- Integrate the different datasets into one single dataset
- Apply different data preprocessing techniques, before you start with business data visualization
- Derive at least five hypotheses to explain the bad quality of some shifts.

4.8.3 Remarks

Use the `productionlog_sample` and `productionlog_extension` dataset for your analysis.

4.8.4 Solutions

You find the solutions of this exercise on the course website.

5 Modeling

5.1.1 Introduction

Congratulations, you have now come very far. You have navigated the perilous shoals of data understanding and fought your way through the churning seas of data preprocessing. Land in sight! We are approaching the modeling phase!

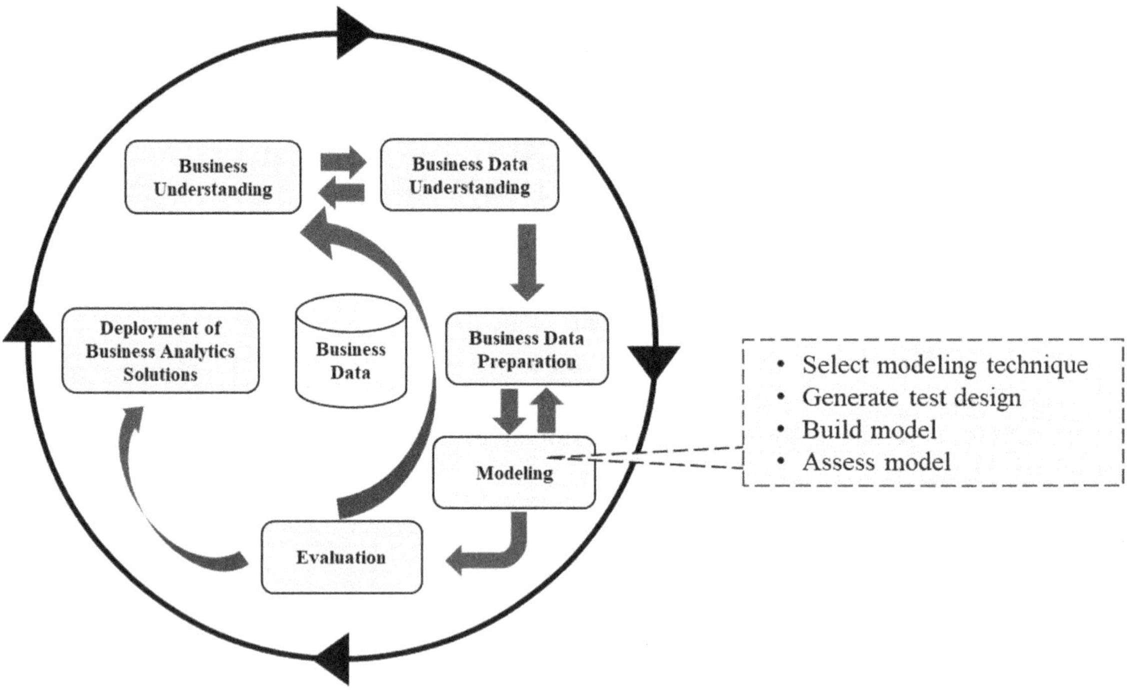

Figure 5-1 The generic tasks of the modeling phase.

As illustrated in Figure 5-1, the modeling phase is building on the data from the business data preparation. However, there might be further data preparations due to the special needs of the modeling algorithm, which is represented by the two arrows in the figure.

The main tasks of the modeling phase are:

- **Select modeling technique**: Decide which problem type you have and which business perspective is relevant for you to choose an appropriate modeling technique
- **Generate test design**: Make a short workflow to test your models on the business problem
- **Build model**: Define parameter settings to configure your model(s), make a short model description to better understand the different approaches
- **Assess model**: Assess the performance of the different models and revise your parameter and model configurations

Initiating the modeling process involves opting for the precise modeling technique to employ. While you might have previously chosen a tool in the business understanding phase, this step

D. Jung, *The Modern Business Data Analyst*, https://doi.org/10.1007/978-3-031-59907-1_5

pertains to the specific method itself, like building a decision tree with version 5.0 or generating different clusters using k-means.

Prior to running a modeling algorithm and constructing a model, it's essential to formulate a strategy or approach to assess the model's effectiveness and authenticity. For instance, to predict classes with classification algorithms, error rates are commonly employed as indicators of model quality. Consequently, the dataset is usually divided into training and testing sets. The model is built on the training set, while its performance is tested on a distinct test set.

If you are ready to construct your model, you utilize the chosen modeling tool on the preprocessed dataset to generate one or more models. In this step numerous hyperparameters can be fine-tuned. Note these parameters along with the selected values, accompanied by an explanation for the rationale behind the chosen parameter configurations in a documentation. Also, the comprehension of these models is crucial to dive deeper in the business and data understanding, highlight any challenges faced while interpreting their implications.

Finally, you as a business data analyst interpret the models together with the domain experts based on your knowledge and expertise and the test design. It is crucial to emphasize that in the modeling phase you concentrate on models, whereas the evaluation phase encompasses all other project-generated outputs (chapter 6). Provide an overview of the outcomes of this step, outlining the performance of the created models (such as accuracy) and establishing a hierarchy of their performance in comparison to one another. Based on the assessment of the models, fine-tune and adjust hyperparameter configurations to build further models. In most cases, you will engage in a repetitive process of constructing models and evaluating them until a satisfying degree of confidence is reached.

5.2 Modeling Methods in Analytics

Every business data problem and related solution modeling is unique and different from the previous one. On the other hand, there are general problem types in this phase that occur again and again. To help you get started at your specific modeling problem, we will look at these common business data analytics problems together. The problem categories are very general but they will help you to organize the huge number of business data analytics modeling techniques and scenarios.

Figure 5-2 is an overview of typical business problems, related questions, and the techniques we will use in this book to solve them. Use the overview presented in Figure 5-2 to get started fast if you face one of the most common problem types. The algorithms are established for these kinds of problems, and you will produce very fast very usable results which can be used as a starting point for your further investigations.

One very popular problem is to compare two or more business actions with each other. For instance, you want to know if a new label influences the sales or if two datasets are significantly different from each other. These kinds of problems are solved with A/B-tests.

Some other problem might be that we have several observations like customers information or production logs and want to know if they can be gathered together into groups.

Problem Type	Business Analytics Perspective	Techniques
Compare Business Decisions	▪ Does a new whisky label increase the sales? ▪ Is there a difference between two customer campaigns in the online shop?	▪ Hypothesis Tests ▪ A/B-Tests
Find Clusters and Outliers	▪ What is the most similar whisky compared to Junglivet? ▪ Can we put our customers together into distinct/different groups for marketing issues?	▪ Clustering ▪ Nearest Neighbour Analysis
Find Rules and Relationships	▪ If a customer buys our 10-year Junglivet whisky, what whisky does he buy next? ▪ Which products are normally bought together?	▪ Association Analysis
Predict and Explain Categorial Values	▪ Is the customer of our shop solvent or not? ▪ Will this customer send back this shipping or not?	▪ Decision Trees ▪ Logistic Regression
Predict and Explain Numeric Values	▪ How much does a new label increase the sales? ▪ Compare the influence of different factors to sales volume	▪ Regression Analysis
Predict and Analyse Developments	▪ How will the number of sales of our products develop? ▪ What future developments are likely?	▪ Time Series Forecasting

Figure 5-2. Typical business data analytics problems and techniques to solve them.

Or we want to get a general overview of the similarity of our observations. This can be used to support decisions on a more general level or be seen as a complement to other techniques like classification. For instance, if we have a set of whiskies that is very similar in their characteristics, we can make conclusions if a whisky with similar characteristics will fit in our collection or not.

If we want to understand or describe the relationships of the data points association analysis might be helpful. This technique is very useful to find hidden patterns that can be used to derive decision rules. The focus is generating such many useful rules instead of predicting. For instance, association rules describe if a specific product is often bought together with a certain set of products.

In other scenarios you want probably predict a value (for instance sales or prices) or a class (whisky type). For these kinds of problems decision trees and regression models have been established in business data analytics. They are good to understand and produce very solid results.

If you are not only interested in the future outcome but in the whole development time series forecasting is the method of your choice. Time series methods can be used to describe how a specific value will develop over time and gives more sophisticated predictions than regression analyses.

Please note that we will focus in this book only on one or two techniques for each problem type. Up to now analytics research has developed endless approaches and algorithms for each problem type. Each method and algorithm have different strengths and weaknesses. A discussion and comparison of all possible techniques will go wide over the scope of this book. Hence, we will discuss the most popular and general method and algorithm for each problem type. In real-life you will come very far with that. However, if your model does not produce good enough results, I want to push you forward to read books about the specific problem type (e.g., clustering or decision tree), to learn further approaches which may help you.

I also want to take a chance and give you a further recommendation. Don't start too early with the sophisticated steps like modeling tuning or feature engineering. Remember, that our main objective is to build analytics solutions to give decision support. Often, most problems can be solved quite well using simple methods instead of complicated solutions. For instance, a simple and well designed A/B test is in most cases much better than using complex modeling techniques just because you know how to use them.

5.2.1 Modeling Workflow and Test Design

The most relevant task in modeling is to design a short workflow to test your models on the business problem. This process should be comprehensible and simple. In this section, we will look how such a process might look like.

A straight forward modeling workflow is illustrated in Figure 5-3. You start with a preprocessed dataset and simply apply your modeling algorithm on it. The result will be one or multiple models that you use to make predictions or compare different business decisions.

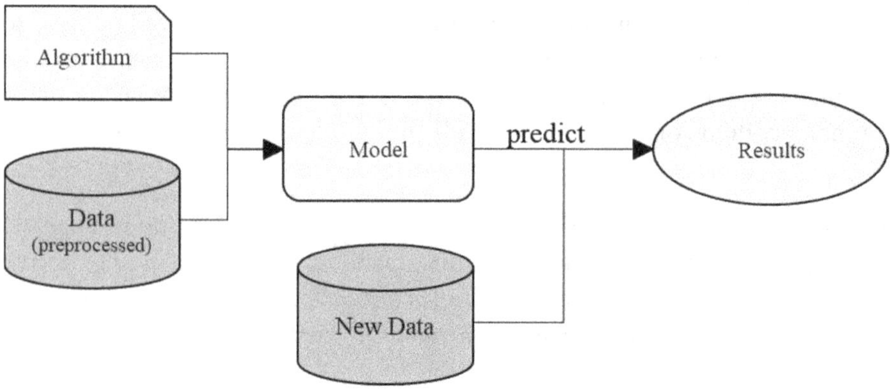

Figure 5-3 Simple workflow to find patterns and compare datasets.

However, this process has some pitfalls. The most relevant is, that your model may become too specific to the training data, capturing noise and irrelevant patterns rather than true relationships. You will choose the model with the best performance, but this will likely be the model that is only the best in the past and not the future. Another issue is, that you cannot use this process to evaluate different model parameters and find the combination that works best.

Therefore, a test design allows you to assess the quality of your analytical models. It helps determine how well your model performs on new, unseen data and whether it produces the desired results. Only if you have such a clean test design, you can use metrics such as accuracy, precision, recall, and F1-score to evaluate your model performance and ensure it meets the project's requirements and objectives.

The most common test design in business data analytics is a training-test split as illustrated in the following Figure 5-4:

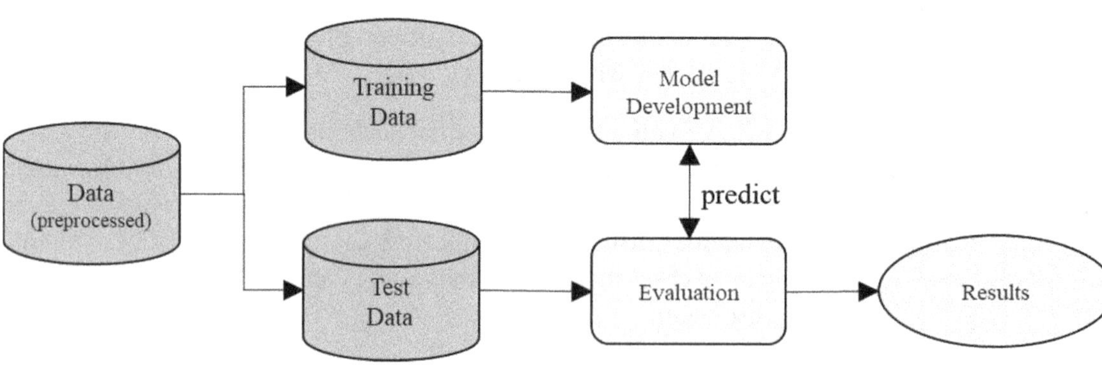

Figure 5-4 Workflow to develop an analytics model to predict future outcomes.

Using the design from Figure 5-4, you start by splitting your data into two subsets. One the training set is only used for model development, while the second one is only used to test your models. A best practice is to split your data into `80%/20%`, while the training set will be `80%` of the data. Because your data might be ordered or contain some structural information, you should do this split randomly in most cases. If you use special techniques like time series, you can do this split based on a specific date in your dataset (e.g., $t <= 01.04.2023$) but should not split it randomly, which would make your time series useless.

For instance, such split of a data frame named `df` can be done in the following manner:

```
ids = sample(c(TRUE, FALSE), nrow(df), replace=TRUE, prob=c(0.8,0.2))
train = df[ids, ]
test = df[!ids, ]
```

5.2.2 The Junglivet Online Shop Dataset

To demonstrate the different models of the following chapter, we'll use again a specific dataset. In most cases we will refer to a sample of the Junglivet `onlineshop` dataset that can be accessed in the `dstools` package:

```
library(dstools)
data("onlineshop")
```

Or you can download it from the course website and load it with the read function:

```
read_excel("onlineshop.xlsx")
```

The dataset is from Lindsey from the marketing team. She and her colleagues gathered every information about the Junglivet customers in the database of the online shop they could find. Then, they extracted it from the Junglivet online shop.

Table 5-1 The `online shop` *dataset from Lindsey*

DATE	USER_ID	AGE	GENDER	TYPE	CREDIT_SCORE	...
2023-12-30	rose_leslie	56	female	NA	4	...
2023-12-30	david-niven	53	male	1	4	...
2023-12-30	usurpgam	48	male	1	5	...
...	...	...	...	...	...	...

As you can see in Table 5-1, the dataset is an excerpt of the different transactions in the online shop. Each row represents one transaction. For example, the first row belongs to a user with an `USER_ID` named `rose_leslie`. The user is `56` and `female`. Based on the column `DATE`, we know that she bought something in the online shop by end of December in 2023. She has a `CREDIT_SCORE` of `4` and there is no further information about her user `TYPE`.

Here is a full list of the variables in the dataset for you:

- `DATE`: Timestamp indicating date of login to online shop
- `USER_ID`: Unique identifier for the online shop user
- `AGE`: Age of the user at time of visiting the website
- `GENDER`: Gender of the user

- TYPE: Internal customer type based on phone interviews (1: Standard Customer, 2: Whisky Connoisseur, NA: no phone interview)
- CREDIT_SCORE: External scoring of the financial solvency of the customer from 1 (very bad) to 5 (excellent)
- LIFETIME: Lifetime of the account of the customers in years after initial registration
- PAYMENT_METHOD: Payment method of the customer like BNPL (Buy now, pay later), credit card, bank transfer or paypal
- TURNOVER: Turnover of the current purchase in EUR
- SENDBACK: Dummy to flag if parts of the purchase were returned within 2 weeks. 1 or TRUE means that it has been send back by the user
- VIDEO_AD: Dummy to indicate the type of daily commercial on the landing page. While 1 or TRUE indicates that the user had a video commercial, 0 or FALSE indicate otherwise like a text prompt or article
- CONVERTED: Dummy to indicate that the user put the whisky or merchandise from the daily commercial in the shopping basket

The dataset contains a lot of information, so don't be confused if you do not understand everything at the moment. We will investigate this dataset step by step in the remaining chapter.

5.3 Compare Business Decisions

Business decision-making often involves assessing various options, strategies, or even design choices. But how can we be sure that the choices we make are truly effective, rather than based on assumptions or intuition? Especially in the areas of web design and online marketing with Youtube and Instagram, analyzing and validating decisions has become a huge thing in business data analytics. In many digital or mobile companies, the main job of business data analytics experts is to evaluate the success or failure of business decisions.

Figure 5-5 Example of two different web designs to compare with an A/B test.

For instance, the product owner of the Junglivet online shop would like to know if the new design of the mobile landing page will increase the sales. Or the marketing manager of the *Junglivet Whisky Company* might be interested in the acceptance of his new marketing campaign on Youtube with influencers compared to a traditional campaign.

In the next section, we'll demystify the art of comparing and testing different decisions with business data analytics methods. We explore the different approaches to make informed decisions based on solid empirical evidence.

5.3.1 Hypothesis Test

But before we come to A/B testing, I would like to give you a short introduction into hypothesis testing in general. With that in mind, let's turn to a current task that we have just received! Or our friend Barney Gumble from the production team has ordered a new strain of barley from his suspicious friend Mr. Burns. The materials from `Burns Best` caused problems in the past (see our introduction case in the preface) and hence Gumble has a serious interest if the different cadmium pollutant levels in the new strain of barley are just random or not.

Testing whether a certain assumption (for instance, Mr. Burns' barley has more pollution than other suppliers) is likely to be true or not is one central topic of traditional statistics and of modern business data analytics. The first and probably most popular hypothesis tests were proposed by our friend Ronal Fisher (Fisher, 1936), we talked before in the introduction of chapter 4. As most (business) data analyst, he had a huge sense of humor and a good portion of irony towards himself. So he instantly used his new developed A/B-testing technique to challenge his guest Lady Blanche Muriel Bristol (Fisher, 1956) during tea time. She had claimed that if you change the filling order of milk and tea in a cup it would influence the flavor. The end of the story: After many cups of tea, he proofed that Muriel Bristol was surprisingly right, and published the findings in a scientific journal as "lady tasting tea experiment".

In the classical statistics and analytics setup, assumptions are called hypotheses. We have a null hypothesis H_0 that represents the default. And we have one or more alternative hypotheses H_1, H_2 $etc.$ that represent what we would like to compare.

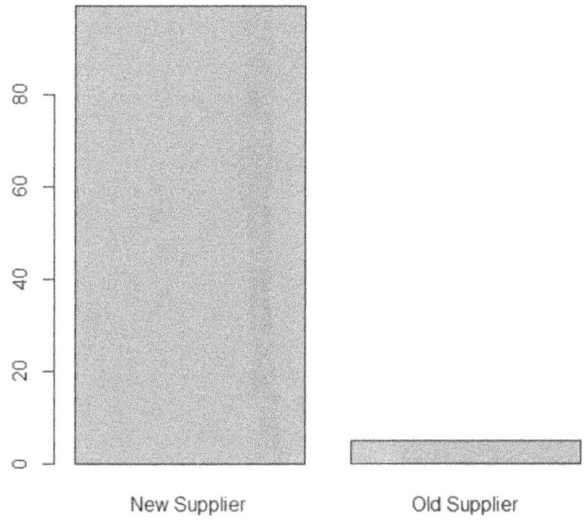

Figure 5-6 Comparison of the different supplier's performance.

For instance, we would like to test the cadmium in a new barley. Then our H_0 would be "There is no difference in cadmium compared to our traditional barley", and our H_1 would be "There is significantly more cadmium". Let us now take a look at a working example. Given the following results from the Junglivet whisky laboratory we have one dataset with the minerals and pollution values from our new barley and our old barley supplier:

If 99% of the observations of the new supplier have critical pollution values, while only 5 % of our other sample do not, we can be pretty confident that our old supplier is better than our new supplier. However, what if only 50 % or 10 % do have critical values? When can we be sure that the differences are real differences. And that the differences are not due to random fluctuations or measurement uncertainties?

Best practice is to use a statistical test to decide if our null hypothesis is likely or unlikely and hence should be rejected. A very popular collection of statistical tests are t-tests with its most popular member the one sample t-test. The group of tests is implemented in base R as `t.test()`.

After this general introduction let us come back to business problems and assume we want to test if there is a significant difference in the turnover of our online shop customers compared to the local shop. We know that the average turnover in the local whisky shop is about `50` EUR. Hence, the related H_1 would be "Online shop customers buy significantly more than 50 EUR".

But before we can test our hypothesis, you should know that the t-test can be used only, when the data is normally distributed and there are at least `30` observations per group. We can check the number of rows with `nrow()` and the normal distribution using Shapiro-Wilk test in the following manner.

```
> shapiro.test(onlineshop$TURNOVER)
       Shapiro-Wilk normality test
data:  onlineshop$TURNOVER
W = 0.75779, p-value < 2.2e-16
```

Our test output shows the p-value is below the significance level `0.05` implying that the distribution of the data is not significantly different from normal distribution. In other words, we can assume that our turnover in the online shop is normally distributed, which makes the t-test the suitable test in this case.

Finally, we can run our one sample t-test with:

```
> t.test(onlineshop$TURNOVER, mu=50)
       One Sample t-test
data:  onlineshop$TURNOVER
t = 14.754, df = 462, p-value < 2.2e-16

alternative hypothesis: true mean is not equal to 50
95 percent confidence interval:
 164.5080 199.6994
sample estimates:
mean of x
 182.1037
```

The results show that the t-test has a p-value below our significance level of `0.05`. Thus, we can conclude that the mean turnover of online shop customers is significantly different from the `50` EUR turnover (`mu = 50`) of local shoppers. As you already might have noticed, the `mu` parameter specifies the value of the null hypothesis being tested. In our example, a one-tailed t-test is performed to check if the mean of the data in `onlineshop$TURNOVER` is greater than `mu=50`.

If you want to test whether the mean turnover is less than `50` EUR you have to run the test with a specified alternative hypothesis `alternative="less"`, and if you want to test if it is significant more you have to set the `alternative="greater"` like this:

```
t.test(onlineshop$TURNOVER, mu=50, alternative="greater")
```

Besides this one-sample t-test there exists further t-tests you should know. To give you an overview where you can come back, I summarized the most important characteristics for you:

- **one-sample t-test**, `t.test(x,y)`: Can be used to figure out if the average (mean) of a sample is significantly different from a known or expected value. E.g., The *Junglivet Whisky Company* produces a popular whisky called "20 years distiller edition," and the advertised alcohol by volume (ABV) is 40%. However, they suspect that the actual ABV might be different due to variations in the production process
- **two-sample t-test**, `t.test(x, y, var.equal=TRUE)`: Can be used to compare the means of two different groups or sets of data. Both groups have to be randomly selected, independent (one group's data doesn't affect the other), and follow a normal distribution. We also assume their variances (how spread out the data is) are equal but unknown. E.g., The *Junglivet Whisky Company* is considering sourcing barley from two different suppliers, Supplier A and Supplier B. They want to know if there's a significant difference in the average barley quality between the two suppliers
- **paired t-test**, `t.test(x, y, paired=TRUE)`: You're comparing the means of two sets of data points that are somehow related or "paired". These pairs come from two different populations. E.g., the *Junglivet Whisky Company* is experimenting with a new fermentation process to improve whisky flavor. To assess its effectiveness, they take a group of whisky barrels and age half of them using the new process, while the other half uses the traditional process.

As you can see t-tests are a powerful statistical tool set used to determine if there's a meaningful difference between groups or if a sample's mean is significantly different from a known value. Each type of t-test has specific applications based on the nature of the data and the question you want to answer.

However, sometimes the t-test will be not the right choice or other tests are more suitable. At this point I can strongly recommend the decision assistant from university Zürich. They developed an online tool that asks you some questions about your analytics problem and helps you to select the correct test.

You can find it at:

https://www.methodenberatung.uzh.ch/de/datenanalyse_spss/entscheidassistent.html

5.3.2 A/B Test

Building on the concept of hypothesis testing, the basic idea of A/B testing is to systematically prepare a data sample measuring the outcome of two competing business decisions and test these two different alternatives against each other. As illustrated before, there are many cases for this in web design and online marketing. But for now, we want to compare two different mobile versions of the same landing pages in the Junglivet online shop to decide which one does better.

In terms of hypothesis testing our question is "Will using video advertisements on the landing page results in more whisky buys?" and the related H_1 would be "Using videos in our advertisements on the landing page will result in more whisky sells". We have the variable `CONVERTED` as our dependent variable which indicates if the whisky from the commercial has been put in the shopping basket after watching the commercial on the landing page and `VIDEO_AD` as independent variable indicating if a video was integrated in the advertisement on the landing page or not. Let us have a short look at our dataset:

```
onlineshop = read_excel("onlineshop.xlsx") %>%
  select(USER_ID, DATE, VIDEO_AD, CONVERTED)

onlineshop = onlineshop %>%
  mutate(MONTH = months(DATE)) %>%
  group_by(MONTH, VIDEO_AD) %>%
  summarize(WHISKY_CONVERSION_RATE = mean(CONVERTED))
onlineshop

ggplot(onlineshop) +
  aes(x = MONTH,
      y = WHISKY_CONVERSION_RATE,
      color = VIDEO_AD,
      group = VIDEO_AD) +
  geom_point() +
  geom_line()
```

The output is illustrated in Figure 5-7 and shows the different whisky conversation rates across different months.

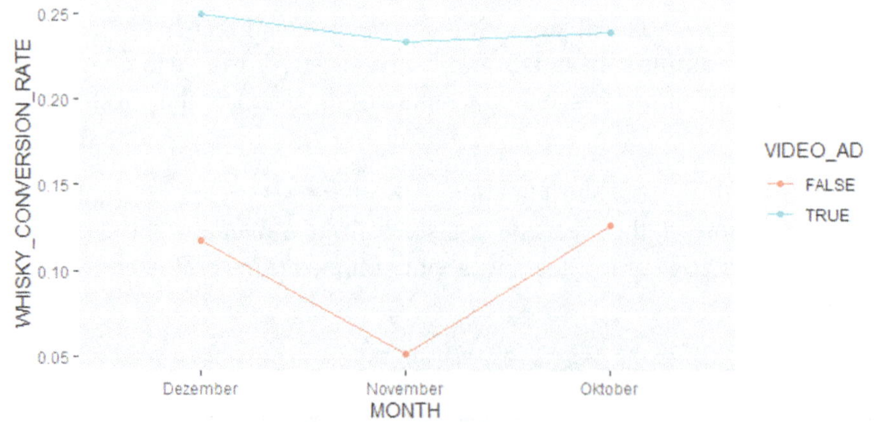

Figure 5-7 Conversion rates across the last quartal.

As you can see, we have a kind of seasonal effect, the whisky sales are dropping in November but increase towards Christmas. However, the conversation rate increases in the video condition compared to the control condition without video.

A very common job would be that the management asks you to make such plots and then test if a new design feature like our video feature works. This is usually done with an A/B test in terms of conversion rate, indicating how many people used your feature to buy new products or recommended a product to their friends after the feature has been introduced.

Before you can start with A/B testing, you have to make some experimental design decisions. First you have to create two groups of data, for instance two groups of users in the Junglivet online shop. One group will be the control group and the other will be the treatment group. We randomly assign the users to one of the two groups. This is also called a randomized-control trial.

Now for the actual A/B test: conveniently, R has the `prop.test()` function, which tests whether two proportions are significantly different (using a Pearson's chi-squared test under the hood).

All we have to do is feed our dataset as a table into the function, and R will do the rest of the work for us:

```
read_excel("onlineshop.xlsx") %>%
  select(VIDEO_AD, CONVERTED) %>%
  table() %>%
  prop.test()
```

Resulting in:

```
2-sample test for equality of proportions with continuity correction
data:  .
X-squared = 23.842, df = 1, p-value = 1.046e-06
alternative hypothesis: two.sided
95 percent confidence interval:
 0.3204785 0.7017437
sample estimates:
   prop 1    prop 2
0.7777778 0.2666667
```

Because the p-value was well below the usual threshold of `0.05`, the difference was highly significant and the null hypothesis (that the difference was by chance) could be rejected. Consequently, we would definitely choose the video advertisement presented to the test group in the future.

An alternative to running this test would be to run a regression with dummies (a column indicating if the user is the control or treatment group). We will learn more about regressions later in this chapter.

If you remove outliers or filter specific observations from your dataset, you can probably bring many of your real-world datasets below the significance level. This is called p-hacking and there should be no doubt that this must not be driven too far. My rule of thumb is that you should clean your data without your alternative hypothesis in mind. Your data cleaning should be traceable, without doing something unusual. Then test your assumptions and do not adjust your cleaning if your hypothesis does not hold.

5.4 Find Clusters

Your colleague Lindsey Naegle from the marketing team contacts you because she and her team would like to design a new marketing campaign. She would like to know which type of customers drink Junglivet whisky. Are there specific groups of customers that share characteristics like favorite local football clubs that Junglivet could sponsor? Are there preferred super markets or do they come from the same area, so that she can setup billboard marketing? Do they share other characteristics like age, gender or income that could help to design the marketing campaign?

Now imagine a new user named `ron_swanson76` makes an account in the Junglivet online shop. If Ron is a `whisky enthusiast`, it's very likely that if he likes what he gets, he will spend a lot of money in the future. So, you could send a special gift in the first order, or if you are very confident send him the whisky for free. However, if you do it for every customer your earnings will drop and you will probably lose money in the long run.

What could you do? If you know more about Ron than just his username, for instance residence, age, gender, or more specific things like how much friends he has who are also *Junglivet whisky enthusiasts* or how much income he has (based on transactions) you could compare his profile with the profile of the other users. Looking at users who are similar in the information you have about Ron, could help you to make a good decision to which type of customer he belongs. And finally, if you should invest some money and add a special Junglivet whisky sample.

Mathematical speaking, this means that we can classify a customer by a majority vote of its neighbors. The new customer is assigned to the class that is most common among its k-nearest neighbors. The distance is computed by comparing the different customer features e.g., x and y with n observations by the Euclidian distance $d(x, y) = |y - x| = \sqrt{\sum_{i=1}^{n}(y_i - x_i)^2}$.

If you know nothing about your customers you should start gathering data about them and use it to understand how your customer behave. Otherwise, you cannot compute such distances. Customers living in the neighborhood of a popular football club and are passionate about football, could be easily convinced to buy Junglivet whisky by sponsoring the local team. If you have different types or clusters of customers, you can design your business actions accordingly.

However, what do you do if you do not find clusters on a first look. Or if you don't want to rely on the human decision or if you don't have any marketing background research about your customers to build the user types from scratch. These methods are called clustering methods.

Let us start with a simple case. Lindsey and her colleagues from the marketing team want you to find some clusters in the transaction data from the Junglivet online shop. They already had a look at the data and recommend you to start with a selection of the variables. In particular, they want you to investigate possible clusters in the following set of variables: `AGE`, `GENDER`, number of buys (as `INTERACTIONS`) and average expenditure in the shop based on `TURNOVER`.

Therefore, we will load the dataset and compute the new variables in the following manner:

```
onlineshop = read_excel("onlineshop.xlsx") %>%
  group_by(USER_ID) %>%
  mutate(EXPENDITURE = round(mean(TURNOVER)), INTERACTIONS = n()) %>%
  select(AGE, GENDER, EXPENDITURE, INTERACTIONS) %>%
  distinct()
```

Our final dataset looks now like this:

```
> onlineshop
   USER_ID               AGE GENDER EXPENDITURE INTERACTIONS
   <chr>               <dbl> <chr>        <dbl>        <int>
 1 cardiB                 27 female       189.            12
 2 natalie_horohlag       42 female       185.            10
 3 babe_ruth              48 male         166.            10
 4 homlesspistachio       60 male         500              1
 5 chrstian_hauff         93 male          79.9            9
 6 marc_sinclair          36 male         184.            11
 7 ron_swanson76          56 male         433.            15
 8 fred_firestone         35 male          99.9           10
 9 amanda_missinger       24 female       228.             9
10 fred-johnson           37 male          73.1            9
```

As you can see, there are different customers of different ages. `ron_swanson76` bought `15` times this year in the online shop, while `homlesspistachio` only once. All the other customers bought around `10` times in this year. Whenever you look at data, it is very common that you intuitively detect clusters like this in your dataset.

For instance, based on marketing research it seems likely that the female shoppers spend more money in the online shop. In our data extract most female spend above 100 EUR which is more than the most males spend. There might be also a group of very old whisky enthusiasts that also spend a lot of money in the shop. However, this is just a guess at the moment and we have to check it.

Furthermore, unlike other analytic problems there is no correct cluster. Clusters are qualitative descriptions of your data that gather your objects into groups that make somehow sense, but could also gather together in another way. We made some educated guesses about possible clusters in our dataset but there might be further clusters we didn't see because we didn't expect them to be there. Or as the famous decision scientist Daniel Kahneman said "We focus on what *we know* and neglect what *we* do not *know*, which makes us overly confident in our beliefs" (Kahneman, 2011). Therefore, it might be helpful to have some analytics methods which could challenge our clusters and make some suggestions for clusters by their own.

5.4.1 k-means

In a first step we want to run a very robust and popular clustering algorithm: k-means. K-means clustering (MacQueen, 1967) is one of the most commonly used algorithms to divide a given business data set into k clusters, where k is the number of groups predetermined by the business analyst. It classifies objects into multiple clusters such that objects within the same cluster are as similar as possible. We say that they have a high intra-class similarity, while objects from other clusters are as dissimilar as possible. In k-means clustering, each cluster is defined by its centroids (the geometric center), which is the mean of the points assigned to the cluster.

Let us have a short look at the mathematical idea behind this. Given n observations $(x_1, x_2, \dots x_n)$, the k-means clustering method partitions our n observations into $k \leq n$ distinct clusters $S = (S_1, S_2, \dots, S_k)$. This is done by minimizing the distance of our observations in each cluster S, measured by the within-cluster sum of squares:

$$\arg\min_{S} \sum_{j=1}^{k} \sum_{x_i \in S_j} \left\| x_i - \overline{m}_j \right\|^2$$

The cluster mean $\overline{m}_j$ is the mean vector of the features averaged over all observations in the cluster S_j. The distance is the (squared) Euclidian distance $\left\| x_i - \overline{m}_j \right\|^2$, which sums the squared differences over all features. Hence, we must assure that our features have comparable or even equal scales and there are no correlations between them. The number of k must be given.

If we go now back to our example. We know that based on the most bought whisky bottle, our marketing team assumes that there are three ($k = 3$) types of customers: *General whisky shopper* (they buy a selected bottle or a present for a friend), *whisky drinker* (they buy randomly low budget whiskies at high frequency), and *Junglivet enthusiasts* (they spend a lot of money in the shop mostly for Junglivet, but do not buy often). Lindsey would be interesting if your analytics methods find a similar cluster characteristic or find other clusters. Please note that clustering algorithms do not label clusters, you will have to do that by looking at the objects in the data and trying to understand the underlying relationships. But by talking to your domain experts, you get a good guess for the number of clusters k-means should search.

In our case we run the k-means algorithm on our online shopping data and investigate the results. But before we can do this, you should know that k-means requires continuous features as input, as it is based on computing distances. So, we have to remove the `USER_ID` because it contains unique character values, otherwise the k-means algorithm would not compute distances in a very meaningful way. Let us therefore mutate our double columns into integer columns, which seems a good time to use the `across()` function inside `mutate()` to mutate all `double` values to `numerics` (check the last chapter if this is new for you):

```
onlineshop_pre = onlineshop %>%
  mutate(across(c(is.double), as.integer)) %>%
  ungroup() %>%
  select(AGE, EXPENDITURE, INTERACTIONS)

fit = kmeans(onlineshop_pre, centers = 3, nstart = 25)
```

The final clustering model is stored in our fit variable.

```
> fit
K-means clustering with 3 clusters of sizes 42, 35, 23

Cluster means:
       AGE EXPENDITURE INTERACTIONS
1 40.04762    59.21429     4.095238
2 42.71429   194.00000     6.542857
3 43.00000   431.08696     2.695652

Clustering vector:
  [1] 2 2 2 3 1 2 3 1 2 1 1 2 1 1 2 1 2 3 2 3 2 1 1 1 2 1 3 1 2 2 1 2
1 2 1 2 1 2 3 1 2 1 1 2 2 1 1 1 1 1 1 2 1 1 1 2 3 3 1 1 1 2 3 1 3 1 2
1 2 2 3 3 1 1 1 3
 [77] 3 3 3 1 2 2 1 2 3 2 1 2 3 2 3 1 2 1 2 3 3 3 2 3

Within cluster sum of squares by cluster:
[1]  47750.60  71389.83 246934.70
 (between SS / total_SS =  84.9 %)

Available components:
[1]  "cluster"           "centers"           "totss"              "withinss"
"tot.withinss" "betweenss"     "size"          "iter"           "ifault"
```

This output tells us that the data has been grouped into three clusters, and it provides the sizes of each cluster (i.e. the number of data points in each cluster).

The next part `Cluster means` provides the mean values of the features within each cluster. For example, in `Cluster 1`, the mean values for the variables `AGE`, `EXPENDITURE`, and `INTERACTIONS` are approximately `40`, `59`, and `4`.

Then, we see the clustering vector, showing the cluster assignments for each data customer. Each number corresponds to a cluster. For instance, if a data point is labeled with `2` it belongs to `Cluster 2`.

Moreover, we get some information how tightly the data points are clustered within each cluster using sum of squares.

And finally, the model consists also of different components like `cluster` or `centers` which you can call directly if you want them to process in your code. For instance, the `cluster` variable can give us the locations of the centroids of the three clusters. We can access them directly via `fit$centers`:

```
> fit$centers
       AGE EXPENDITURE INTERACTIONS
1 40.04762    59.21429     4.095238
2 42.71429   194.00000     6.542857
3 43.00000   431.08696     2.695652
```

Let us now make a plot to investigate our results. Therefore, we add the cluster vector `fit$cluster` to our online shop data frame and then plot it.

Coloring by cluster and using different symbol shapes to highlight the gender or age for comparison.

```
onlineshop$CLUSTER = as.factor(fit$cluster)
ggplot(onlineshop) +
  aes(x=INTERACTIONS, y=EXPENDITURE, color=CLUSTER, shape = GENDER) +
  geom_point() +
  labs(x="Number of interactions", y="Expenditure in EUR") +
  ggtitle("Customer types in the Junglivet onlineshop")
```

The final results are illustrated in the following Figure 5-8. As you can see, the results show no useful clusters. By eye you can see that there are definitely two major clusters below `2` interactions and above `8` interactions.

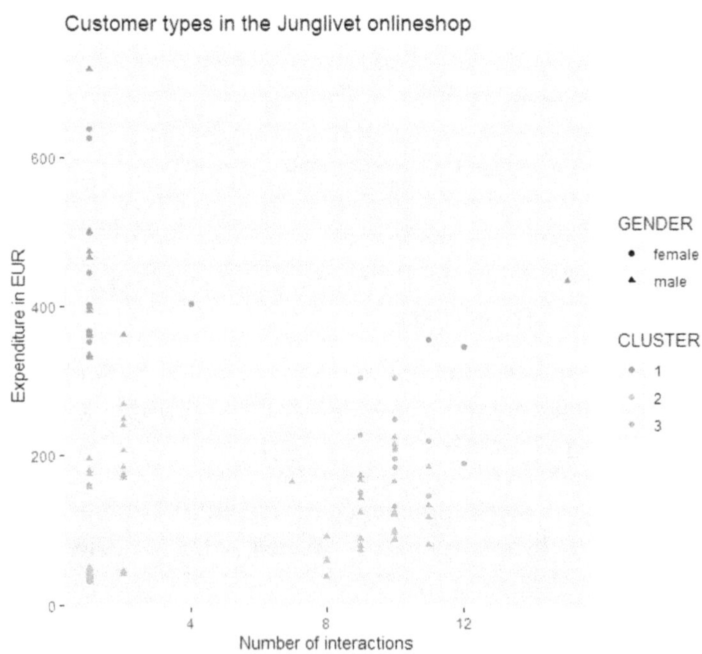

Figure 5-8 Customer types based on k-means clustering without scaling.

However, the algorithm does not detect them. The reason is that your features use different scales. `AGE` is a feature representing values between 21 and 100, while `EXPENDITURE` is a value much higher. Our model cannot understand the reason for this and hence will have difficulties to interpret them. Remembering the chapter 4 about data preprocessing, we rerun our code with scaled features.

You can also normalize them, which will lead to similar results:

```
onlineshop_pre = onlineshop %>%
  ungroup() %>%
  select(AGE, EXPENDITURE, INTERACTIONS) %>%
  mutate(across(c(is.double), as.integer)) %>%
  mutate(across(everything(), scale))
fit = kmeans(onlineshop_pre, centers = 3, nstart = 25)
onlineshop$CLUSTER = as.factor(fit$cluster)

ggplot(onlineshop) +
  aes(x=INTERACTIONS, y=EXPENDITURE, color=CLUSTER, shape = GENDER) +
  geom_point() +
  labs(x="Number of interactions", y="Expenditure in EUR") +
  ggtitle("Customer types in the Junglivet onlineshop")
```

And tada! There are well-formatted features as illustrated in the following Figure 5-9:

Figure 5-9 Better clusters with k-means and scaled features.

I hope this example illustrated the relevance of data preprocessing for the modeling part. Furthermore, we can improve the clustering results by removing noise and outliers. But for now, we want to continue.

A very common error in clustering is that analysts forget to scale their data. If you do not scale your features k-means can still produces somehow useful clusters, but if you start to add other features with other scales the clusters start to become useless. The reason in our example is that because the range of the numbers of `AGE` and `EXPENDITURE` are much bigger than number of interactions, which will make it difficult for k-means to consider the number of interactions correctly.

While we started building our clusters fairly quickly in the previous pages, we deliberately skipped a key question. As a business data analyst, how do I actually know what number of clusters makes sense?

A rather simple way to answer this question is to look at the development of the error values (weighted sum of squares) of a cluster. We assume that a suitable number of clusters has to be between `2` and `10`, thus we plot the within-groups sum of squares against the number of clusters and look for a characteristic slowdown in the decline:

```
# prepare dataset for k-means
onlineshop_pre = onlineshop %>%
  mutate(across(c(is.double), as.integer)) %>%
  ungroup() %>%
  select(AGE, EXPENDITURE, INTERACTIONS)

# weighted sum squares (wss) for all cluster numbers from 2 to 15
wss = c()
for(k in 2:10){
  wss[k-1] = sum(
    kmeans(
      onlineshop_pre,
      nstart=1000,
      centers=i
    )$withinss
  )
}

wss_clusters = data.frame(centers = 2:10, wss)

# make plot
ggplot(wss_clusters, aes(x = centers, y = wss)) +
  geom_point() +
  geom_line() +
  xlab("Number of clusters") +
  ylab("Within groups sum of squares")
```

The result is a plot that will somehow look like this (I write somehow because k-means starts with random cluster centers and hence every run will be slightly different).

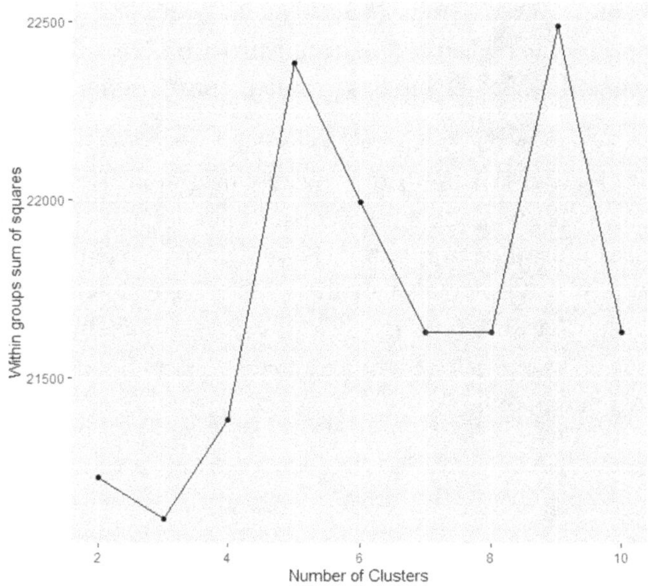

Figure 5.10 Development of the errors across cluster numbers in k-means.

As you can see if we increase the number of clusters from 2 to 3 the error declines and then increases with each further cluster. We can conclude that a suitable number of clusters would be 2 or 3 clusters.

5.4.2 k-nearest-neighbor

Another very common use case of clustering is the assignment of new points to an existing clustering model or to find similar points that are comparable to an existing group represented by an existing cluster. To illustrate this, let's assume that after a discussion of our new clustering-model with the marketing department, we want to send out personalized promotional flyers with vouchers and a Junglivet whisky sample to a selected set of customers of the online shop to promote Junglivet whisky. The idea is to increase the image of Junglivet for these customers and motivate them to buy Junglivet more frequently or at least try it in the future.

The k-nearest-neighbors algorithm (or in business data analytics slang just KNN) can help you decide which customers to send the flyers to, based on customer clusters in our data. In our case the marketing team made some market research and contacted some of the customers personally. As a result, they could label a subset of the customers and you know that you have a group of whisky connoisseurs in the online shop. Customers of this cluster of whisky connoisseurs tend to buy premium alcoholics. They are mostly foodies and liquor fans and would be open to spend a lot on high quality whisky like Junglivet. The idea is that you could use KNN to detect further customers who are similar to this cluster in our online shop dataset.

Let's say you have prepared 100 flyers and whisky samples and want to send them to 100 customers who are most similar to our whisky connoisseur group (represented by our clustering-model). KNN would look at the data of all our customers and find the 100 customers who are closest in terms of age and spending habits to the group of customers who are mostly foodies.

You would then send the flyers, vouchers and whisky samples to these 100 customers, hoping that they would also be interested in buying Junglivet whisky. You could also use KNN to decide if a new customer who registered in the online shop belongs to this group of connoisseurs or not.

KNN helps you make marketing decisions based on the behavior of similar customers, just like sorting balls based on their colors using the rulebook. It finds the "nearest neighbor(s)" in your customer data and uses their behavior to make predictions or decisions for your marketing strategy. So, when a new data point enters, as illustrated in Figure 5-11, and the nearest neighbor (`k=1`) is in cluster green, we assign the new customer also to this cluster. To increase the reliability further, we could also compare the nearest 2 (`k=2`), 3 (`k=3`) or 5 (`k=5`) neighbors or all neighbors in a specific range. And if out of 5 neighbors, 3 are green and 2 are red, we assign the new customer to the category with most neighbors.

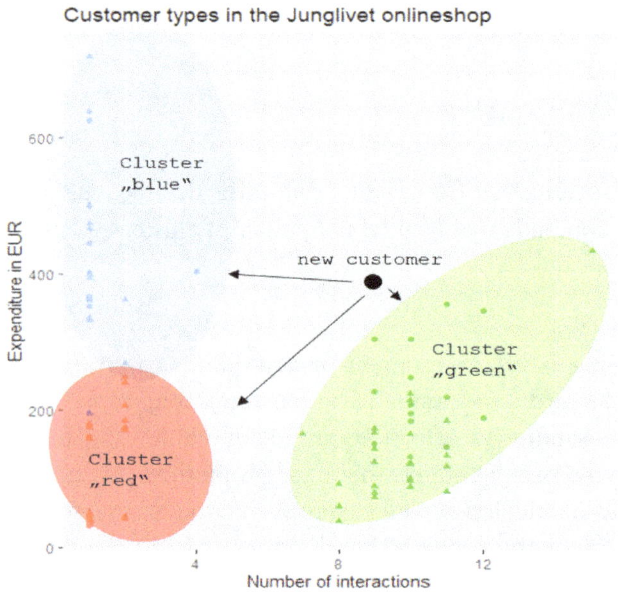

Figure 5-11 KNN helps to find similar points.

As you have seen, the algorithm is simple but is still very useful to predict data points in a for human understandable and comprehensible way. The only requirement is that the variables or columns in your dataset represent distances. That means that the observations or data points in your dataset that are close to each other have also values that are close to each other. For that purpose, you should avoid none-quantitative variables like names or other nominal and ordinal variables. But if you used k-means to build your clusters before, this should be fine. If you use other methods or existing clustering models than check if they fulfill these requirements.

Let us now take a look at our online shop data and implement our first nearest-neighbor classifier. We will use it to estimate the missing `TYPE` values for customers that have not been investigated by the marketing campaign.

But before we start to code, we have to decide which columns we want to consider. Best practice is to talk to the domain experts in your business data analytics project or rely on the reports of the data understanding phase. In our case, we will focus on the following numeric columns to predict the `TYPE` of a customer:

- `AGE`: This is a demographic variable which helps to describe the user even further. In business scenarios you would even have much more demographics like home address, friends (based on referrals), or hobbies (based on other visited sites).
- `LIFETIME`: This is an account specific feature. It describes the lifetime of the account in years after initial registration. This can be useful to analyze whether the customer is a returning customer or not.
- `CREDIT_SCORE`: Information like credit card scoring can be used to offer different shipping and payment options. Customers that have a low rating should pay before the shipment process is started.
- `EXPENDITURE`: As business data analyst you use such features to understand if this is an important customer or not. If the customer spends a lot of time in the shop, you can design special offers explicitly for the customer.
- `INTERACTIONS`: How often the customer buys in the online shop. Together with `EXPENDITURE` it is useful to better understand if the customer generates high or low profit.
- `TYPE`: This feature represents if the marketing team could contact this customer and labeled them based on a telephone interview in one of two groups (standard and connoisseur). We want to use clustering to label the missing customers based on their user characteristics.

Now we are ready to prepare our variables and reduce the dataset to the variables of interest. Furthermore, we will have to load and install the `class` package, which is necessary, because it has a very popular KNN-implementation, which we want to use later in this process.

```
if (!require(class)) install.packages("class")
library(class)
```

Then we start to prepare our dataset. In this scenario, we have to separate our dataset of customers in two subsets of customers with known `TYPE` and customers with unknown `TYPE`.

Then we will build our model on the dataset with labels based on the phone interview and use it to predict the missing `TYPE` of the other dataset:

```
onlineshop = read_excel("onlineshop.xlsx") %>%
  group_by(USER_ID) %>%
  mutate(EXPENDITURE = round(mean(TURNOVER)), INTERACTIONS = n()) %>%
  select(AGE, LIFETIME, CREDIT_SCORE, EXPENDITURE, INTERACTIONS, TYPE)
%>%
  distinct()

onlineshop$AGE = normalize(onlineshop$AGE)
onlineshop$EXPENDITURE = normalize(onlineshop$EXPENDITURE)
onlineshop$INTERACTIONS = normalize(onlineshop$INTERACTIONS)
onlineshop$TYPE = as.factor(onlineshop$TYPE)

customers_unknown = filter(onlineshop, is.na(TYPE))
customers_known = filter(onlineshop, !is.na(TYPE))
```

The default KNN implementation uses Euclidian distance, which makes it like the k-means algorithm sensitive to magnitudes. Hence, we normalize all features with `normalize()` from the `dstools` package to weigh features equally before we build the KNN model. Scaling and normalization are often used interchangeably, but if you prefer a fix range of values you should use normalization. Finally, we have to transform the `TYPE` column as factor value, because our KNN algorithm needs the target feature as factor type.

To build our model, we rely on the modeling procedure in Figure 5-4 from the previous section about modeling methods. This means we have to split our dataset in a test and a train set. We can do that in the following manner:

```
split = sample(c(TRUE, FALSE), nrow(customers_known), replace=TRUE,
prob=c(0.7,0.3))
customers_train_types = customers_known[split, ]$TYPE
customers_train = customers_known[split, ] %>% select(., -TYPE)
customers_test_types = customers_known[!split, ]$TYPE
customers_test = customers_known[!split, ] %>% select(.,-TYPE)
```

Then we build our KNN model, using the next customer in the feature space as a reference to decide the `TYPE`:

```
fit = knn(train=customers_train,
          test=customers_test,
          cl=customers_train_types,
          k=1)
```

Then we evaluate the results and compute some typical error measures based on a confusion matrix like accuracy. A confusion matrix is a very popular measure used in classification problems to assess the performance of a model. A template for such a confusion matrix is illustrated in Figure 5-12.

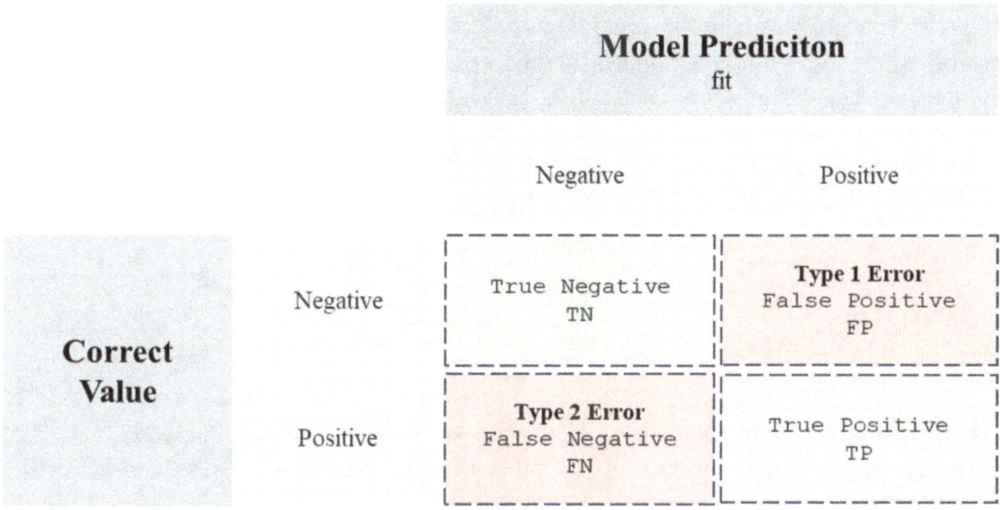

Figure 5-12 Confusion matrix to compute performance of classification models.

It provides a summary of how well the model's predictions align with the actual outcomes or ground truth. It's called a "confusion" matrix because it can help you understand where the model is getting confused. The matrix consists of four values:

- `True Positives (TP)`: Which are the cases where the model predicted the positive class correctly, and they were actually positive according to the ground truth.
- `True Negatives (TN)`: Where the model predicted the negative class correctly, and they were actually negative according to the ground truth.
- `False Positives (FP)`: Cases where the model predicted the positive class, but they were actually negative. This is also known as a Type I error.
- `False Negatives (FN)`: Cases where the model predicted the negative class, but they were actually positive. This is also known as a Type II error.

Based on the matrix we can compute the following popular error measures:

- **Accuracy**: Measures how many predictions the model got correct out of all predictions. It's the mean error rate across all observations calculated as (TP + TN) / (TP + TN + FP + FN).
- **Precision**: Measures the proportion of true positive predictions among all positive predictions made by the model. It's calculated as TP / (TP + FP). Precision is a measure of the model's ability to avoid false positives.
- **Recall** (also sensitivity): Measures the proportion of true positive predictions among all actual positive cases. It's calculated as TP / (TP + FN). Recall is a measure of the model's ability to capture all positive cases.
- **Specificity**: Measures the proportion of true negative predictions among all actual negative cases. It's calculated as TN / (TN + FP). Specificity is a measure of the model's ability to avoid false negatives.

- **F1-Score**: Harmonic mean of precision and recall. It provides a balance between precision and recall, which can be useful when there's an imbalance between the classes in the dataset.

To compute these performance or error measures in our example, we have to build a confusion matrix for our model with R. We can do that with the `table()` function in the following manner:

```
> table(customers_test_types, fit)
                    fit
customers_test_types  1  2
                   1 24  5
                   2  2  0
```

Please note that the matrix might look different in your case. The reason is the same as before in the k-means example. We split our dataset randomly in different initial subsets, and hence the split is different every time you run your code. Nevertheless, your results might be slightly different you can continue and compute e.g., the accuracy to validate the performance of our model:

```
misClassError = mean(fit != customers_test_types)
print(paste('Accuracy =', 1-misClassError))
```

Regardless, of the random splits your model should have an accuracy of around 70%, which means it predicts round about `70` of `100` customers correctly. Let us now use the model to predict the missing `NA` values in the `TYPE` feature of the unknown customers dataset (`customers_unknow`). Therefore, we have to replace the test dataset with our `customers_unknow` dataset:

```
customers_unknown = customers_unknown %>% select(.,-TYPE)
fit = knn(train=customers_train,
    test=customers_unknown,
    cl=customers_train_types,
    k=1)
```

Finally, we can access the predictions for our unknown customers with:

```
> fit
 [1] 2 1 1 1 2 2 1 1 1 1 1 1 1 1 1 1 1 1 1 1 2 1 1 1 1 1 1 1 1 2
Levels: 1 2
```

In our case we just used the next neighbor to decide for a customer `TYPE`. However, you can easily increase the neighbors the algorithm should look at to decide for a customer `TYPE`. Try to figure out how to do this before you continue with the next chapter!

5.5 Find Rules and Relationships

As you have seen, there are quite a few business situations where finding clusters in your business data is very useful. However, there are also many business scenarios where it is not enough just to know that such clusters exist. Your manager or the marketing department might be interested in finding further insights in the business data. They might want to find relationships between the different customers.

For instance, they want to know if there is a relationship between creditworthiness and send backs. Or if there are rules which help to identify very profitable customers. In the following section we will take a look at two methods to taggle these issues: correlation analysis and association analysis.

5.5.1 Correlation Analysis

Correlation analysis is a business data analytics method that is used to determine the strength and direction of the relationship between two or more variables in your dataset. Therefore, you will compute a correlation coefficient. The value of the coefficient ranges from `-1` to `1`, with `0` indicating no correlation, `-1` indicating a perfect negative correlation, and `1` indicating a perfect positive correlation.

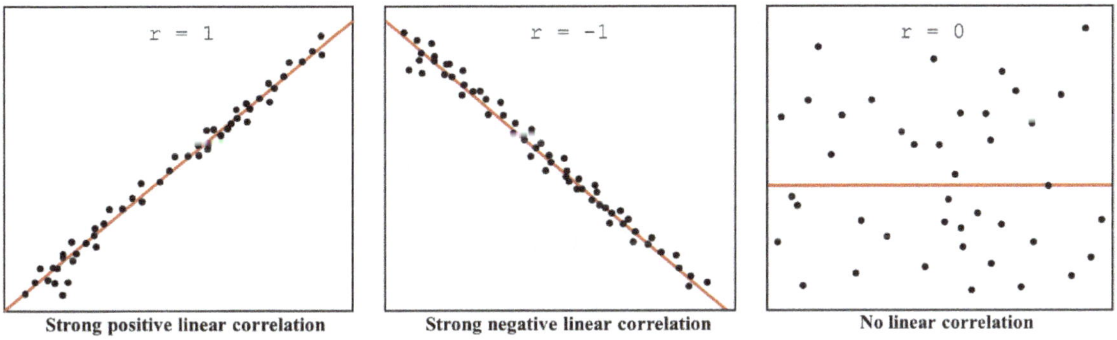

Figure 5-13 Correlation coefficients and their interpretation.

Suppose Lindsey Maegle from the marketing team wants to understand the relationship between the average temperature and your daily sales revenue in the online shop. Therefore, you write a web scraper and collect data on the average temperature and join it with your daily sales revenue in the Junglivet online shop over a period of one year.

After collecting the data, you can perform correlation analysis to determine if there is a relationship between temperature and sales revenue. Let's assume that you find a positive correlation coefficient of `-0.75` between average temperature and sales in the online shop.

This negative correlation coefficient of `-0.75` indicates that there is a moderate negative relationship between temperature and sales. As the temperature decreases, the sales revenue tends to increase, and vice versa. As a consequence, you can assume that in the cold winter months more customers buy whisky than in summer.

Even if you might think this information is trivial, it can be useful for many departments of the *Junglivet Whisky Company*. Lindsey Maegle can create marketing campaigns or promotions that are tailored to weather conditions. She can now run promotions on winter-themed products when the temperature is low to capitalize on increased consumer demand. Or Ted Gumble from the production team can use this information to adjust stock and production levels based on temperature trends. While Alice Gleck from the management can adjust Junglivet pricing strategy based on temperature trends.

Besides this example it's important to note that correlation does not imply causation. A strong correlation between two variables does not necessarily mean that one variable causes the other. There may be other underlying factors or variables that are driving the observed correlation. Therefore, it's important to be cautious and consider other factors in addition to correlation analysis when making business decisions. Correlation analysis is just one tool in the business data analytics toolbox that can provide insight into the relationships between variables, but further analysis and consideration of other factors is often necessary to make informed business decisions.

Let us now compute a simple correlation analysis with the whole `onlineshop` dataset:

```
> onlineshop = read_excel("onlineshop.xlsx")
> cor(onlineshop$AGE, onlineshop$LIFETIME)
[1] 0.5141696
```

This means that there is correlation of `0.5` between `AGE` and `LIFETIME`. This indicates that the older the account the older the customer who belongs to this account. But also, the other way round, the older the customer the longer he or she is customer of the online shop. This shows one relevant problem of correlation analysis. Correlation is not causation. You can find relationships, but domain knowledge and further investigations are necessary to come to causal conclusions.

Another challenge of correlation analysis is to find the right level of aggregation of your data. One could argue that our dataset contains one row for each purchase. Hence, if we compute the correlation analysis like we did in our example, customer with multiple purchases will also count multiple times in the correlation computation. If you are interested in understanding a relationship between the things represented by the features, it might be necessary to summarize your dataset and then compute the correlation.

However, to detect such correlations in your variables is an important step in data preprocessing and data analysis. It might be helpful to find features that might measure the same thing and can be removed. Or it might help you to find patterns and relationships.

Therefore, it is very convenient that there is a way in R to directly visualize all correlations in a data set: The correlation matrix and its visualization the correlation plot! The correlation matrix is very helpful because it is a matrix storing all correlation coefficients of all numeric variables, we are interested in. If you have built such a correlation matrix you can visualize it as correlation plot to see correlations between your features easily.

Let us know build such a correlation matrix and plot. We will focus on the four numeric features `AGE`, `LIFETIME`, `CREDIT_SCORE` and `TURNOVER` to start:

```
onlineshop_selection = onlineshop %>%
  select(AGE, LIFETIME, CREDIT_SCORE, TURNOVER)
correlation_matrix = round(cor(onlineshop_selection), 2)
```

The result is a correlation matrix storing rounded correlation coefficients:

```
> correlation_matrix
              AGE LIFETIME CREDIT_SCORE TURNOVER
AGE          1.00     0.51         0.09     0.01
LIFETIME     0.51     1.00         0.29     0.02
CREDIT_SCORE 0.09     0.29         1.00    -0.05
TURNOVER     0.01     0.02        -0.05     1.00
```

The first row shows that between the features `AGE` and `AGE` is obviously 1, which equals 100% correlation (they are the same feature). Between `AGE` and `LIFETIME` we have our `0.51` correlation from before, between `AGE` and `CREDIT_SCORE` we have no correlation (`0.09`) as between `AGE` and `TURNOVER` and so on. Now that we have our correlation matrix, we can create our correlation plot with the `ggcorrplot` package:

```
install.packages(c("ggcorrplot"), dependencies = TRUE)
library(ggcorrplot)
ggcorrplot(correlation_matrix)
```

Resulting in the following correlation plot, represented in Figure 5-14.

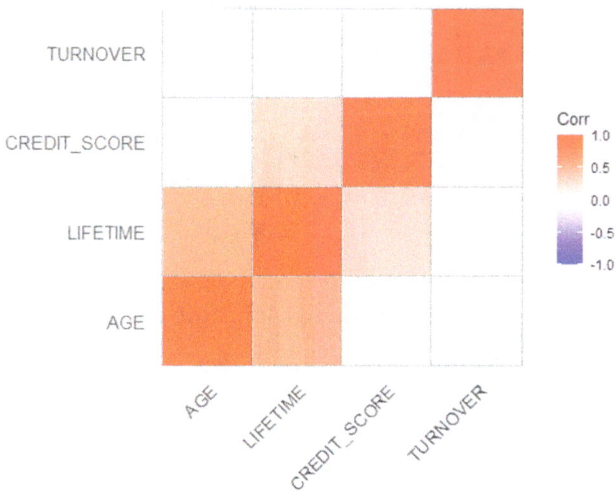

Figure 5-14 Correlation plot.

You can see that the plot is a direct representation of the matrix. Red cells indicate a strong positive correlation, while blue ones a negative correlation. In contrast to a matrix, correlations can be uncovered and recognized much more quickly for humans. However, to a certain extent it is also a matter of taste whether a business data analyst chooses a matrix or a plot to compare correlations.

5.5.2 Association Analysis

Association analysis, also known as market basket analysis or affinity analysis, is a business data analytics technique used to discover rules in the form of relationships or patterns among items. The idea is to find associations between items that are often bought together. As a consequence, association analysis is commonly used in retail and e-commerce settings to analyze customer purchase data and identify item combinations.

The method is also useful for the marketing team of our online shop. It can help to identify items that fit to each other. For instance, Lindsey Maegle from the marketing team may use the result of our analysis to discover patterns like that customers who buy Glenmorangie are also likely to buy Junglivet and the fancy whisky glasses set, leading to cross-selling opportunities.

The primary output of association analysis is a set of rules, known as association rules, which express the relationships between items in the form of "if-then" statements. The rules consist of a left-hand side and a right-hand side, separated by an arrow. The left-hand side represents the items that are found to be associated or correlated, while the right-hand represents the items that tend to occur together with the left-hand items.

Association rules are typically evaluated based on metrics such as support, confidence, and lift. Support measures the frequency of occurrence of an itemset or rule in the dataset, confidence measures the conditional probability of the consequent given the antecedent, and lift measures the strength of association between the antecedent and consequent, taking into account the frequency of occurrence of both items independently.

To perform the association analysis in R, we use the `arules` and `arulesViz` packages, which we install with:

```
if(!require(arules)) install.packages("arules")
if(!require(arulesViz)) install.packages("arulesViz")

library(arules)
library(arulesViz)
```

Let us now load the `supermarket` dataset from the `dstools` package consisting of transactions of a Scottish supermarket next to the Junglivet distillery:

```
data("supermarket")
```

Or if you downloaded the file, you can load the transactions with:

```
read.transactions(file = "supermarket.csv", sep = ",")
```

The resulting dataset is no normal data frame, which you know from the other chapters. This dataset is a so called "transaction" dataset which represents different shopping baskets. You can check it with `class()` to validate that it has been loaded correctly as transaction dataset.

```
> class(supermarket)
[1] "transactions"
attr(,"package")
[1] "arules"
```

This kind of format is based on the so called "market basket" model of data. It represents a many-many relationship between items. It consists of different customers in a store buying different items and putting them in their basket. After they finish shopping the checkout and the result is a transaction, which is one line of data in our dataset.

Table 5-2 Example transactions with items based on the Scottish supermarket dataset from the `dstools` *package*

TRANSACTION	ITEMS
1	coffee, tropical fruit, yogurt
2	whole milk
3	cream cheese, meat spreads, pip fruit, yogurt
...	...

As illustrated in Figure 5-2, each basket includes so called item sets (like `{coffee, tropical fruit, yogurt}`). We see that there is no number of purchases. So, the first customer might have bought 10 packs of coffee or just one – we do not know it. And the second relevant aspect is that we do not know which customer belongs to which transaction. For instance, Sherry Young could have quickly gone to the supermarket during her lunch break and bought some coffee, tropical fruits and yogurt (`transaction 1`). Then she realized that she had forgotten milk for her coffee and quickly went back in and bought it (`transaction 2`). Transactions `1` and `2` could therefore be the same but also different people. Keep that in mind when you do association analysis.

You should also note that an alternative representation in many packages is in the form of a tidy frame. In this case the rows represent the different users or baskets and the columns represent the different products of the store. The values of the product columns are Boolean indicating whether or not the product is purchased by the user. In this format the dimensions of the data frame are very large, and consequently the transaction data format is preferred over the tidy format in this case.

If you want to investigate the transactions dataset yourself you can do that with:

```
> inspect(head(supermarket))
    items
[1] {coffee, tropical fruit, yogurt}
[2] {whole milk}
[3] {cream cheese, meat spreads, pip fruit, yogurt}
[4] {condensed milk, long life bakery product, other vegetables, whole
milk}
[5] {abrasive cleaner, butter, rice, whole milk, yogurt}
[6] {rolls/buns}
```

Our goal in the association analysis is to analyze these transactions to find frequent item sets, which are sets of items that appear in many of the baskets together. This information is often presented as a collection of association rules.

There are different algorithms for finding frequent item-sets. In this book we will use the a-priori algorithm. It works by eliminating item sets by looking first at smaller sets and recognizing that a large set cannot be frequent unless all its subsets are. Or in other words, the algorithm assumes that if an itemset is not frequent, then all its supersets must also be not frequent.

Let us now investigate the transaction in our dataset:

```
> dim(supermarket)
[1] 9834  170
```

This means we have `9834` transactions and `170` different items in our dataset. If you want to list the distinct items in the dataset you can run:

```
> itemLabels(supermarket)
[1] "abrasive cleaner" "artif. sweetener" "baby cosmetics" "baby food"
```

Or you can use the `summary()` function you know from the previous chapters:

```
> summary(supermarket)
transactions as itemMatrix in sparse format with
 9834 rows (elements/itemsets/transactions) and
 170 columns (items) and a density of 0.02598907

most frequent items:
whole milk   other vegetables  rolls/buns  soda  yogurt  (Other)
   2513          1901                1808       1714  1372    34140

element (itemset/transaction) length distribution:
sizes
   1    2    3    4    5    6    7    8    9   10   11   12   13
14   15   16   17   18   19   20   21   22   23   24   26   27
28   29   32
2133 1656 1294 1013  857  647  548  439  349  248  181  117   79
77   55   46   29   14   14    9   11    4    6    1    1    1
1    3    1

   Min. 1st Qu.  Median    Mean 3rd Qu.    Max.
  1.000   2.000   3.000   4.418   6.000  32.000

includes extended item information - examples:
            labels
1 abrasive cleaner
2 artif. sweetener
3   baby cosmetics
```

The output shows the number of transactions and items (which we already know), but also the basket sizes (or transaction lengths) and more information about their distribution.

You can also plot the item frequencies of the top `10` items with:

```
itemFrequencyPlot(supermarket, topN=10,  cex.names=1)
```

Resulting in the plot in Figure 5-15.

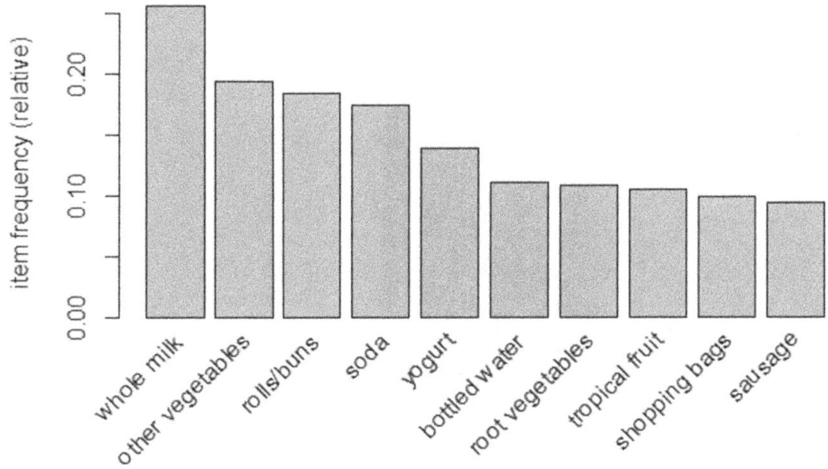

Figure 5-15 Top 10 item frequency in the `Scottish supermaket` *dataset.*

As you can see `milk` is the most frequent item, followed by "`other vegetables`" which is probably a placeholder for vegetables that are not in the regular assortment of the supermarket, other items like `buns`, `soda` or `sausages` are also in the top `10`.

The next step is to analyze the transactions using the a-priori algorithm with the function `apriori()`. This function requires both a minimum support and a minimum confidence constraint at the same time. Support represents how popular a set of items is, as measured by the proportion of transactions in which the itemset appears.

While confidence tells us how likely an item is purchased given that another specific item is purchased. Let us now find all rules with a `support` of `0.001` and a `confidence` of `0.001` (considering that we have about 10000 items in our dataset) to predict if a customer will buy whisky or not.

```
rules = apriori(data=supermarket,
                parameter=list(supp=0.001,conf = 0.001),
                appearance=list(rhs="whisky"),
                control=list(verbose=F))
```

Which results in a set of `11` rules:

```
> rules
set of 11 rules
```

We can also sort the rules based on their confidence score and print them with:

```
> rules = sort(rules, decreasing=TRUE,by="confidence")
> inspect(rules)
        lhs            rhs          support      confidence  coverage     lift
          count
[1] {steak} => {whisky} 0.002542201 0.694444444 0.003660769 84.3106996
25
[2] {beef}   => {whisky} 0.001728696 0.032442748 0.053284523  3.9387899
17
[3]  {bottled  beer}  =>  {whisky}  0.001220256  0.01511335  0.080740289
1.83487 12  ...
```

The result shows us that whisky is often bought together with steak or beef. In particular rule `[1]` has a confidence of `69`%, which means that 69% of the times a customer buys steak, a whisky is bought as well. Furthermore, rule `[1]` has support of `0.0025` which means it has been found in `0.25` percent of all transactions (which is ok because most customer in grocery stores do not buy whisky).

As a consequence, we can report this finding to Lindsey Maegle which can now contact the supermarket to offer Junglivet coupons for customers buying steak or other promotional discounts. An alternative is to launch advertisements to target customers who are planning a barbecue and are still looking for alcoholic drinks to go with their steak. Or as Carl Leonard from the production team proposes, the Junglivet company could think about launching a steak sauce with whisky flavor.

As the team brainstorming really picks up steam, Lindsey asks if there is no function to visualize all the rules we found in the dataset. With a big grin, you open your laptop and run the following functions which are part of the `arulesViz` package:

```
ruleExplorer(rules)
```

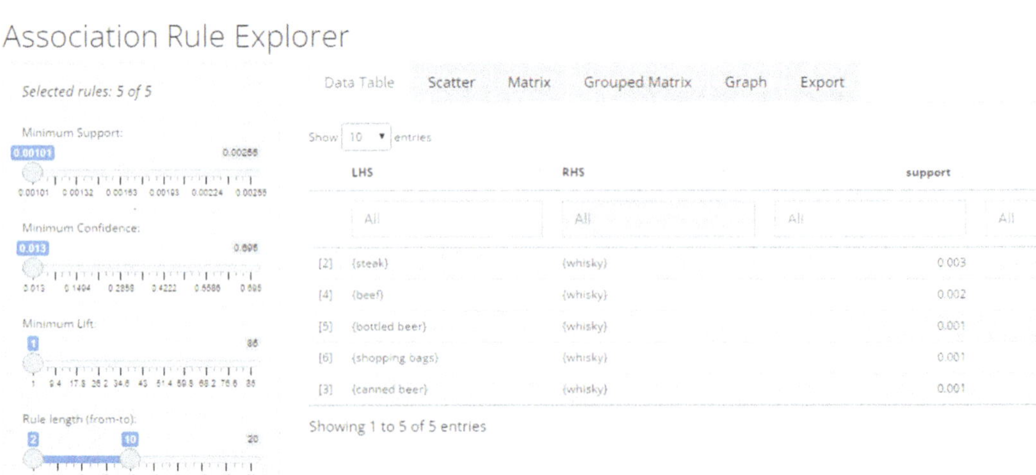

Figure 5-16 `Arules` explorer allows you to investigate the association rules interactively.

Which will open an interactive dashboard as illustrated in Figure 5-16 to investigate all the different association rules we could find with the a-priori algorithm:

In the rules explorer you can select different rules by their support, confidence, lift or length. Furthermore, you can filter your rules or investigate the most relevant plots directly. Finally, there is a CSV export functionality to save your results.

Association analysis can provide valuable insights for businesses by revealing hidden patterns and relationships in data, which can be used to support business decision-making like marketing strategies, product recommendations, and personalized offers to customers. However, it's important to interpret the association rules carefully, as correlation does not necessarily imply causation, and other factors may be at play in the observed associations.

5.6 Predict Categorial Values

In the previous section we used k-nearest-neighbor to predict consumer behavior based on previous shopping data. We used it to compare `ron_swanson` to the other customer types and identified him as whisky enthusiast. In this section we want to go even further. We want to build models that explicitly predict if a specific data point will belong to a specific class or category. And even better allows us to understand which attributes really influence the probability of class.

These algorithms are called classification algorithms. Classification in this context means that we want to predict a categorial feature (also known as “target feature” or "class label"). We develop a model based on input features that can accurately classify new, previously unseen data into one of several predefined categories.

The input features can be numerical or categorical and can represent various types of information, such as demographic information, medical test results, or text data. The target feature is typically a binary (e.g., true/false) or a multi-class label (e.g., categories such as "whisky fan" vs. "whisky enthusiast").

There are many different algorithms and techniques used for classification, including decision trees, random forests, some regression techniques, and neural networks. The choice of algorithm depends on the specific problem and the characteristics of the data.

The most popular classification algorithms in business data analytics are decision trees. Decision trees are models that use a tree structure to represent decision paths and their related outcomes. Hence, decision trees can be visualized easily and are easy to understand and interpret, even for non-experts. Which makes them a very solid choice for business data analytics.

If you investigate your decision tree you can identify the most relevant features for a given task, which can help to reduce overfitting and improve generalization performance. Another benefit is that decision trees make no assumptions about the distribution of the data or the functional form of the relationship between the input features and the output variable. This makes them flexible and able to capture complex patterns in the data.

Decision trees in general, consist of nodes, branches and leaves. Nodes represent decision rules or tests if a specific attribute is bigger or smaller a specific value. The branches represent these values and the leaves stand for the outcome of the decision process.

To train a classification model, we typically use a labeled dataset, where the input features and corresponding output labels are known. The model is trained to learn the relationship between the input features and the output labels by minimizing the difference between its predictions and the true labels in the training data.

Once the model is trained, it can be used to predict the output labels for new, unseen data. The accuracy of the model is typically evaluated using metrics such as accuracy, precision or recall, which reflect how well the model is able to correctly classify instances from each class.

5.6.1 C5.0 Tree

One of the most popular decision tree algorithms is the C5.0 algorithm. The algorithm is an extension of previous decision tree algorithms like ID3 or C4.5 algorithms (Quinlan, 2014), and it uses an entropy-based criterion to select the best feature to split the data at each node of the tree.

To run the algorithm, we have to install and load the `C50` package:

```
install.packages("C50", dependencies = TRUE)
library(C50)
```

To further extend our knowledge of decision trees, let's help Lindsey Maegle. Lindsey is the online store manager and has found that determining a customer's `CREDIT_SCORE` costs a relatively large amount of money. The scoring is created by an external company and is requested on demand (via an API) for each purchase. However, the benefit of the `CREDIT_SCORE` for the online store is relatively small. It is mainly used to select the available `PAYMENT_METHOD`s. If a customer has a very bad rating, he will only be offered `bank transfer` as payment option (in most cases), for example. Lindsey would be greatly helped if we could build a model to predict the `CREDIT_SCORE` so she has not to buy it. No problem for us as business data analysts!

The `CREDIT_SCORE` feature in the dataset is therefore our target variable. It indicates whether a customer is a candidate for payment default or not. We now want to build a model that predicts an individual scoring based on the order information such as `TURNOVER` but also the customer characteristics like `AGE` or `GENDER`.

To start, we have to prepare our dataset for the algorithm and keep the features we think that are likely associated with a `CREDIT_SCORE`. Furthermore, you should know that our target feature `CREDIT_SCORE` can be seen as a categorial feature but also as a numerical feature (see next section).

In this case, we want to predict it with a decision tree, which requires our target feature to be of the type `factor`, which we can do easily by converting it with `as.factor()`:

```
onlineshop_sample = onlineshop %>%
  mutate(CREDIT_SCORE = as.factor(CREDIT_SCORE)) %>%
  distinct() %>%
  select(AGE, GENDER, TURNOVER, CREDIT_SCORE)
```

We have to reduce the online shop data to an individual level, because otherwise we have customers with many transactions counting multiple times in our prediction model. Then we split the data into training and testing set, before we feed our training dataset to our C5.0 algorithm to predict the `SENDBACK` of customers:

```
set.seed(1)
ids  =  sample(c(TRUE,  FALSE),  nrow(onlineshop_sample),  replace=TRUE,
prob=c(0.7,0.3))
train = onlineshop_sample[ids,]
test = onlineshop_sample[!ids,]

fit = C5.0(x=select(train,-CREDIT_SCORE), y=train$CREDIT_SCORE)
```

Our final result is the model stored in the variable fit. We can investigate it with:

```
> summary(fit)
Call:
C5.0.default(x = select(train, -CREDIT_SCORE), y = train$CREDIT_SCORE)
C5.0 [Release 2.07 GPL Edition]     Sat Sep 16 18:58:06 2023
-------------------------------
Class specified by attribute `outcome'
Read 329 cases (4 attributes) from undefined.data

Decision tree:
AGE <= 27:
:...GENDER = male:
:   :...AGE <= 23: 3 (9/3)
:   :   AGE > 23: 2 (8/1)
:   GENDER = female:
:   :...AGE <= 24: 5 (11/2)
:       AGE > 24:
:       :...AGE <= 26: 1 (14/1)
:           AGE > 26: 3 (11/2)
...
```

```
Evaluation on training data (329 cases):
          Decision Tree
        ----------------
        Size       Errors
          22    62(18.8%)    <<
         (a)    (b)    (c)    (d)    (e)     <-classified as
        ----   ----   ----   ----   ----
          13                    2            (a): class 1
                 21             3     12     (b): class 2
                  1     48      2      3     (c): class 3
           1            12    116      4     (d): class 4
                  3     11      8     69     (e): class 5

      Attribute usage:
      100.00%      AGE
       51.98%      GENDER
Time: 0.0 secs
```

The summary provides information about the decision tree structure, evaluation on the training data, and attribute usage in building the tree.

The decision tree is printed out in a hierarchical format, with each internal node representing a condition based on an attribute, and each leaf node representing a class prediction. As you can see in the summary output, we have a huge decision tree (which is reduced in the print version of this book, but your output should be much longer) and further information of our model.

The conditions of our model are represented by the attribute name, followed by the threshold value, and the number of cases that satisfy the condition. For example, the first condition is " `AGE <= 27`", which splits the tree into two branches. The next split is based on whether the value of `GENDER` is `male` or `female`. If you follow one of this decision paths, you get the "decision" of the decision tree at the end of the path in the leaf.

The evaluation on the training data shows that the size of the decision tree is `22`, and the model makes errors on `62` cases, which is a classification error rate of `18.8%`.

Finally, the attribute usage table shows the percentage of times each attribute was used in the decision tree, with `AGE` used `100`% of the time, and `GENDER` used `51.98%` of the time. The order related feature `TURNOVER` has no influence on the `CREDIT_SCORE`, because it is not used. Hence, we can remove it from the model and are able to provide our credit scoring regardless of the customer order.

An more suitable way to represent the tree is the graphical method by the `plot()` method as shown in Figure 5-17.

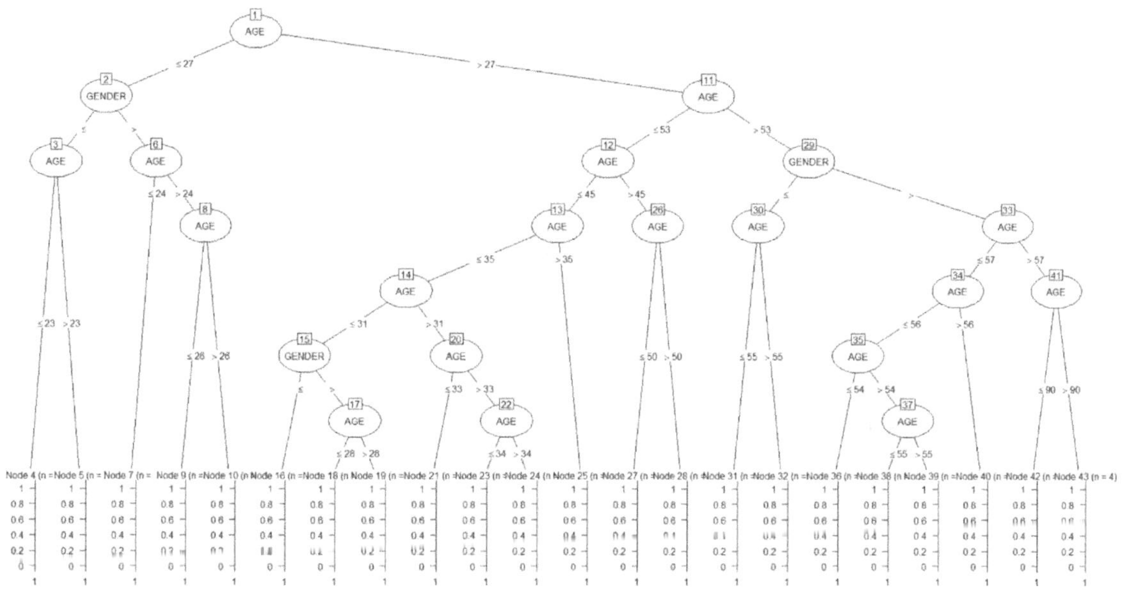

Figure 5-17 A graphical representation of the decision tree.

Furthermore, the C5.0 algorithm can create an initial tree model and then decompose the tree structure into a set of mutually exclusive rules. These rules can then be pruned and modified into a smaller set of potentially overlapping rules. The rules can be created using the rules option:

```
fit_rules = C5.0(x=train[,-4], y=train$CREDIT_SCORE, rules=TRUE)
```

However there is no `plot()` functionality for rule-based decision trees at the moment.

The final models, both the classical C5.0 tree in `fit` and the rule-based tree in `fit_rules` can be used to label new data. For instance, you can use them to predict the target in the test dataset with:

```
predict(fit, newdata=test)
```

Or:

```
predict(fit_rules, newdata=test)
```

Which will return a vector with predictions:

```
> predict(fit, newdata=test)
  [1] 2 4 5 5 1 3 3 ...
```

If you are instead interested in probabilities you can set the `type="prob"`, which will return the probability of each class of our target variable:

```
> predict(fit, newdata=test, type="prob")
            1           2           3          4          5
1 0.005065856 0.567713609 0.129348193 0.04491726 0.25295508
2 0.015376363 0.031020919 0.031825496 0.80006258 0.12171464
3 0.003507131 0.008417115 0.012625672 0.03109656 0.94435352
4 0.001112017 0.295351783 0.004003262 0.05864038 0.64089256
5 0.869706174 0.007294833 0.010942250 0.09361703 0.01843972
```

This is very useful if you want to better understand the model and the scoring. It can be also helpful if you want to work with a kind of probabilistic values to score your customer.

5.6.2 Random Forest

One technique to combine different decision trees is the random forest method. The random forest algorithm builds multiple decision trees and combines their outputs by voting, where each tree is trained on a random subset of the data and a random subset of the features.

One of the most relevant benefits of random forest is that it can provide accurate predictions even when working with complex datasets with a large number of variables. By design it is very good in reducing overfitting by using a combination of multiple decision trees with random subsets of variables, which makes it less sensitive to outliers compared to the C5.0 tree.

Let us have a look at it and install and load the `randomForest` package in R:

```
install.packages(randomForest)
library(randomForest)
```

Then we prepare our data. Random forest can handle both numerical and categorical variables, but as before in the C5.0 example categorical variables should be converted to factors.

```
onlineshop_sample = onlineshop %>%
  mutate(CREDIT_SCORE=as.factor(CREDIT_SCORE)) %>%
  distinct() %>%
  select(AGE, GENDER, LIFETIME, CREDIT_SCORE)
```

Then, we split the data into training and testing sets:

```
set.seed(1)
ids = sample(c(TRUE, FALSE), nrow(onlineshop_sample), replace=TRUE,
prob=c(0.7,0.3))
train = onlineshop_sample[ids,]
test = onlineshop_sample[!ids,]
```

Then we train the random forest model using the `randomForest` function. Therefore, we specify the predictor variables (`x`), the response variable (`y`), and the number of trees to grow (`ntree`).

```
fit = randomForest(x=train[,-4], y=train$CREDIT_SCORE, ntree = 500)
```

We use the predict function to make predictions on the testing data.

```
predictions = predict(fit, newdata=test)
```

Which can be investigated with the `summary()` and visualized with the `plot()` function like a C5.0 tree in the previous example.

5.6.3 XGBoost

Extreme gradient boosting is an advanced implementation of gradient boosting algorithm that has become popular among business analysts for solving classification problems (and regression problems!). Good news that it is implemented in many programming languages including R. In R, the `xgboost` package provides a set of functions to train and predict using the XGBoost algorithm.

As before, let us have a short look how to use `xgboost` in R. First, you need to install and load the `xgboost` package using the following commands:

```
install.packages("xgboost", dependencies = TRUE)
library(xgboost)
```

Then we prepare our data by splitting it into training and testing sets. Furthermore, XGBoost only accepts data in numeric format and as a matrix. Therefore, we have to convert our data frame to the matrix format and our feature values to numeric values.

```
onlineshop_sample = onlineshop %>%
  select(AGE, GENDER, LIFETIME, CREDIT_SCORE) %>%
  distinct() %>%
  mutate(GENDER = ifelse(GENDER == "male",1,0)) %>%
  mutate(across(everything(), as.numeric)) %>%
  as.matrix()

set.seed(1)
ids = sample(c(TRUE, FALSE), nrow(onlineshop_sample), replace=TRUE,
prob=c(0.7,0.3))
train = onlineshop_sample[ids,]
test = onlineshop_sample[!ids,]
```

Now we can train the model using the `xgboost()` function with the training data.

```
xgboost(data=train[, -4], label=train[, 4], nrounds = 10)
```

And use the trained model to make predictions on the test data using the `predict()` function.

```
predictions = predict(fit, newdata = test[, -4])
```

Even if XGBoost can only use numerical values, it has become popular due to its ability to handle large datasets and high-dimensional features, as well as its high accuracy and speed.

Additionally, it has built-in handling of missing values and can automatically handle feature selection which saves you a lot of time and speeding up your analytics workflow.

5.7 Predict Numeric Values

In the previous chapter we successfully used classification algorithms to make predictions about numeric values. But sometimes you do have a numeric and not a class variable as target. In this case you have to use other algorithms that can handle this type of data. One of the most popular approaches to model numeric target variables is regression analysis.

Regression analysis is a statistical technique that is used to analyze the relationship between one or more independent variables and a single dependent variable. As before, we want to build a model that explains the relationship between the variables and to make predictions based on that model.

There are several types of regression models, including linear regression, generalized linear regression, polynomial regression, logistic regression, or mixed-effects models and more. Linear regression is the most commonly used regression model and is used to analyze the relationship between a continuous dependent variable and one or more continuous independent variables. Logistic regression, on the other hand, is used for classification modeling, which means that we have a binary dependent variable (e.g., yes or no) and one or more independent variables. The other regression types are used for more complex modeling scenarios.

In business data analytics regression modeling is often used to make predictions or to understand the relationships between different business variables. For example, the *Junglivet Whisky Company* may use regression models to predict online sales based on advertising spending or to understand the factors that influence customer satisfaction of the online shop. Another very popular scenario is that you can use regression modeling to test hypotheses about the relationships between variables (instead of selecting and running a hypothesis test). For example, you could use linear regression modeling to test whether the supplier of whisky grain has a significant effect on the outcome of your whisky production!

5.7.1 Linear Regression

The most straight forward regression analysis is with a linear regression. In linear regression, we assume that our numeric target variable can be predicted by a single predictor variable. And that there is a realistic chance that they have a linear relationship. Therefore, the model fits a single regression line that describes how the outcome changes in response to changes in the predictor.

In a standard linear regression model, we have two central components. A target variable y which represents the column we are interested in to predict. This variable must be numeric and continuous. And secondly, we have one or multiple (numeric) predictor variables $x_1, x_2, \dots x_n$ that are used together with the model coefficients $\beta_1, \beta_2, \dots \beta_n$ to predict our target variable y. A linear regression with one single predictor variable has the form:

$$y = f(X) = \beta_0 + \beta_1 x_1 + \varepsilon$$

This regression function you see above describes a line defined by two important parameters (regression coefficients):

- **Intercept (β_0)**: Where the regression line crosses the origin on the y-axis for $\varepsilon = 0$
- **Slope (β_1)**: How much does the outcome change in response to a change in the predictor variable. This can be positive, as x_1 goes up or negative vise-versa.

As we said before our regression model is represented by this regression function $f(\cdot)$, which is linear in the n predictor variables (with $n = 1$ in the standard linear regression). Our data is represented by a data frame or data matrix $X = [x_1, x_2, \dots, x_n]$. Which are also the input of our regression function $f(\cdot)$. The last term in the model ε represents a so-called error term that captures all other factors which influence our target variable. It follows a normal distribution assumption with mean zero. This means that we assume that there is no linear relationship between the errors and the independent variables of our model, and on the other hand that we can model our target variable as a linear function of the predictor variables. That's why sometimes the term ε is also called noise.

The regression line will probably not predict each data point exactly. In most cases, there will be some error in the model. The differences between our observed values and the regression line are called regression error or residuals.

For example, imagine you want to predict the numerical rating of a whisky between 0 and 5 based on its price. Then rating will be our target variable, and price will be our predictor variable. Such regression models are called standard linear regressions. We will have a look at this kind of regression in this chapter. However, in business scenarios it makes sense to use further characteristics such as flavor, age, number of sales etc. In this case the rating will still be our target variable, but we now have multiple characteristics which we can use as predictor variables.

As a consequence, we can expand our model to a linear multiple regression function with the form:

$$y = f(X) = \beta_1 x_1 + \beta_2 x_2 + \dots + \beta_n x_n + \varepsilon$$

As before, we can also expand this model by adding an intercept β_0 which represents the value of the target variable when all predictor variables are zero. You can think of it as the point where the regression line crosses the y-axis. But this is not always necessary.

To find the optimal values for our parameter β_i we must find a way to minimize the differences between the observations and the predictions of our model. And since our model can also make negative predictions, we have to consider that these values could cancel out positive values in the summation. Thus, best practice is to use the least square estimates that minimizes the sum of the squared differences between the values and the predictions of our model:

$$\widehat{D} = \sum_{i=1}^{m} (y_i - \widehat{y_i})^2$$

or

$$\hat{D} = \sum_{i=1}^{m} [y_i - (\beta_1 x_{1,i} + \beta_2 x_{2,i} + \dots + \beta_n x_{n,i})]^2$$

The outcomes are m estimates for our model parameter. We can use them to measure the performance of our model. $\hat{y}_i$

The R-square is calculated by:

$$R^2 = 1 - \left[\frac{\hat{D}}{\sum_{i=1}^{m}(y_i - \bar{y})^2}\right]$$

It expresses the proportion of variation that can be explained by our regression model.

The F-statistic is used to compute the overall significance of our model with k predictor variables:

$$F = \frac{\frac{\left[\sum_{i=1}^{n}(y_i - \bar{y})^2 - \hat{D}\right]}{k}}{\frac{\hat{D}}{(n-k-1)}},$$

A high value means that none of our predictor variables have an influence.

Other error measures are the mean error

$$ME = \frac{1}{m} \sum_{i=1}^{m} (y_i - \hat{y}_i)^2$$

The root mean square error:

$$RMSE = \sqrt{\frac{1}{m} \sum_{i=1}^{m} (y_i - \hat{y}_i)^2}$$

And mean absolute percent error:

$$MAPE = \frac{100}{m} \sum_{i=1}^{m} \frac{|y_i - \hat{y}_i|}{y_i}$$

The benefit of a linear regression is that, if all other variables are fixed, a change in one of the predictor variables $x_1, x_2, \dots, x_n$ changes the expected mean target variable by β_i units.

We will come back to this information later. But let us now first help Lindsey Maegle again with a pressing problem. She would like to have a prediction for each customer how much money he or she will spend in the online shop. To help her out, we could use a linear regression in R to investigate if we can predict the turnover of a customer by using the turnover of the last order.

Therefore, we have to compute a new variable `LAST_TURNOVER` for each customer by its `USER_ID`. We load the online shop dataset and compute the `LAST_TURNOVER` with the `lag()` function:

```
onlineshop = read_excel("onlineshop.xlsx")
head(onlineshop)

onlineshop = onlineshop %>%
  group_by(USER_ID) %>%
  arrange(DATE) %>%
  mutate(LAST_TURNOVER = lag(TURNOVER)) %>%
  ungroup()
```

In the next step, we can check if it worked with `tail()`. The result should look somehow like this:

```
> tail(onlineshop_sample)
# A tibble: 6 x 4
# Groups:   USER_ID [6]
  DATE                USER_ID             AGE  ...  LAST_TURNOVER
  <dttm>              <chr>             <dbl>  ...          <dbl>
1 2023-12-30 00:00:00 rose_leslie          56  ...             39
2 2023-12-30 00:00:00 david-niven          53  ...             32
3 2023-12-30 00:00:00 usurpgam             48  ...             47
4 2023-12-30 00:00:00 mr.smith             37  ...            316
5 2023-12-30 00:00:00 amanda_missinger     24  ...             33
6 2023-12-30 00:00:00 TH_fouche            57  ...             NA
```

As you can see, we computed a new column `LAST_TURNOVER` with the turnover of the last order for each customer. However, if we look at the last row, or at the first rows of our dataset with `head()`, we notice that we have some missing values in the `LAST_TURNOVER` column:

```
> head(onlineshop_sample)
# A tibble: 6 x 6
# Groups:   USER_ID [6]
  DATE                USER_ID              AGE GENDER TURNOVER LAST_TURNOVER
  <dttm>              <chr>              <dbl> <chr>     <dbl>         <dbl>
1 2023-10-02 00:00:00 cardiB                27 female       44            NA
2 2023-10-02 00:00:00 natalie_hershlag      42 female       34            NA
3 2023-10-02 00:00:00 babe_ruth             48 male         34            NA
4 2023-10-02 00:00:00 homlesspistachio      60 male        500            NA
5 2023-10-02 00:00:00 chrstian_hauff        93 male         43            NA
6 2023-10-02 00:00:00 marc_sinclair         36 male         47            NA
```

The reason for the missing values is that the oldest shopping transaction has no other or previous shopping transaction that we can use to compute the `LAST_TURNOVER` variable. Therefore, there is a `NA` value.

To continue with modeling, we remove these rows from the dataset with:

```
onlineshop_sample = onlineshop_sample %>%
  filter(!is.na(LAST_TURNOVER))
```

Now we can build a linear regression with:

```
fit = lm(TURNOVER ~ LAST_TURNOVER, data= onlineshop_sample)
```

As in our other examples, we can now use the `summary()` function as a diagnostic tool to assess the fit and assumptions of our regression model. This can help us to identify any potential issues with our model and ensure that it is accurately capturing the relationship between the variables.

```
> summary(fit)
Call:
lm(formula = TURNOVER ~ LAST_TURNOVER, data = onlineshop_sample)

Residuals:
   Min      1Q Median      3Q     Max
-189.6 -137.9 -125.4  165.6  511.5

Coefficients:
               Estimate Std. Error t value Pr(>|t|)
(Intercept)   168.16971   13.97512  12.034   <2e-16 ***
LAST_TURNOVER   0.08151    0.05273   1.546    0.123
---
Signif. codes:  0 '***' 0.001 '**' 0.01 '*' 0.05 '.' 0.1 ' ' 1

Residual standard error: 192.5 on 361 degrees of freedom
Multiple R-squared:  0.006576,     Adjusted R-squared:  0.003824
F-statistic:  2.39 on 1 and 361 DF,  p-value: 0.123
```

Let us now have a short look at the results of the `summary()` function. In the first row, we see our model, which tries to predict the `TURNOVER` based on the `LAST_TURNOVER`. In the next rows, we see the residuals (we mentioned before). The residuals are the differences between the actual values of the dependent variable (`TURNOVER`) and the predicted values by the model. Our model makes guesses in `50%` of the cases between `-137.0` too low and `165.6` too high. In one case it overestimates the `TURNOVER` by `511.5` EUR.

The next part about coefficients is the most relevant one. It provides information about the coefficients of our linear regression model. The coefficients represent the relationship between the predictor variable (`LAST_TURNOVER`) and the response variable (`TURNOVER`). There are four columns: the estimate, which is the estimated coefficient for `LAST_TURNOVER`. In this case, it's approximately `0.08151`. Second, the standard error associated with the coefficient estimate. Third, the t-value, which measures how many standard errors the coefficient estimate is away from zero. In this case, it's `1.546`. And the p-value in the last column is associated with the significance of the coefficient. It indicates the probability of observing a t-statistic as extreme as the one computed if the true coefficient were zero. In this case, it's `0.123`, which is greater than `0.05`, suggesting that the coefficient is not statistically significant at the `0.05` significance level.

Which means, we can use this model but it is probably not suitable and will return mostly incorrect predictions.

This is represented by the error values in the next part. The residual standard error, which is an estimate of the standard deviation of the residuals is approximately `192.5`. Another very popular performance measure is the multiple R-squared. This is a measure of how well the independent variable (`LAST_TURNOVER`) explains the variation in the dependent variable (`TURNOVER`). In this case, it's `0.006576`, indicating that the independent variable explains a very small proportion of the variance in the dependent variable. The adjusted R-squared is a variation of R-squared that takes into account the number of predictors in the model. In this case, it's `0.003824`, which is also very bad.

The last row is the F-statistic. This statistic tests whether the overall regression model is statistically significant. In this case, it's `2.39`, and it is associated with a p-value of `0.123`. Since this p-value is greater than `0.05`, it suggests that the model as a whole is not statistically significant which means we have to invest some more effort and work on our model.

As an alternative to the summary function, we can also plot the results with:

```
ggplot(fit, aes(x = LAST_TURNOVER, y = TURNOVER)) +
  geom_point() +
  geom_smooth(method = "lm", se = FALSE, fullrange = TRUE)
```

Which results in Figure 5-18:

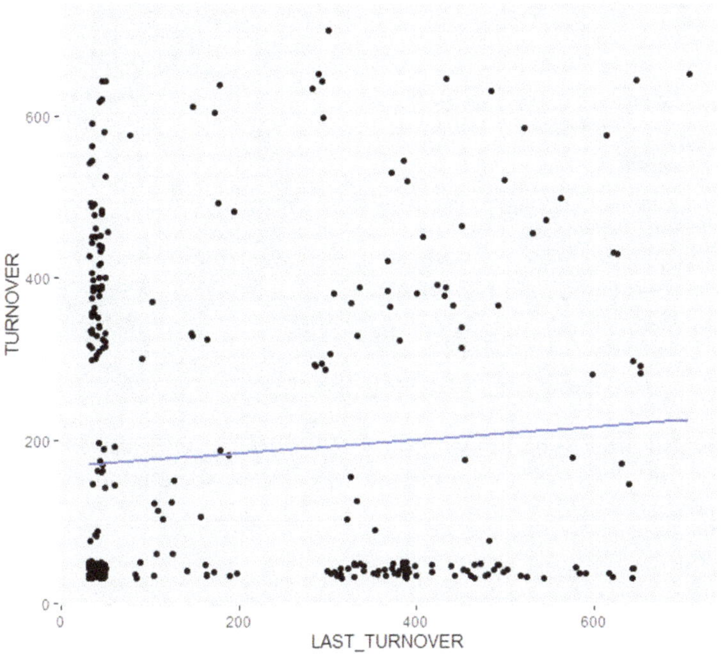

Figure 5-18 Regression plot with `ggplot2`.

5.7.2 Multiple Linear Regression

In the last section, we unfortunately found that our model did not perform particularly well. The reason was that it more or less just missed the usual significance level of `.10` or `.05`. This is not surprising, however, since we also used only one variable `LAST_TURNOVER`. In most real-life scenarios you have more than just one variable that influences your target variable. And if you are lucky, you also have data about it. Also, when we look at the regression plot in Figure 5-18, the question arises whether the relationship between `LAST_TURNOVER` and `TURNOVER` is really linear.

A common solution is to use a multiple linear (or even polynomial) regression instead of a simple linear regression. Which means we can now extend our model with all our features available. We can do that by adding a "`.`" to our model, which means use all features available. We remove the `USER_ID` because this would result in a feature per customer, which is not helpful in our case. Accordingly, our regression model changes to:

```
fit = lm(TURNOVER ~ ., data=select(onlineshop_sample, -USER_ID))
summary(fit)
```

If we now investigate our model again, we can see a significant influence of many features on the `TURNOVER` indicated by the stars (`*`) in the column `Pr(>|t|)`:

```
Coefficients:
                             Estimate Std. Error t value Pr(>|t|)
(Intercept)                 4.527e+03  8.235e+03   0.550   0.5830
DATE                       -2.537e-06  4.845e-06  -0.524   0.6010
AGE                         1.018e+00  8.277e-01   1.230   0.2199
GENDERmale                 -1.341e+02  2.681e+01  -5.001 1.06e-06 ***
TYPE                        9.245e+01  4.360e+01   2.120   0.0349 *
CREDIT_SCORE               -5.571e+00  1.078e+01  -0.517   0.6058
LIFETIME                   -7.755e+00  3.910e+00  -1.983   0.0484 *
PAYMENT_METHODBNPL          7.118e+01  3.151e+01   2.259   0.0247 *
PAYMENT_METHODcredit card -4.690e+01  3.816e+01  -1.229   0.2202
PAYMENT_METHODpaypal       -1.744e+01  3.349e+01  -0.521   0.6030
SENDBACKTRUE                4.571e+01  4.038e+01   1.132   0.2588
VIDEO_ADTRUE               -9.713e+00  2.136e+01  -0.455   0.6497
CONVERTED                  -1.186e+01  2.932e+01  -0.404   0.6862
LAST_TURNOVER              -6.572e-02  6.118e-02  -1.074   0.2837
```

Based on these results we can conclude that we should build a new model using the features `GENDER`, `TYPE`, `LIFETIME` and `PAYMENT_METHOD` to predict the `TURNOVER` of a customer. Here we go:

```
fit = lm(TURNOVER ~ GENDER + TYPE + LIFETIME + PAYMENT_METHOD,
data=onlineshop_sample)
```

You can now continue this process and deep dive in your model until you reach a point where you have only significant variables, use the variables the management wants you to use or you are happy with the error measures like R-squared or F-statistic. The best model depends on your business understanding and project objective.

5.7.3 GLM

The next level of regression models are GLMs, which is short for generalized linear models. These models are generalizations of the ordinary linear regression you met in the previous section. The benefit of GLMs is that they allow for response variables to have other error distribution models than a normal distribution like Gaussian distribution. This means that we can use them to make a logistic regression to predict a binary output like yes or no, or true or false (see the chapter about classification models).

GLMs are fit with the function `glm()`. Like our ordinary linear models, generalized linear models have formulas and data as inputs, but additionally they also have a family input. It would clearly far exceed the focus of this book, if you want to understand all the different types and approaches in GLM. Nevertheless, let's briefly take a look at how you can use one of these models in R, the logistic regression.

In the standard linear regression model, our target variable is numeric and continuous such as the rating or age of whisky. Our linear regression model has the form:

$$y = f(X) = \beta_0 + \beta_1 x_1 + \beta_2 x_2 + \dots + \beta_n x_n + \varepsilon$$

In logistic regression models the target variable is binary. A binary variable can be either `TRUE` or `FALSE`. This is usually coded as `1` and `0`. For example a whisky can be labeled “single-malt” or “not single malt” or be on the personal wish list of a customer of the Junglivet online shop or not. In some business analytics scenarios, it is often useful to reduce a complex target variable or problem to a simple binary or yes-or-no decision. For instance, it reduces the complexity of a prediction problem, if we focus to predict, if a customer will pay the bill or not. Or if a website user is a potential customer or not.

The logistic regression relies on maximum likelihood estimation, which differs from the method of least squares that we used in our linear regression model.

While linear regression attempts to find a linear function $f(\cdot)$ that best fits the data, logistic regression uses a different type of function known as the logistic or sigmoid function. This kind of function is shaped like a "S", symmetric, and asymptotically approaches $y = 0$ and $y = 1$. This means that the logistic function only yields values between 0 and 1. That is necessary because our binary target variable must have values between 0 and 1.

Therefore, the values produced by our logistic model can be interpreted as probabilities. Specifically, they represent the probability that the dependent variable (y) takes on the value 1, given the values of the independent variables (x). In logistic regression modeling, the aim is not to predict the values of the dependent variable but rather to estimate the likelihood of its occurrence. A value close to 0 suggests that the occurrence of y ($y = 1$) is highly unlikely, whereas a value close to 1 indicates a high likelihood of y occurring. This results in the following regression model:

$$P(y = 1) = \frac{1}{1 + e^{-(\beta_0 + \beta_1 x_1 + \beta_2 x_2 + \dots + \beta_n x_n + \varepsilon)}}$$

Compared to the linear regression the interpretation of the regression output and its coefficients is a bit different. Logistic regression deals with predicting probabilities of a binary outcome rather than directly predicting numerical values. Here, the effect of a one-unit change in regressor variables is not constant but varies based on the current value of the regressor and the parameters of the logistic function. A common approach to make the models better understandable would be to compute the average partial effects, which is not necessary in our example.

For instance, in logistic regression predicting the probability of a customer buying a product based on their age, a one-year increase in age might not have a constant impact. It could have a larger or smaller effect depending on the age range in which it occurs and the nature of the relationship between age and the probability of buying the product. To illustrate this let say our regression has the coefficient $\beta_1 = -0.25$ and the related factor in our model is $e^{\beta_1} = e^{-0.25} = 0.78$. And let x_1 be the age of the customer. If everything else stays the same, and the age increases by one year from x_1 to $x_1 + 1$, then the log odds that the product is actually bought changes by the amount of β_1. Which is a reduction of $100 \cdot (1 - 0.78) = 22\%$. Or generally speaked, if β_1 is greater than 1, a change from x_1 to $x_1 + 1$ increases the odds, while β_1 smaller than 1 will decrease the odds.

To illustrate this with an example, we can try to predict a binary variable like `SENDBACK`, which flags if the customer sends parts of the order back with a logistic regression model:

```
fit = glm(SENDBACK ~ GENDER + TURNOVER, data = onlineshop, family =
binomial)
```

5.8 Predict Developments

While sitting relaxed in your recliner in the courtyard of the production hall and comfortably reading the chapter about regression analysis, Ted Gumble contacts you with a pressing problem. He collected a dataset consisting of the whisky production costs of the last years. Could you, he asks, use this data to predict future costs? So that he won't waste money by ordering too much or too few whisky packages and other materials?

This seems like a good case for time series analysis, another very mighty tool in the analytics toolbox! Time series prediction or forecasting is designed specifically for data that is collected over time, such as daily stock prices, weekly sales figures, or monthly production data. The forecasting takes into account the temporal structure of the data and is able to capture the patterns and trends that are unique to time series data. This makes it well-suited for making predictions about future values of a time series, such as predicting tomorrow's sales or next month's production levels.

In the last section about regression modeling, we focused on analyzing relationship between two or more variables, such as the relationship between a company's advertising spending and its sales revenue. While regression models can include time series data as predictor variables, they are not specifically designed to analyze time series data and may not be able to capture the temporal patterns and trends that are unique to time series data.

In general, you would use time series forecasting when you are interested in making predictions about future values of a time series, and regression modeling when you are interested in understanding the relationship between variables. Hence, if you have time series data and want to make predictions about future values, then time series forecasting would be the more appropriate technique and if you have cross-sectional data and want to analyze the relationship between variables, then regression modeling would be the more appropriate technique.

The first step in time series analysis is to see how a data object behaves over a period of time. In R, it can be easily done by the `ts()` function with some parameters and the `plot()` or `ggplot()` function. The format of `ts()` is `ts(data, start=, end=, frequency=)` where `start` and `end` are the first and last observation and frequency is the number of observations per unit time (e.g. `1` for annual, or `12` for monthly).

Let's take the example dataset `costs` from the `dstools` package. The dataset consists of `5` features representing the costs and the date:

- `ID`: Unique identifier of the production costs
- `YEAR`: Year of the costs entry
- `MONTH`: Month of the costs entry
- `DAY`: Day of the month of the costs entry
- `COSTS`: Production costs in Euro of the *Junglivet Whisky Company*

We load this dataset in our R-environment and check if it's loaded correctly with the `head()` function:

```
> head(costs)
  ID YEAR MONTH DAY COSTS
1  1 2015     1  31 33014
2  2 2015     2  28 31631
3  3 2015     3  31 32136
4  4 2015     4  30 32715
5  5 2015     5  31 32005
6  6 2015     6  30 34788
```

As you can see the dataset consists of different columns for each month per year and its related costs. The day column has always the last day of the related month as value, indicating that each row represents the costs at the end of each month. The column with the feature `ID` represents the identification number of each row.

To work with time series, we have to compute a new column in the `Date` format that represents the x-axis in the plot. This new column `DATE` has to be formatted explicitly `as.Date()` in the International format month-day-year to avoid problems with other packages. The value in the `COSTS` column will be plotted on the y-axis.

```
costs = costs %>%
  mutate(DATE = paste(MONTH, DAY, YEAR, sep="/")) %>%
  mutate(DATE = as.Date(DATE, format="%m/%d/%Y"))
```

Finally, we can plot it in base R with the `plot()` function. But we have to format it explicitly as time series with the `ts()` command:

```
costs_ts = ts(costs$COSTS, frequency=12, start=c(2015,1))
plot(costs_ts)
```

Or we can plot the `costs` dataset directly in `ggplot2`, where we do not have to format it before:

```
ggplot(costs) +
  aes(x=DATE, y=COSTS) +
  geom_line() +
  xlab("")
```

Resulting in the plots in Figure 5-19.

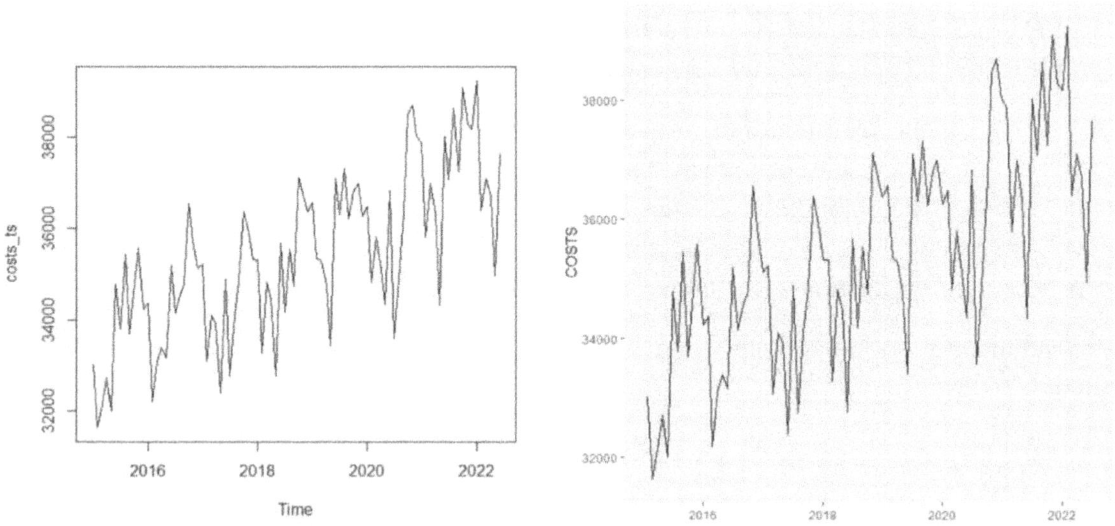

Figure 5-19 Time series plots of the `costs` dataset with base R and `ggplot2`.

If you need a more detailed view you can change the x-axis format in another representation by setting a new `scale_x_date()` format:

```
ggplot(costs) +
  aes(x=DATE, y=COSTS) +
  geom_line() +
  xlab("") +
  scale_x_date(date_labels="%m-%Y")
```

Which you can also use to set a limit in your visualization:

```
scale_x_date(limit=c(as.Date("2018-01-01"),as.Date("2020-06-30")))
```

There we go, we will now use this time series to predict the future costs for Ted Gumble!

5.8.1 ARIMA

One very popular method in time series analysis is the "Auto Regressive Integrated Moving Average" technique (ARIMA), which is actually a class of models that can be used for forecasting time series based on its own past values. The term “auto regressive” in ARIMA means it is a linear regression model (which we know already from the previous chapter) that uses its own lags as predictors. You should know that linear regression models, and hence ARIMA models, work best when the predictors are not correlated and are independent of each other.

To forecast we have to install and load the package:

```
install.packages("forecast", dependencies = TRUE)
library(forecast)
```

And use the ARIMA model to predict the missing months of the last year in our dataset with:

```
costs_ts = ts(costs$COSTS, frequency=12, start=c(2015,1))
fit = auto.arima(costs_ts)
costs_prediction = forecast(fit, 6)
```

Returning the predictions and the confidence intervals:

```
> costs_prediction
         Point Forecast    Lo 80    Hi 80    Lo 95    Hi 95
Jul 2022       36517.53 35784.52 37250.54 35396.49 37638.57
Aug 2022       37827.91 36995.11 38660.72 36554.24 39101.58
Sep 2022       37572.64 36712.98 38432.30 36257.90 38887.37
Oct 2022       38969.57 38102.26 39836.88 37643.14 40296.01
Nov 2022       38993.65 38124.13 39863.17 37663.83 40323.47
Dec 2022       38345.94 37475.79 39216.09 37015.16 39676.72
```

5.8.2 Prophet

For business data analysts, time series forecasts are a very helpful tool. The problem, however, is that forecasting with ARIMA quickly reaches its limits and the alternative methods are relatively complex (if you are working with data that has multiple seasonality). Therefore, in 2017, Facebook's Data Science team published a paper called, "Forecasting at Scale" (Taylor & Letham, 2018) which introduced an open-source project to provide quick, powerful, and accessible time-series modeling to business data analysts: the Prophet algorithm.

The Prophet algorithm uses a decomposable time-series model with three main model components: Trend, Seasonality and Holidays. It automates the tuning of the time series model and provides hence an easy-to-use time series algorithm that can be used in most business scenarios.

Before we can start to use it, we have to install the `prophet` package and load it in our workspace:

```
install.packages("prophet", dependencies = TRUE)
library(prophet)
```

Sometimes the installation of rstan which is necessary to use prophet makes problems. I recommend to check the official help on: *https://github.com/stan-dev/rstan/wiki/RStan-Getting-Started*

In the next step we have to prepare our data for the prophet algorithm. Prophet expects input data to have 2 columns with the name `ds` and `y`. Hence, let us rename our columns to match that with:

```
names(costs) = c("ds", "y")
```

Then we use the prophet function to fit our new time series model. The first argument is the training dataset as data frame. The additional arguments control how Prophet fits the data:

```
> fit = prophet(costs)
Disabling weekly seasonality. Run prophet with weekly.seasonality=TRUE
to override this.
Disabling daily seasonality. Run prophet with daily.seasonality=TRUE to
override this.
```

The warnings indicate that our dataset has no weekly or daily information. Which is fine because we have monthly data. In the next step we want to use our model to make a prediction. To do so we have to make an empty data frame with one single column which contains the dates where we want to predict values. We can then use the `predict()` function, which we have used before, to predict these values in the data frame:

```
future = make_future_dataframe(fit, periods = 36)
prediction = predict(fit, future)
```

Which we can use to plot the data and the forecast together with:

```
plot(fit, prediction)
```

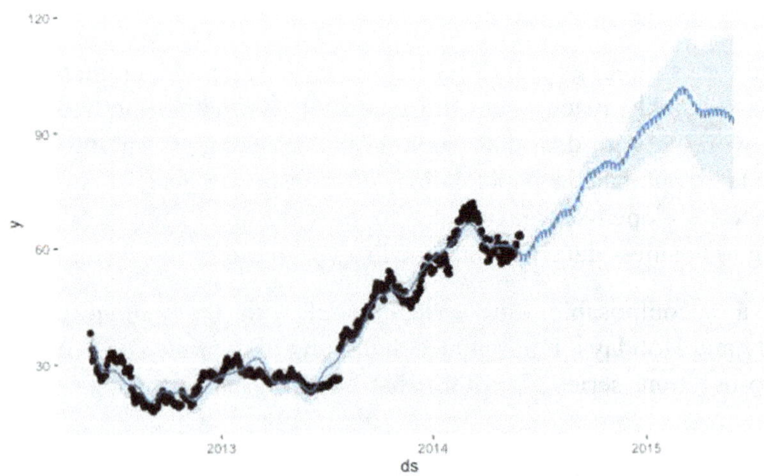

Figure 5-20 Prediction of the future of the black time series with confidence intervals (blue).

In the plot in Figure 5-20, the black dots represent actual measurements, while the blue line displays Prophet's forecast. The light blue window indicates uncertainty intervals.

We can also plot the different components of the model with:

```
prophet_plot_components(fit, prediction)
```

Resulting in the visualization in Figure 5-21:

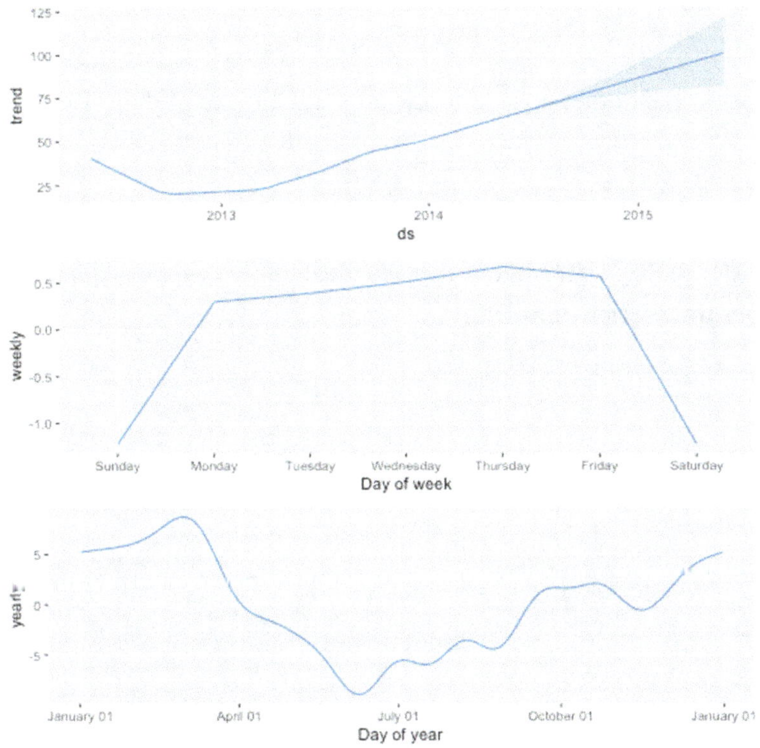

Figure 5-21 The different components of the model.

In both cases, we see a direct influence of our components on the target feature `COSTS`. Furthermore, we can now use these results to make predictions about future production costs and help Ted Gumble to optimize the shipping and logistics of the *Junglivet Whisky Company*.

5.9 Useful R Functions for Everyday Business Data Analytics

In this section I would like to present some useful functions that might help you in preparing business data. They are very useful, but the situations in which you might want to use them are rather rare. So, you are free to just skip this section and come back to it at a later time.

In the previous section you have meet the `seed()` function. The `seed()` function is mostly used to set the seed for generating random numbers. In our examples it has been used to ensure the reproducibility of random processes in our algorithms like k-means. When you set a seed using `seed()`, it initializes the internal random number generator algorithm in such a way that subsequent calls to random number functions will produce the same sequence of random numbers every time you run your code.

Furthermore, I want to point out that R has many different implementations of data structures for representing time series. I have not tried them all, but I can say that the `zoo` and `xts` packages are excellent starting points to continue working with time series data.

Sometimes you might want to select one or more elements from a time series to make samples. The naïve approach would be to just index the time series by date with:

```
ts[as.Date("yyyy-mm-dd")]
```

But this will be a lot of manual work. Best practice is to index it by a sequence of dates to build a sample of your dataset:

```
dates = seq(startdate, enddate, i)
ts[dates]
```

Where `i` is the increment. As an alternative, you can use the `window()` function to select a range by start and end date of your time series sample:

```
window(ts, start = startdate, end = enddate)
```

5.10 Checklist

1. Phase: Business Understanding			
1.1	Determine business objectives	❯	• Background • Business objectives • Business success criteria
1.2	Assess situation	❯	• Inventory of resources and capabilities • Requirements, assumptions and constraints • Risks and contingencies • Terminology • Costs and benefits
1.3	Determine analytics goals	❯	• Business data analytics goals • Business data analytics success criteria
1.4	Produce project plan	❯	• Project plan • Initial assessment of tools and techniques
2. Phase: Business Data Understanding			
2.1	Collect initial data	❯	• Initial data collection report
2.2	Describe data	❯	• Data description report
2.3	Explore data	❯	• Data exploration report
2.4	Verify data quality	❯	• Data quality report
3. Phase: Business Data Preparation			
3.1	Select data	❯	• Rationale for inclusion/exclusion
3.2	Clean data	❯	• Data cleaning report
3.3	Construct data	❯	• Derived features • Generated records
3.4	Integrate data	❯	• Merged data
3.5	Format data		• Reformatted data

		❯	• Dataset • Dataset description
4. Phase: Modeling			
4.1	Select modeling techniques	❯	• Modeling technique • Modeling assumptions
4.2	Generate test design	❯	• Test design
4.3	Build model	❯	• Parameter settings • Models • Model descriptions
4.4	Assess model	❯	• Model assessment • Revised parameter settings
5. Phase: Evaluation			
5.1	Evaluate results	❯	• Assessment of results w.r.t. success criteria • Approved models
5.2	Review process	❯	• Review of process
5.3	Determine next steps	❯	• List of possible actions decisions
6. Phase: Deployment			

5.11 Business Case Exercise: Finding Clusters in the Whisky Market

5.11.1 Scenario

It is a brisk morning when Lindsey Maegle, the indomitable marketing lead, bursts into your office. Her eyes ablaze with a fervor matched only by the fiery depths of the Junglivet Premium Selection Cask. You need no words to realize that a task of utmost importance rests upon her shoulders, one that promises to unravel the secrets of the whisky market's geographical and spatial distribution in this storied land.

The air is thick with anticipation as Lindsey unveils the grand plan. The Junglivet Whisky Company is poised to release their crowning achievement – the new 10-year Junglivet whisky. It's described as "body and smoky," a tantalizing symphony of flavors that speaks to the very essence of the Scottish Highlands. But here's the challenge: How can they ensure that this elixir of delight reaches the eager palates of those who appreciate its unique characteristics?

With a map of Scotland spread before you, Lindsey's fervent voice fills the room as she reveals the mission at hand. The task is twofold: to cluster the whiskies and unearth the most "body and smoky" group, if such a cluster exists. It's a quest for the essence of flavor, a journey into the heart of whisky craftsmanship. The fate of the Junglivet 10-year whisky hangs in the balance, its destiny intertwined with the artistry of data analysis.

5.11.2 Task

Your task is to:

- structure your work according the process model from the lecture
- understand the data and document it before you start clustering
- cluster your data with k-means
- plot the results on a map to illustrate your findings

5.11.3 Remarks

Please take a look if it is true that the smokiest and most body whisky distilleries are from Islay. Use the whiskies.csv file for your analysis.

5.11.4 Solutions

You find the solutions of this exercise on the course website.

6 Business Data Products

6.1 Introduction

In the last chapters we used many methods and approaches to generate insights with the motivation to help business decision-makers to make better decisions. For most practice problems, however, simply finding a good model to make predictions is not enough. It is important to communicate this solution to the decision-makers and make it available to them. If the customer can't do anything with our solution - we as business data analytics experts have failed.

There are numerous challenges to overcome in the process. Either the results have to be pitched and explained in front of the management, for this the results have to be prepared and translated into simple management language. Or there are different models and approaches and these must be weighed against each other and compared according to the use case. Furthermore, it is always important to provide the results as efficiently as possible: Sometimes your customers want that you compute an actual prediction every morning, or that you provide insights for them with their specific problem every day. Wouldn't it be awesome to provide the insights from business data analytics in a more general way?

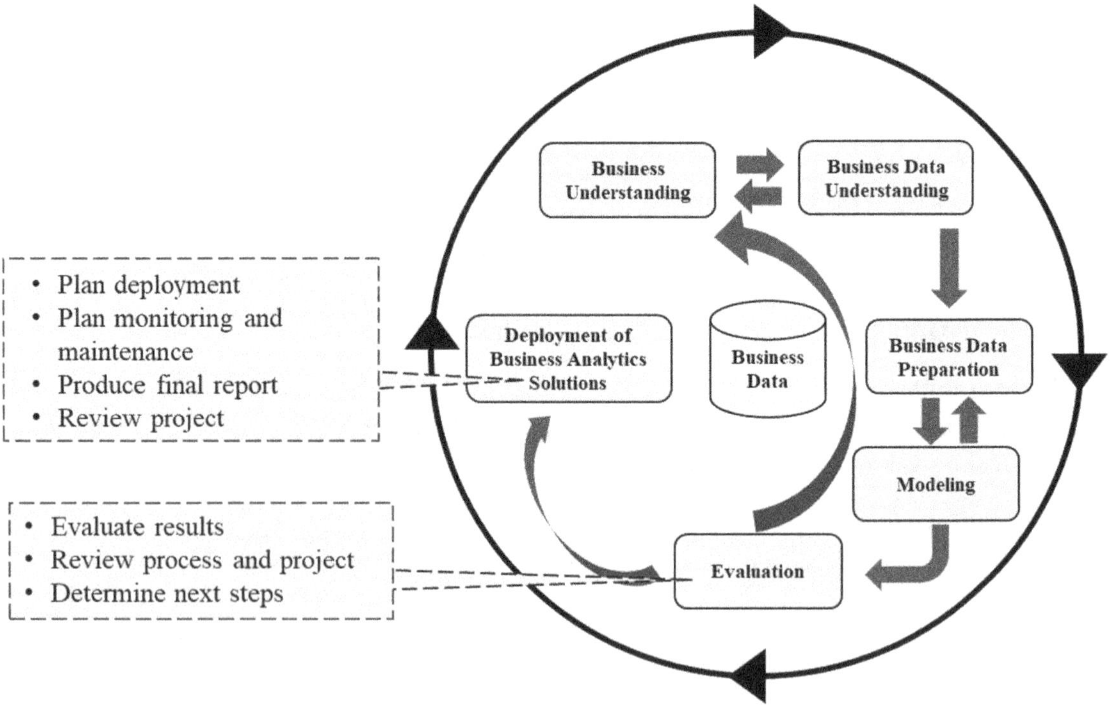

Figure 6-1 The generic tasks of the deployment phase.

In this chapter we will now discuss how we can evaluate and prepare our results from the previous steps so that we can make them available to our customers in a long-term solution. For that purpose, you must think about the following to-dos:

D. Jung, *The Modern Business Data Analyst*, https://doi.org/10.1007/978-3-031-59907-1_6

- **Evaluate results**: Assess your outputs with respect to the business success criteria you defined in the beginning
- **Review process**: Review the project and process
- **Determine next steps**: Make a list of possible actions and decisions for or against different types of analytics solutions

In the last chapter, we developed models and focused on inherent aspects like model accuracy and applicability. In the next phase, our goal is to evaluate how well the model and our findings align with the business objectives we derived at the beginning of the project. And subsequently we want to identify any potential shortcomings that might arise due to business-specific requirements. Most relevant to mention are time and budget, which allow us (or not) to test alternative approaches involving testing the model(s) on real-world applications or scenarios. Let me illustrate this with an example. I once had a business data analytics project focused on predicting customers beforehand. By the project's end, it turned out that the available data only allowed this to a limited extent. We were then faced with a choice: either redo the project extensively (with data that was extremely hard to obtain) or accept that our current data wouldn't allow us to proceed and, consequently, we couldn't identify customers in advance. The damage caused by not being able to predict was significantly less than the costs of the subsequent project. Therefore, we decided not to pursue a solution.

To make such difficult decisions you have to communicate your findings from your analysis that are directly linked to the initial business aims, as well as any other discoveries that might reveal further insights, challenges, or opportunities for future projects. Condense these findings concerning the business success benchmarks, and conclude with an affirmation on whether your business data analytics project aligns with the original business goals you identified at the beginning. In the example above, we aimed to proactively approach a specific group of customers facing a particular product issue. Through this, we gained extensive insights into the problem and its challenges. Consequently, we didn't just generate models but also gained valuable insights into the actual issue. This, in turn, allowed us to adapt processes within the company to prevent the problem from occurring.

Based on your business success criteria, you have to select the final models you want to use for your final data product like a dashboard or reporting. For that purpose, you have to face many quality assurance considerations, such as: Was the model constructed accurately? Were solely usable and available features employed for the analysis and future use? Provide a concise overview of the evaluation, emphasizing any overlooked activities that need to be addressed or repeated in future projects.

Based on these considerations, you have to identify subsequent actions. I recommend that you decide together with the project team the appropriate course of action. The team discusses whether to conclude the current project and its transition to deployment, if additional iterations are needed (CRISP-DM cycle), or if you have to initiate another business analytics project. Consider the existing and available resources and budget in your ultimate decisions. As a result, you will generate different course of actions (for instance, 1. "low budget", 2. "medium budget", and 3. "high budget" or A: "add more data sources", B: "roll out with current data").

Best practice is that you prepare then a short PowerPoint presentation for the management that enlists the potential actions, providing the reasoning behind the different decisions and outlining the pros and cons of each alternative. The management can then decide together with the team on the chosen course of action.

If your findings and models are promising and you pitched your findings successfully before management, the next likely step is to deploy the knowledge you generated during the project in form of a business data analytics solution. As illustrated in Figure 6-1, you have then to clarify the following aspects:

- **Plan deployment**: Decide which type of solution you want to deploy
- **Plan monitoring and maintenance**: Clarify monitoring and maintenance for your solution
- **Produce final report**: Make a final report and presentation for the management and project stakeholders
- **Review project**: Make a short experience documentation for yourself

In this phase, you want to leverage the findings to formulate a deployment strategy.

If a comprehensive procedure has been identified to generate the model(s), this method is documented at this stage for future deployment purposes.

Integrating the outcomes of the business data analytics project into your day-to-day business operations requires monitoring and maintenance of the solution. This is important to ensure that your findings are used correctly and consistently over time. A well-thought-out maintenance strategy is crucial to avoid prolonged misuse of your results. To effectively oversee the implementation of these findings, your project needs a carefully crafted monitoring plan tailored to fit your specific deployment scenario in the landscape of your organization.

Finally, evaluate the business data analytics project by analyzing both its successes and shortcomings, identifying strengths and areas requiring enhancement. The scope of this report depends on the needs of the management and the resources you used in your project. It could be a brief summary of the project and the insights gained if they haven't been consistently documented. Alternatively, it might serve as an exhaustive representation of the business data analytics results, depending on our project's requirements.

6.2 Evaluation and Deployment of Business Data Products

Dhanurjay Patil, former United States chief data scientist, defines "data products" as a product that facilitate an end goal through the use of data (Patil, 2012). As a consequence, in business data analytics, business data products help to manage, organize and make sense of the vast amount of business data and findings we generated in the business data analytics process.

Our customer will use business data products to access the results of our data analysis in a usable and understandable manner. For instance, Lindsey Maegle from the marketing team could use our findings from the customer clustering to start a marketing campaign or Ted Gumble needs your models to find the right supplier for the whisky production, while Alice Gleck from the management team could use your time series models for churn prediction and planning.

In business data analytics, business data analysts build data products to help these people by using the results and models of their business data analytics process. The related data products can have different forms and types depending on the business data analytics problem and users. Lindsey will be happy if we provide her a simple list with the different customers and their related cluster, Ted would like to have a report illustrating the different suppliers and their related quality issues, while Alice would need a kind of planning systems to make quarterly forecasts.

As you can see, for simple problems a single number might be enough to answer the business problem, while complex business problems might require a more sophisticated solution. With that in mind it's not surprising that the different data products range from very simple catchy ones like KPIs over interactive reporting solutions to complex analytics systems like dashboards. A short overview of the different business data products is provided in Figure 6-2.

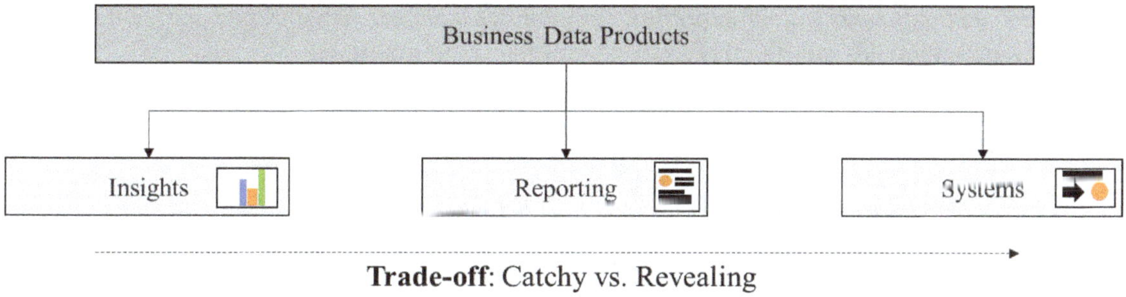

Figure 6-2 Business data products from catchy to revealing with example products.

If your project was a one-time project and the focus was very project-specific, you are probably only interested in gaining one or more simple insights. Overall, business data analytics insights are findings that give business decision-makers a deeper understanding of their operations, customers or market dynamics. By using data-driven insights, companies can make informed decisions, improve performance and gain a competitive advantage in today's data-rich business environment. In the previous chapters, we learned how to communicate simple insights with charts and descriptive summaries, so we do not go into this further in this chapter. Furthermore, it is also very rare that we only need to provide insights. In most projects, the requirement to transfer these into a reporting or a system arises relatively quickly.

Business data reporting stands for solutions that combine multiple insights and interpret them in a report or illustrate them interactively in a dashboard or notebook. The purpose is every time to provide the findings from the business data analytics project as actionable information that stakeholders can understand and utilize to monitor performance, assess trends, identify opportunities, or address challenges. Business data analytics reports typically include key performance indicators (KPIs), metrics, visualizations, and summaries that facilitate a clear and concise understanding of the business data.

Business analytics systems, also known as business intelligence (BI) systems, are software applications or platforms that enable organizations to collect, analyze, and interpret data to gain valuable insights for decision-making (Michalczyk et al., 2021). Therefore, business data analytics systems bring together data and insights from different areas and sources to provide users with information beyond the usual reports and summaries. Sometimes your business data analytics solution has to be integrated in an existing information system or the management wants you to build an information system to build on the findings from your project. As a consequence, you have to setup a pipeline or provide your model as an API for other systems.

6.3 Business Data Reporting

If simple numbers, plots and descriptive summaries are not enough, business data analysts build more complex solutions which I summarized under the category "Business Data Reporting" in Figure 6-18. We said that business data reporting involves gathering and analyzing data to create reporting solutions which help people to make decisions. These reporting solutions show how a business is doing, what trends are happening, and what actions should be taken. When we build these reporting solutions, we will use charts and graphs to make the information clear and easy to see. By using business data reporting, companies can see what's working, what needs improvement, and make smarter choices based on facts and numbers.

In general, there are four types of business data reporting, which we will now investigate further.

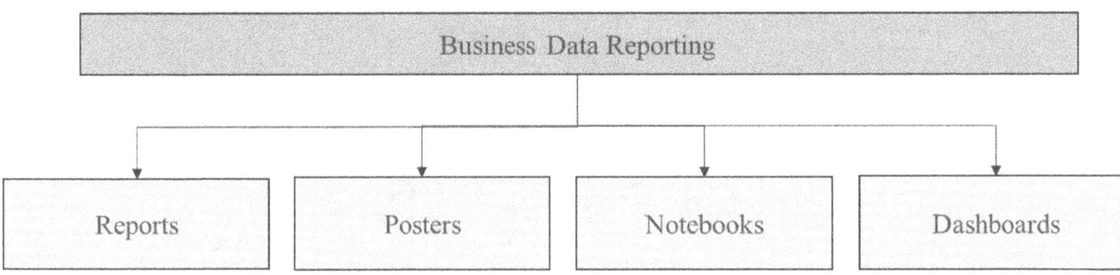

Figure 6-3. Different types of business data reporting solutions.

The first and most popular type is a traditional report. By definition a report is a structured document where you summarize information, your analysis, findings, or recommendations about a specific business analytics problem or project. In most cases this will be a simple PowerPoint presentation you will have to pitch at the management, a detailed email or word document with information about the problem or even a short one-pager you share with your stakeholders.

Another more scientific way of reporting are posters. You can think of them as visual representations of a report. In the past they were commonly used in conferences, symposiums, or academic settings to present complex information in a concise and visually appealing format. But more and more companies have started to organize internal events and workshops for which posters have to be prepared.

Another form of reporting are notebooks. They are interactive documents that combine code, results, text, and visualizations. It's a powerful tool used to share insights with other R developers in a project, because R Notebooks allow you to integrate R code chunks with narrative text, equations, visualizations, and explanatory content in a single document. This format is the best way to go if you have made a complex analysis and want to explain it to someone. But is not the preferred way if you want to share insights with the management or users that cannot use R.

Finally, dashboards are reporting apps to show key metrics, data, visualizations and insights from various sources or systems in a consolidated and easily understandable format. It allows your users to investigate the dataset interactively and by themselves. Which helps them to understand the problem more deeply.

Before starting building one of these solutions, I recommend that you ask yourself:

- Who is my audience / user group?
- What questions do they have?
- What answers do I have for them?
- What other questions will my visualizations inspire?

6.3.1 Automated Reports and Presentations with officer

Creating charts that can be inserted into PowerPoint presentations to share results with other teams, partners or clients is one of the most recurring tasks of a business data analyst. The difficulty with the creation of PowerPoint presentations is that it is a manual and time-consuming process. Of course, there are also cases in which you can pass on the code directly or at least have it prepared in a notebook, but it does not always work in collaborative environments where PowerPoint is the primary tool and your partners are not familiar with R.

When working with PowerPoint using the `officer` package, we need to prepare a PowerPoint deck for that. I recommend that you use the `Projectslides Officer Example.pptx`, which I prepared to work straight away. Alternatively, you can also use the slidepool `CRISP-DM Ready-to-use Projectslides.pptx`, but it contains many templates and slides that will just make this tutorial unnecessary complex. You can download that PowerPoint template on git or the course homepage.

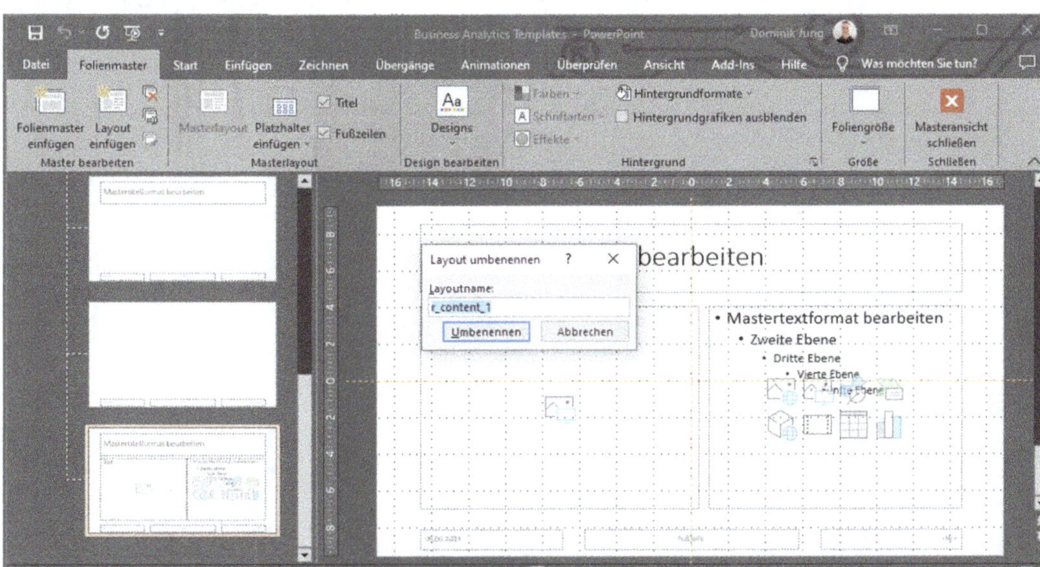

Figure 6-4 I made a new template with the name "`r_content_1`" which we want to use to add our plots.

If you want to make your own template, you have to setup slides and templates in PowerPoint. The way to do this in PowerPoint is to use the "Slide Master" view. There, you have to create new layout types with a content box that has the same dimensions as the saved graphs (11 cm x 15 cm) or content you want to add. If the dimensions don't match, you're going to end up with some weird scaling and warping problems. You can also have a look at the template I prepared if you are not sure how to do this.

If you didn't install the package, you can do that with:

```
install.packages("officer", dependencies = TRUE)
```

Then load it and the related PowerPoint template with:

```
library(officer)
template_pptx = read_pptx("Projectslides Officer Example.pptx")
```

Please verify now, that there is a master template with the name "`r_content_1`" by running the following command:

```
> layout_summary(template_pptx)
       layout                   master
1       title business_data_analytics
2   content_1 business_data_analytics
3   content_2 business_data_analytics
4   content_3 business_data_analytics
5 content_r_1 business_data_analytics
...
```

As you can see in line `11`, there is our `r_content_1` layout, which we now want to use to fill with a plot. For that we have to change the different elements of our layout (as illustrated in Figure 6-5) by referring to them.

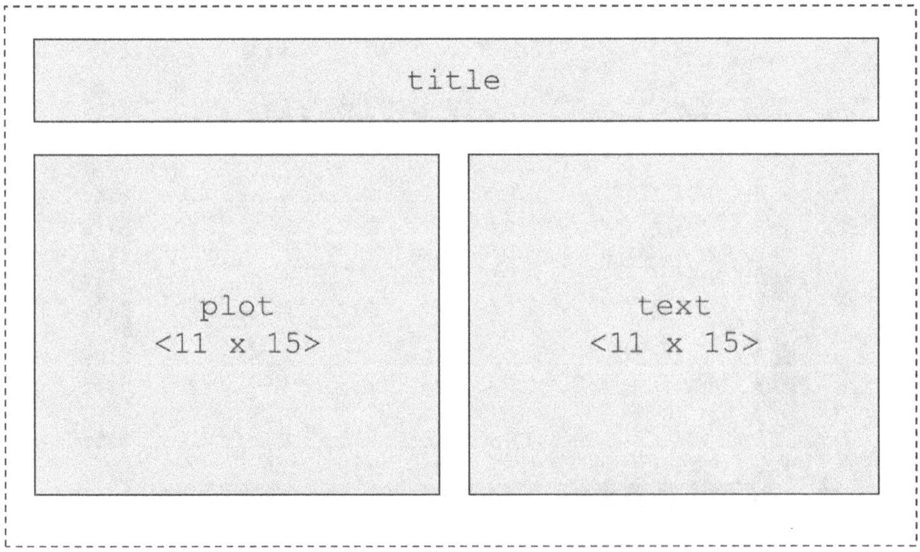

Figure 6-5 The different layout elements of your PowerPoint template.

Let us now add a first slide of our layout and add a title. Then save the new presentation as "`report.pptx`" and check if everything worked fine. We do that with:

```
template_pptx = read_pptx("Projectslides Officer Example.pptx")

template_pptx = add_slide(template_pptx, layout = "r_content_1", master
= "business_data_analytics")

template_pptx = ph_with(
  x = template_pptx, value = "Onlineshop Payment Methods",
  location = ph_location_label(ph_label = "title")
)

print(template_pptx, target = "reporting.pptx")
```

Please remember that the layout and master values in your code must match the values from the initial document. If you are unsecure with the naming, layout names and master layout names can be read easily with the function `layout_summary()`. You can now open the file "`report.pptx`" in your project folder to check if everything worked.

In the next step we want to add a plot related to the topic of this slide in the plot placeholder. For that we have to use the `ph_with_img()` function. Finally, we can build our slide with:

```
template_pptx = read_pptx("Projectslides Officer Example.pptx")

onlineshop = read_excel("onlineshop.xlsx")

plot_payments = ggplot(data=onlineshop) +
  aes(x=PAYMENT_METHOD) +
  geom_bar() +
  labs(x="", y="Frecquency") +
  ggtitle("Prefered Payment Methods in Junglivet Onlineshop")

plot_description = paste("The plot shows the amount of payments per
payment method from ",
min(onlineshop$DATE),
" to ",
max(onlineshop$DATE),
sep="")

template_pptx = add_slide(template_pptx, layout="r_content_1",
master="business_data_analytics")

template_pptx = ph_with(
  x=template_pptx,
  value="Onlineshop Payment Methods",
  location=ph_location_label(ph_label = "title")
)
```

```
template_pptx = ph_with(
  x=template_pptx,
  value=plot_payments,
  location=ph_location_label(ph_label="plot")
)

template_pptx = ph_with(
  x=template_pptx,
  value=plot_description,
  location=ph_location_label(ph_label = "text")
)

print(template_pptx, target = "reporting.pptx")
```

The final slide is illustrated in Figure 6-6. It shows a header "`Onlineshop Payment Methods`" and a related chart on the left side and a description of the chart on the right side. All elements are generated automatically and change if the underlying dataset changes. Furthermore, you can also use your company's template to design the slides more appealing.

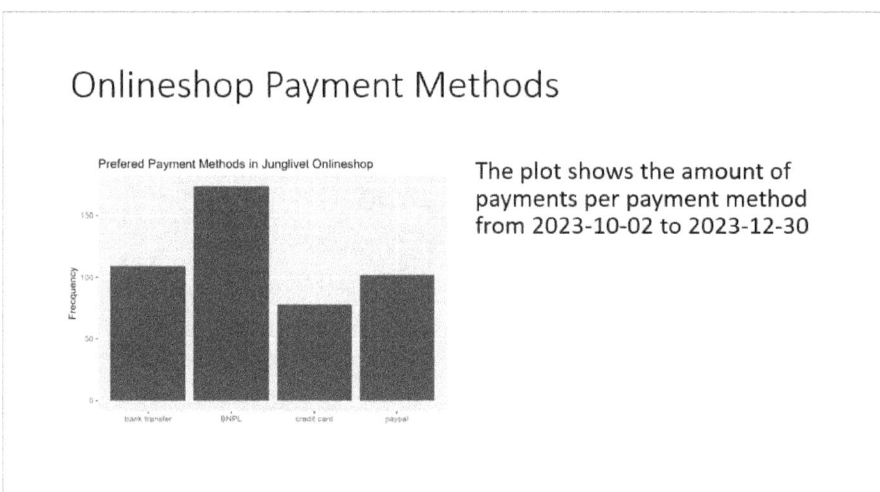

Figure 6-6 The final slide with the different payment methods of the online shop.

The `officer` package is quite powerful, and I want to highlight, it also supports Word documents. If you want to automate parts of your manual written report in Word, you can also rely on it to add plots or text chunks. I further recommend you to take a look at the official documentation, which contains many more functions and hints and is very helpful to streamline your reporting automation.

6.3.2 Poster and Flyers

One of the constants in my professional career is making summaries of my business data analytics project and insights in the form of handouts or posters so that my messages remain in memory and can be presented to other teams, partners, or clients. Handouts are more common in the business context, while posters are very widely spread in the academic context. Designing a handout or scientific poster to present business data analytics results is very similar (despite the size) and involves several key steps, which we will now discuss in detail. Also feel free to use the template `Projectslides Templates.pptx` if you plan to design a flyer or a handout, or the `Poster Template.pptx` as a starting point for your personal poster.

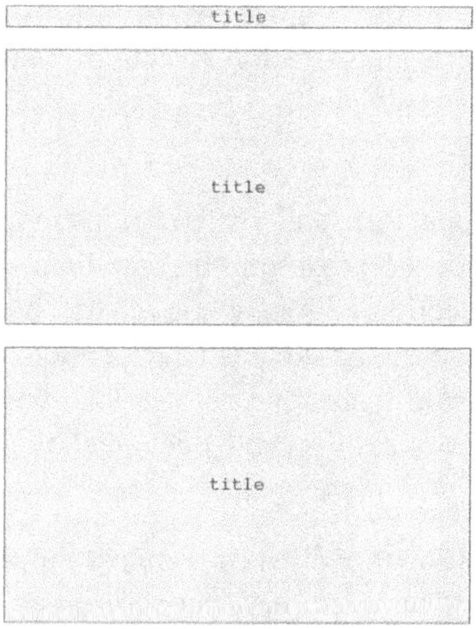

Figure 6-7 Poster template which you can use as a starting point for your project.

I recommend to start by organizing your content logically and determining the main sections of your poster or handout. In most cases, you do fine with the different steps of the CRISP-DM process like *Business Understanding*, *Data Understanding and Preprocessing, Modeling, Evaluation, Solutions and Deployment*. If you want to highlight the project you could organize your poster in the sections *Introduction*, *Method*, *Results*, *Discussion*, and *Conclusion*. Clearly define the purpose and objectives of your project and your findings, so that the message is clear and visible.

If you start designing the content and the different findings, try to use visual hierarchy techniques to guide viewers' attention. Highlight important points using headings, subheadings, and bullet points. Use larger fonts for titles and key findings to make them stand out. Also keep the text on your poster concise and to the point. Use brief statements, bullet points, and phrases instead of lengthy paragraphs. Use clear and straightforward language to communicate your findings effectively.

The most important aspect of your poster are your plots and visualizations. In most cases you will rely on `ggplot2` to present your data in a visually appealing and easily understandable way. Take a look back at chapter 3 to find the right visualization for your problem. Choose appropriate visual representations based on the type of data or insight you are presenting. Make sure that your visualizations have the correct size and ensure that the resolution is high enough for clear printing.

If you work in a business context, incorporate your organization's branding elements, such as logos and colors, into the design. Maintain a consistent layout throughout the poster, with clear sections and a logical flow of information. Probably, there will be a cooperate design guideline where you can rely on.

By following these guidelines, you can design a visually appealing and informative scientific poster to effectively showcase your business data analytics results. Remember to prioritize clarity, simplicity, and visual impact to engage your audience and effectively convey your research findings.

6.3.3 Interactive Notebooks with R Markdown

Interactive notebooks in R refer to documents that combine code, text, and visualizations in an interactive environment. R notebooks are an implementation of the programming paradigm of literate programming (Knuth, 1984) that allows for direct interaction with R, while reading and writing a report. Therefore, notebooks consist of code chunks that can be executed independently and interactively and are enriched by explanatory text written in markdown. Notebooks with markdown allow business data analysts to execute code, view results, and document their workflow in a single document.

The most popular interactive notebook for R is R markdown, which integrates the R programming language with markdown syntax. Any R markdown document can be used as a notebook, and all R notebooks can be rendered to other R markdown document types. A notebook can therefore be thought of as a special execution mode for R markdown documents. An example of such a notebook is illustrated in Figure 6-8:

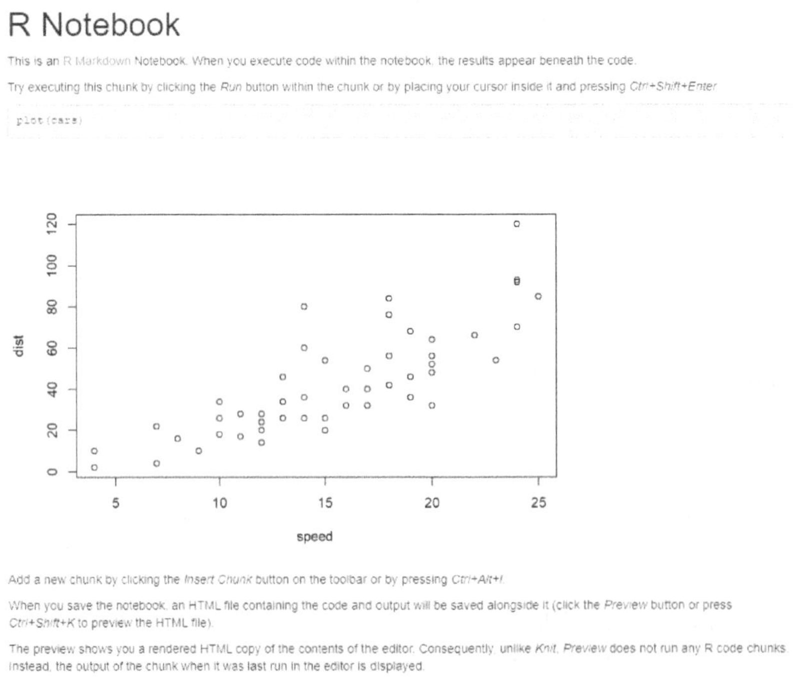

Figure 6-8 Example notebook in R with Markdown.

You can create a new notebook or simple interactive file in RStudio by open an R notebook or a new R markdown in the toolbar as illustrated in Figure 6-9. If you do not use RStudio, you need the `rmarkdown` package, but for now you don't need to explicitly install it, as RStudio automatically does it for you when needed.

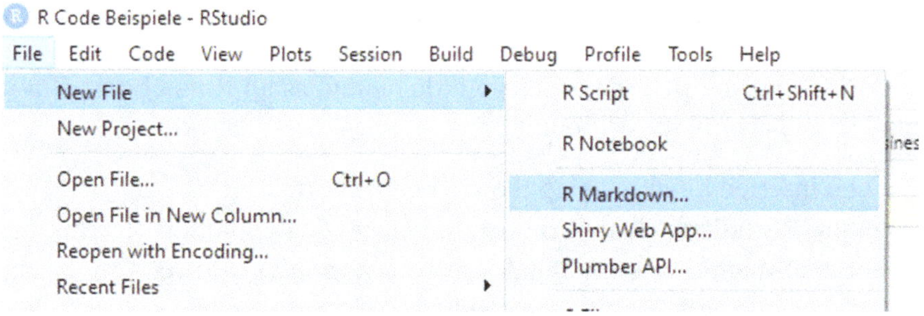

Figure 6-9 Make a new "R Notebook" or "R Markdown" file in RStudio.

Let us now open a new R notebook to write some simple R code and document it. Use the option "R Notebook" in the toolbar for that. You get a notebook interface in which code and output are nested inside each other. The new file will contain the following code:

```
---
title: "R Notebook"
output: html_notebook
---

This is an [R Markdown](http://rmarkdown.RStudio .com) Notebook. When you execute code within the notebook, the results appear beneath the code.

Try executing this chunk by clicking the *Run* button within the chunk or by placing your cursor inside it and pressing *Ctrl+Shift+Enter*.

```{r}
plot(cars)
```

Add a new chunk by clicking the *Insert Chunk* button on the toolbar or by pressing *Ctrl+Alt+I*.

When you save the notebook, an HTML file containing the code and output will be saved alongside it (click the *Preview* button or press *Ctrl+Shift+K* to preview the HTML file).

The preview shows you a rendered HTML copy of the contents of the editor. Consequently, unlike *Knit*, *Preview* does not run any R code chunks. Instead, the output of the chunk when it was last run in the editor is displayed.
```

Congratulation! This is your first R notebook with some dummy code to start. The dummy code consists of three important content sections that are typical for R notebooks:

- A header, surrounded by `---`
- Your code chunks, surrounded by ` ``` `
- And text mixed with some formatting commands like `*` or `_`

The header is a so called YAML header that identifies your document as an R notebook. But it is not necessary and your file will also work without it.

Notebook code chunks can be inserted quickly using the keyboard shortcut `Ctrl + Alt + I`, or via the `Insert` menu in the RStudio toolbar. You can run each code chunk by clicking the run icon (green triangle button at the top of each chunk), or by pressing `Cmd/Ctrl + Shift + Enter`. RStudio executes the code and displays the results inl3ine with the code:

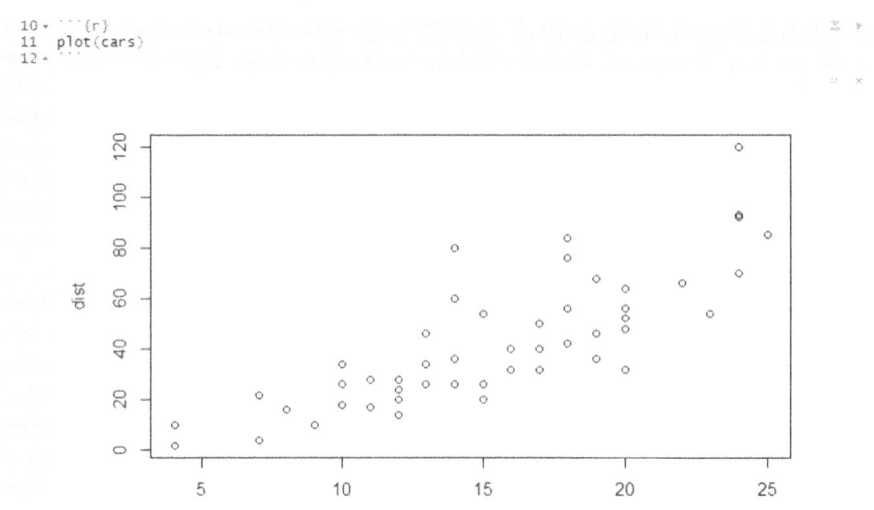

Figure 6-10 Running the plot(cars) command will plot the cars dataset in your notebook.

Code in the notebook is executed with the same manner. The primary difference is that when executing chunks in an R Markdown document, all the code is sent to the console at once, but in a notebook, only one line at a time is sent. This allows execution to stop if a line raises an error.

The last relevant part of the notebook is text in Markdown. Markdown are a set of lightweight conventions to format plain text. The main purpose is readability and simpleness. Therefore, it only consists of a limited number of commands to format your plain text.

The most relevant are:

- `_italic_`
- `__bold__`
- `# 1st level header`
- `## 2st level header`
- `* list`

Besides there exists a minimal number of further commands, which you can look up here:

https://rmarkdown.RStudio.com/lesson-1.html

But they are definitely not necessary to write suitable and understandable notebooks in R. Now, just add some of your own text and try to change the dummy code as you want. If you are finished, we come to the next step!

To render your notebook in a complete report containing all code, text, plots and results, just click the "Preview" or the "Knit" command as illustrated in Figure 6-11 in the toolbar or just press `Cmd/Ctrl + Shift + K`. If you did not select an output file format the command will be "Preview" otherwise it will change to "Knit".

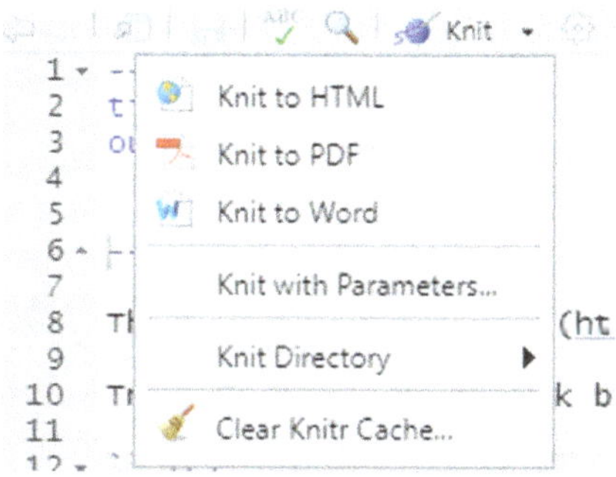

Figure 6-11 There are different options to render an R notebook.

As always, there is also an R command, if you want to automate that or run it programmatically:

```
rmarkdown::render("R Notebook.Rmd")
```

The result will be a self-contained HTML file that you can share with others and contains all the information to run independently. You can look at it in the preview pane on the right in RStudio to convince yourself of the quality.

6.3.4 Dashboards with Flexdashboard

As useful and descriptive as R-notebooks may be, in practice one will be much more drawn to build so-called dashboards. Dashboards provide visual representations of complex insights much more suitable than notebooks, allowing users to quickly grasp key information. More sophisticated user interfaces make it easier to identify trends, patterns, and anomalies in the data. In the best case, they are connected to your organization's database, hence they offer real-time or near-real-time updates on important values. This enables decision-makers to monitor business performance continuously and respond promptly to changing conditions or issues.

If you want to build complex dashboards, you will have to use the `shiny` package (we will come to that in the next chapter). However, if you want to build a simple straight forward dashboard, I recommend to take a look at the `flexdashboard` package. The `flexdashboard` package builds on R Markdown and is the most efficient way to build straight forward, light-weighted dashboards.

To use `flexdashboard` you have to install it:

```
install.packages("flexdashboard", dependencies = TRUE)
```

Furthermore, you have to install shiny which we will use to handle the user input correctly:

```
install.packages("shiny", dependencies = TRUE)
```

Then you will be able to select "Flex Dashboard" in the R Markdown menu, if you make a new R Markdown file and choose the selection "From Template":

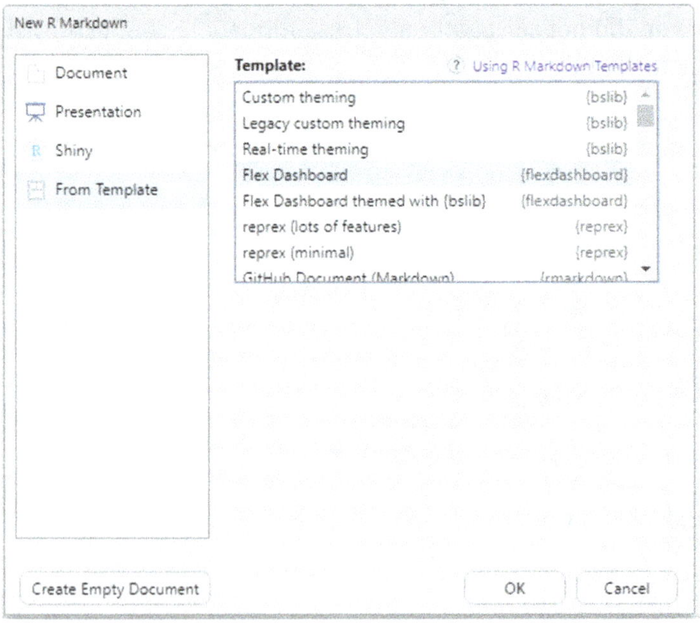

Figure 6-12 Choose "New File", "R Markdown", and then "From Template" to select "Flex Dashboard".

As before, you can also run it programmatically with:

```
rmarkdown::draft("dashboard.Rmd", template = "flex_dashboard",
package="flexdashboard")
```

The resulting file will consist of the following dummy code:

```
---
title: "Untitled"
output:
  flexdashboard::flex_dashboard:
    orientation: columns
    vertical_layout: fill
---

```{r setup, include=FALSE}
library(flexdashboard)
```
```

```
Column {data-width=650}
-----------------------------------------------------------------------
--

### Chart A

```{r}

```

Column {data-width=350}

--

Chart B

```{r}

```

Chart C

```{r}

```
```

Like an R notebook it consists of a header, code chunks and text which are later combined into a final output. If you run it, you get the skeleton of the following dashboard illustrated in Figure 6-13.

Figure 6-13 Skeleton of the `flexdashboard` dashboard in R.

`flexdashboard` is a big package with many functions. The full documentation of `flexdashboard` is online available at:

https://rmarkdown.RStudio .com/flexdashboard

In this section, we will only cover some basic features and usage in this chapter to get you familiar with it before we continue with other tools. Hence, I recommend that you have a look at the documentation if you want to start your full `flexdashboard` project.

Let us now continue to adapt the `flexdashboard` template to our needs. In a first step, we have to add your content to the dashboard - you can think of it as an advanced R notebook. The functionality and interactivity will be inside the ` ```{r} ... ``` ` code chunks.

Let us now customize the code to build a simple dashboard visualizing the `whiskycollection` dataset from the previous chapters. For that purpose, take a look at the sketch in Figure 6-14.

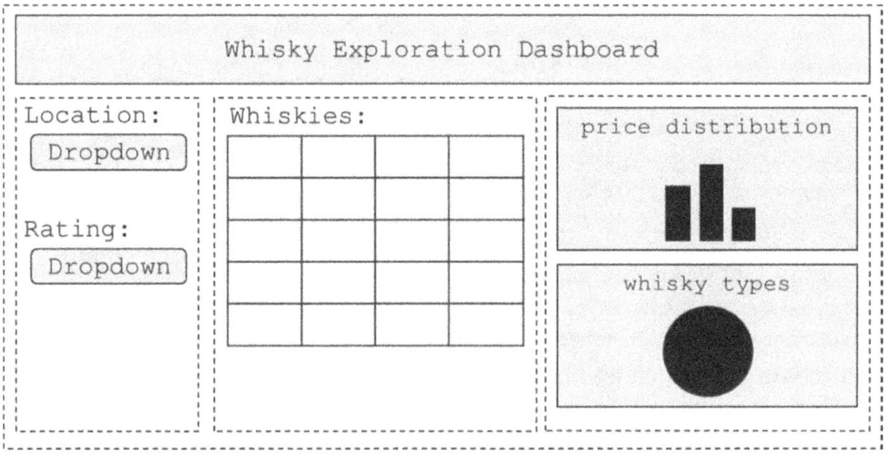

Figure 6-14 A sketch of a possible dashboard to visualize our whisky collection.

Our dashboard will be used to investigate our whisky collection which is stored in the `whiskycollection` dataset. The dashboard will consist of a sidebar on the left. The sidebar will be used for user input. For that purpose, we add there two drop-down menus. Based on the user input, a table will show all the whiskies that have the related rating and location. In the area on the right, we will put our visualizations. There will be a bar plot to illustrate the prices of the whiskies that have been selected, and a pie chart to show the different types in the sample.

In a first step, we will add the shiny runtime to our dashboard allowing reactive content and interactivity (don't worry in the next chapter we will make a deep dive into shiny). We can do that by adding it to the YAML header at the top of the document. Furthermore, we have to add the packages we want to use and load our dataset into the dashboard environment in a new data chunk so that we can use it. Remember to store both the dashboard files and the data in the same working directory otherwise R cannot access it.

Our dashboard code should now look like this:

```
---
title: "Whisky Exploration Dashboard"
output:
  flexdashboard::flex_dashboard:
    orientation: columns
    vertical_layout: fill
  runtime: shiny
---

```{r setup, include=FALSE}
library(flexdashboard)
library(shiny)
library(dplyr)
library(ggplot2)
library(readxl)
```

```{r data}
whiskycollection = read_excel("whiskycollection.xlsx")
```
```

Then we will add a sidebar and adjust the size of the columns to be equal:

```
Column {.sidebar data-width=200}
-----------------------------------------------------------------------
### Sidebar
```{r}
```

Column {data-width=400}

Chart A
```{r}
```

Column {data-width=400}

Chart B
```{r}
```

Chart C
```{r}
```
```

If you copied these steps correctly your layout should look like below if you run your dashboard.

Otherwise, you can copy-paste the code from the `dashboard.Rmd` file on the course homepage.

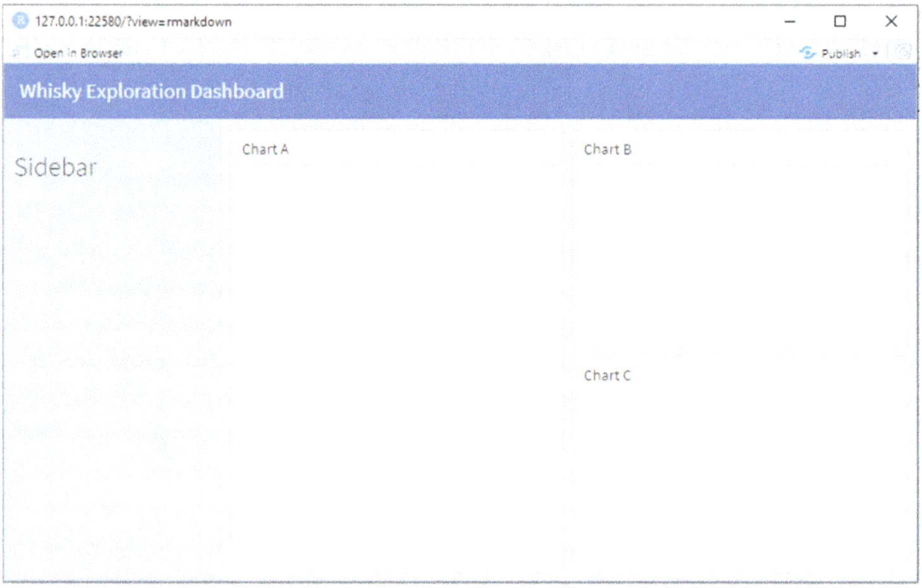

Figure 6-15 Adjusted dashboard with sidebar and new title.

In the next step we want to add some user input elements (in R-jargon termed "widgets"). In our case we will add so called `SelectInput` widgets, which allow the user to select a variable based on its value:

````
### Sidebar
```{r}
locations = unique(whisky_collection$LOCATION)
selectInput("selected_location",          label="Whisky          origin:",
choices=locations)

ratings = unique(whisky_collection$RATING)
selectInput("selected_rating",          label="Whisky          rating:",
choices=ratings)
```
````

The widget has three arguments `name`, `label` and `choices`. The `name` is invisible to the user and is the name of the widget which we can use to access its value in our code. This can be done with the command `input$<name>`. The `label` is the text that will appear above the widget, while `choices` defines the distinct values the user can select from.

The final results are illustrated in the following Figure 6-16.

Figure 6-16 Dashboard with a sidebar and control widgets for user input.

We can now use the values from the widgets as user input for our dashboard. In particular, we will use the input of the widgets for `reactive` outputs. Reactivity is what makes Shiny apps responsive, meaning that they update automatically when changed. And changes to inputs automatically render code and update outputs. We have this functionality already, because we imported the shiny package in the first step. The most relevant render functions that are relevant for dashboards in `flexdashboard` are:

- `renderTable()`
- `renderPlot()`
- `renderText()`

You can think of these functions as functions that continuously monitor their input. If something happens, they get active and render again with the new input. But for this to work properly, we need to make sure that the input is also `reactive`. This can easily be done with the function `reactive({ })`. The combination of braces and curly braces indicates that the content is `reactive`.

In our dashboard we want now add a `reactive` table with all whiskies that suit the filter requirement of the user in the left content box. If the user changes the whisky origin or the rating the table should adjust. For that we will use the `renderTable()` command. In the two boxes on the right, we will show two plots illustrating the whisky rating and prices.

To add the table, we add the following code in the code chunk for the sidebar and change the name from “Chart A” to “Whisky Overview”:

```
### Whisky Overview

```{r}
location = reactive({input$selected_location})
rating = reactive({input$selected_rating})

selection = reactive({whisky_collection %>%
  filter(LOCATION == location() & RATING == rating()) %>%
  select(NAME, DISTILLERY, REGION, TYPE, PRICE)})

renderTable({selection()})
```
```

You can now restart your dashboard to see that the changes in the sidebar will result in different whisky selections in the table. Finally, we also add the two plots in our dashboard:

```
### Price comparison

```{r}
location = reactive({input$selected_location})
rating = reactive({input$selected_rating})

selection_distilleries = reactive({
  whisky_collection %>%
  filter(LOCATION == location() & RATING == rating(), PRICE <= 100) %>%
  select(NAME, DISTILLERY, REGION, TYPE, PRICE)
  })

plot_distilleries = reactive({
  ggplot(data=selection_distilleries()) +
  aes(x=DISTILLERY, y=PRICE) +
  geom_bar(stat="identity") +
  labs(x="Distillery", y="Price in EUR")
  })

renderPlot(plot_distilleries())
```


Whisky types

```{r}
location = reactive({input$selected_location})
rating = reactive({input$selected_rating})

selection_whisky_types = reactive({
  whisky_collection %>%
  filter(LOCATION == location() & RATING == rating()) %>%
  group_by(TYPE) %>%
  summarize(NUM=n())
  })
```
```

~~~
plot_whisky_types = reactive({
    ggplot(data=selection_whisky_types()) +
    aes(x="", y=NUM, fill=TYPE, label=TYPE) +
    geom_bar(stat = "identity", width = 1) +
    geom_text(position=position_stack(vjust = 0.5)) +
    coord_polar("y", start = 0) +
    theme_void()
  })

renderPlot(plot_whisky_types())
```
~~~

If you added everything correctly, you can now restart your dashboard. Congratulation you build your first dashboard in R! If there are some error messages, you can always fall back to the solution on the course homepage.

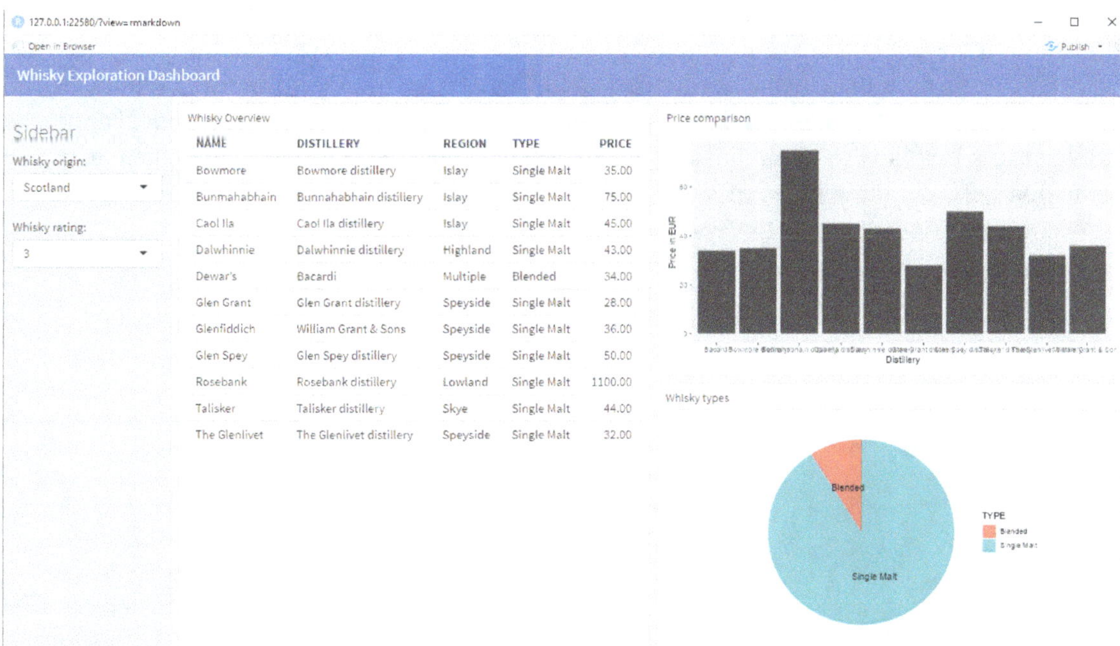

Figure 6-17 Final dashboard to explore the prices and types of whiskies per country.

The good thing about an open-source software solution like R is that you don't have to reinvent the wheel all the time. There is a collection of many dashboards and templates on the web that can be used or from which you can copy the parts that suit you. In my experience, you learn the most the same way, especially if you look closely at the code of the dashboards and solutions and try not only to copy them but also to understand them. Have a look at the `flexdashboard` examples on:

https://pkgs.RStudio .com/flexdashboard/articles/examples.html

Besides, I also strongly recommend checking out the official `flexdashboard` documentation, where you can find further features, widgets and details we could not do in this tutorial. Go and have a look at it, if you start to build your own first dashboard!

6.4 Business Analytics Systems

In the last chapter, I mentioned the `shiny` package to add reactivity to your dashboard with `flexdashboard`. The package `shiny` is a very popular R package to build analytic systems, which have more functionalities than simple dashboards. Data and computer scientists around the world develop such analytic systems to provide the insights from the business data analytics process in an interactive and professional way besides management presentations and reporting.

In this chapter we will first look at business data analytics systems on a generic level. And understand why they are a very popular way to integrate or combine your business data analytics findings in an existing information system.

In business data analytics, we distinguish between three generic types of business analytics systems. Please note that in the academic world there exists many different taxonomies and discussions about system types and characteristics. But in this book, we take a practical approach and want to know how to communicate our findings in practice. Therefore, we make a simple distinction between three types of systems that are relevant for us: *Decision-Support Systems*, *Recommender Systems* and *APIs/Pipelines*.

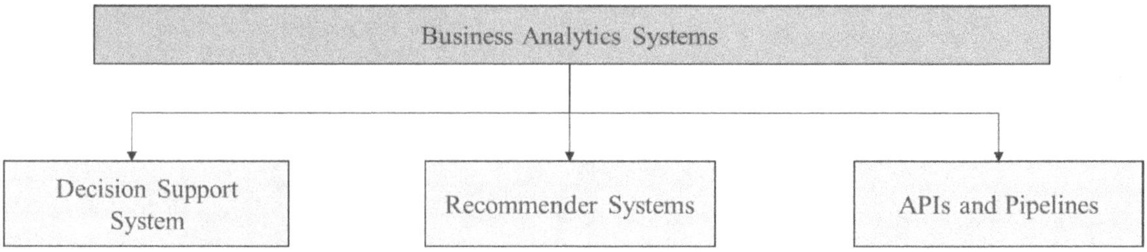

Figure 6-18. Different types of business analytics systems in analytics practice.

A decision support system is an interactive system that analyses large amounts of business data to support informed business decisions. A decision-support system helps the management, operational and planning levels of an organization make better decisions by assessing the importance of uncertainties and the trade-offs associated with one decision over another (Aronson et al., 2005; Keen, 1980). In most cases decision support systems will contain an analytics module or a recommender engine to help the users to make decisions.

Meanwhile, these systems are becoming very versatile and are also being combined with new technological developments. Be it through digital assistants such as robo-advisors that accompany their users in the complete decision-making process (Jung et al., 2018) or services that offer their users detailed explanations via large language models (Microsoft 365 Copilot). The more complex these systems become, the more R comes to an end and it becomes more of a software development project than a business data analytics project. We will focus here on the solutions that can be implemented sensibly and realistically with R.

A direct relative of decision support systems are recommender systems. Recommender systems are information systems to provide personalized recommendations to users, specifically in the context of business-related decisions (Resnick & Varian, 1997). For instance, they help users to find similar products or items in an online shop. Or if you shop at amazon and get a recommendation what to buy, you interact with recommender system. For that purpose, the recommender system analyzes historical data, user preferences, and other relevant factors to suggest suitable actions, strategies, or insights for optimizing business performance. If you want to test one of my favorite recommender systems, you should check out Jester. It's a joke recommender that learns on your preferences and recommends you jokes based on your personal humor: *http://shadow.ieor.berkeley.edu/humor/*.

Finally, sometimes you do not want to build your own system or have to integrate your system into the IT landscape of your organization. For that purpose, it might be likely that you setup an analytics pipeline that feeds other systems or writes results back into a database. Another very common scenario is to provide your results as a service with an API.

6.4.1 Decision Support Systems with Shiny

Once you have created your first dashboards with `flexdashboard`, you are ready to take the next step: We will now look at how we can build slightly more complex systems like advanced dashboards or even decision support systems!

Do not understand me wrong, notebooks and simple dashboards are good and very common tools for business data analysts. But if you have a lot of user input or want to map complex user interactions, you will quickly reach the limits of what is possible with `flexdashboard`. Therefore, the R community has a package called `shiny`, which can help us to make full functional web apps that can be used as decision support systems. Are you now motivated? Good, let's go!

If you didn't install `shiny` in the previous section, then it's now time to install it with:

```
install.packages("shiny", dependencies = TRUE)
```

If everything is installed, you can now start a complete new shiny project by selecting *File – New Project – New Directory – Shiny Web Application*. This allows you to work in a clean project, focusing on your app. If you want that the app is part of another R project, you can just add shiny as a single-file-application as illustrated in Figure 6-19.

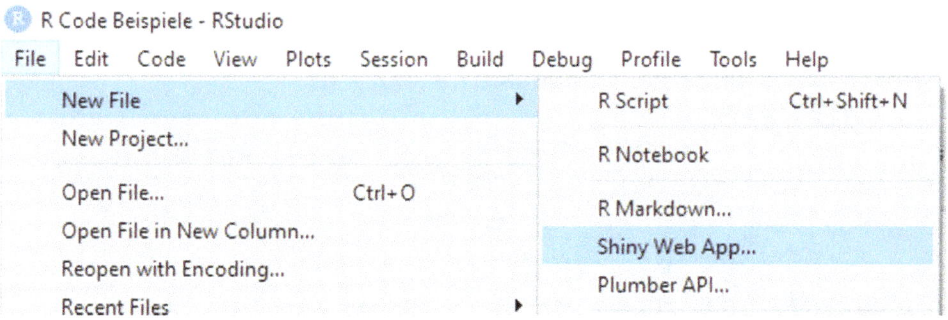

Figure 6-19 Select "Shiny Web App" to start a new shiny app in RStudio.

Depending on what you have chosen, RStudio has now created a file `app.R` in a new or your current project. This file serves as a starting point for our own app and already contains a few demo code. It should look somehow like this:

```
# This is a Shiny web application. You can run the application by
# clicking the 'Run App' button above.
#
# Find out more about building applications with Shiny here:
#
#    http://shiny.RStudio .com/

library(shiny)

# Define UI for application that draws a histogram
ui = fluidPage(

    # Application title
    titlePanel("Old Faithful Geyser Data"),

    # Sidebar with a slider input for number of bins
    sidebarLayout(
        sidebarPanel(
            sliderInput("bins",
                        "Number of bins:",
                        min = 1,
                        max = 50,
                        value = 30)
        ),

        # Show a plot of the generated distribution
        mainPanel(
           plotOutput("distPlot")
        )
    )
)

# Define server logic required to draw a histogram
server = function(input, output) {

    output$distPlot <- renderPlot({
        # generate bins based on input$bins from ui.R
        x    <- faithful[, 2]
        bins <- seq(min(x), max(x), length.out = input$bins + 1)
        # draw the histogram with the specified number of bins
        hist(x, breaks = bins, col = 'darkgray', border = 'white')
    })
}
# Run the application
shinyApp(ui = ui, server = server)
```

Each Shiny app consists of two parts, the user interface and the server, which is responsible for the interactions:

- `ui()`: controls the layout and appearance of your app
- `server()`: contains all server functions to provide interactivity

Besides, the command `shinyApp()` at the end of the file creates a shiny app object from your explicit ui/server pair and is used to start your app. In our example both parts are stored in one file, which makes our app a single-file application. But instead of using a single file you can also store your app in two parts. Therefore create a file "`ui.R`" and a file "`server.R`" and store your code there.

In general, you can organize your shiny project the way you want and give funny names to your functions and files. But if you start a shiny project, it is convention to organize your project in the following folder structure. This is necessary if you deploy your app on a public hosting service or on the on-prem R-server of your organization to ensure that everything works fine:

```
name_of_your_project
├── app.R
├── data
│   └── onlineshop.csv
└── www
    ├── junglivet.jpg
    └── helper.R
```

You can see two folders in this project. The `data` folder is used to store your data that is used by the app. This is necessary if you do not work with live data connections. And the `www` folder contains all files that are otherwise used by the app like images (`junglivet.jpg`) or other R functions (`helper.R`).

To start the shiny app, simply click on "Run App" on the right above the editor. RStudio starts now a local web server to host your app. This is a service that runs on your computer and provides the shiny web app. Then an RStudio browser window opens with the shiny application. If you want, you can also enter the address `http://127.0.0.1:3906` in your web browser of your choice to access your app. `127.0.0.1` is the address (localhost) of your computer and `3906` is the port on which the app can be reached.

The result is illustrated in the following Figure 6-20:

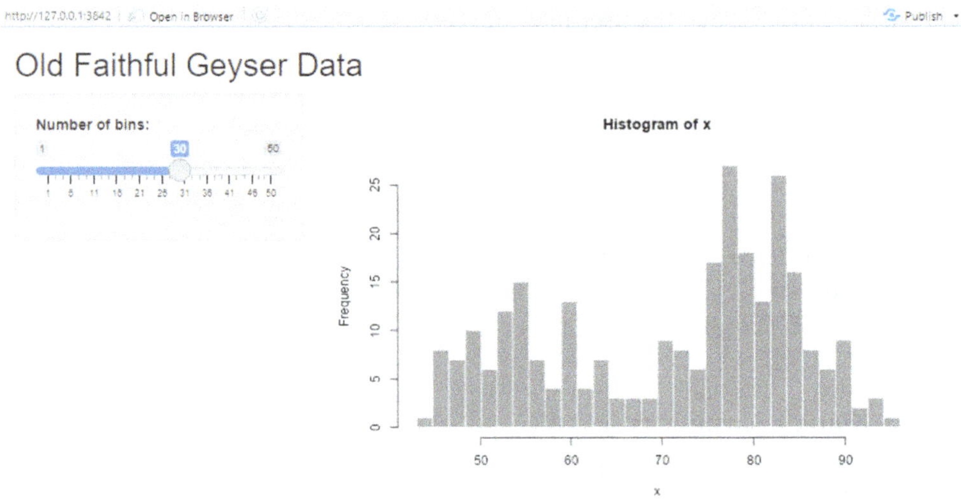

Figure 6-20 The shiny demo app `old faithful geyser` *data.*

Now that you know what the app looks like, let's get back to the code and its two components: the `ui` and the `server`.

As we said before, we specify the layout of our shiny app in the ui part of your app. Shiny apps have a panel structure to layout the content of an app. And panels are a kind of content container which can contain text, plots, tables but also your widgets to control the app. So far, there are many layouts you can use. One very good way to start with is the `fluidPage()` layout, which organize the content in rows which in turn consist of columns. The elements of the rows are there to appear on the same line, while columns exist to define how much horizontal space your elements should occupy within a 12-unit-wide grid.

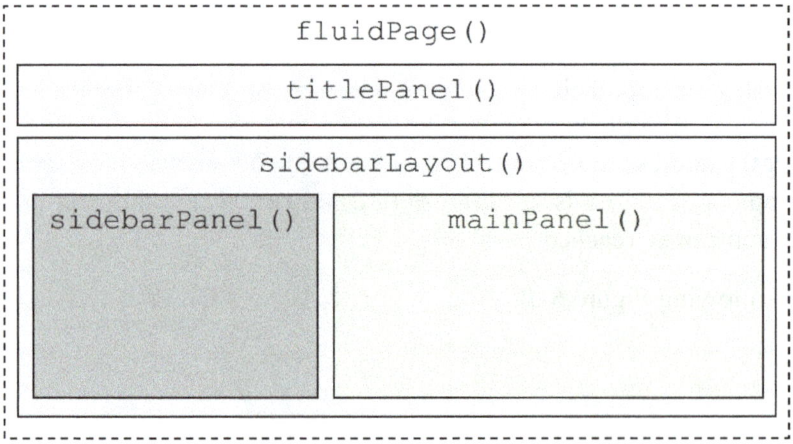

Figure 6-21 Simple illustration of an app layout with `sidebarLayout()`*.*

Alternatively, you can use `fluidRow()` to restructure your app in rows, as illustrated in Figure 6-22.

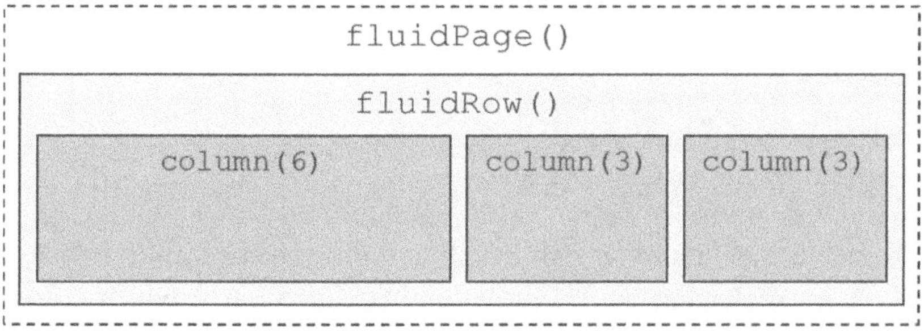

Figure 6-22 Simple illustration of an app layout with `fluidrow()`.

In our example, the first line is the title using `titlePanel()`. In the second row we then define a `sidebarLayout`, which consists of a `sidebarPanel` and a `mainPanel`. The slider is In the `sidebarPanel` and the histogram in the `mainPanel`. All kinds of structures can be realized in this way: multiple pages, tabs, nested rows/columns, etc.

Another very relevant part of your ui are control widgets. In our example, the app has a `sliderInput()` to get the number of bins from the user. Besides the `sliderInput()`, and the `selectInput()` we know from our `flexdashboard` dashboard, there are many more input widgets in shiny. The most relevant are:

- `actionButton(inputId="action", label="Go!")`
- `fileInput("file1", "Choose CSV File", accept=".csv")`
- `numericInput("obs", "Observations:", 10, min=1, max=100)`
- `radioButtons(inputId="radio", label="Radio Buttons", choices=c("A", "B"))`
- `textInput("caption", "Caption", "Data Summary")`

Each widget needs an `inputID`, which you can use to access the value in your server and a `label` argument.

After that, let us now come to the server-side of your shiny app. The server-side of an app is encapsulating the functionalities. In our example form above, it's very simple. It contains a function to generate a random distribution and plots it as histogram with the requested number of bins. As before in our `flexdashboard` example, the code that generates the plot is wrapped in a render function `renderPlot()`.

In shiny, we call these stuff output. Outputs can be in the form of plots, tables, maps or text. In shiny logic outputs are created by placing code in the double brackets `({ })` in the server part of your code:

```
output$distPlot = renderPlot({ … })
```

In most cases, we want that the output is interactive created. If the user changes something in the ui, we expect our app to react to this. Therefore, we have to access the user input, which is stored in the `input` variable in shiny. If the user changes a value in a widget, we can access the value from the `input` variable.

To make the outputs appear on our app, we need to add the rendered object to the `ui` object with an _output() function like `plotOutput()` for plots or `tableOutput()` for tables etc.:

```
mainPanel( plotOutput("distPlot") )
```

Now you are ready to create your own `shiny` app with R!

Like with `flexdashboard`, I recommend that you start benchmarking code examples to better understand how shiny works. A good starting point is the shiny package itself, which comes with further built-in examples that demonstrates how it works. You can access the `11` examples with:

```
library(shiny)
runExample("01_hello")
```

Additionally, there is an excellent beginner tutorial from shiny itself at *https://shiny.posit.co/r/getstarted/shiny-basics/lesson1/index.html*. It contains mostly the same content as in this chapter, but might help you to recap the most relevant aspects.

Finally, there is a famous collection of numerous templates on the web that you can access for free. To take a closer look at the possibilities of shiny, take a look at the shiny use case library: *https://shiny.posit.co/r/gallery*

6.4.2 Recommender Systems

Another very common task in business data analytics is to support business decision-making by means of recommendations. And if we build systems that can give these recommendations automatically, then we speak from recommender systems. The applications of recommender systems are manifold: A user of our whisky shop might be interested in other whiskies that are similar to his interest. If you compare different whisky, you might be interested in other supplier that have comparable goods.

In this chapter we want to implement a collaborative recommender system from scratch to understand the idea behind recommender systems. Collaborative filtering is a technique for predicting the interests of one user based on the preference information of other users. For that purpose, we rely on the package `recommenderlab`. Please install and load the package, before you continue with:

```
install.packages("recommenderlab", dependencies = TRUE)
library(recommenderlab)
```

For the following example, we will use a public dataset from MovieLense to build a quick and dirty recommender system. The website is a very popular community for movie fans and also recommends movies for its users to watch based on the user ratings of the community. The university of Minnesota is the host of this website and uses the data for their research in recommender systems.

The starting point in collaborative filtering is a rating matrix in which rows correspond to users and columns correspond to items (Gorakala & Usuelli, 2015). We can load the matrix from the

MovieLense website (*https://grouplens.org/datasets/movielens/latest/*) or by loading a sample from the `recommenderlab` package with:

```
> data(MovieLense)
> MovieLense
943 x 1664 rating matrix of class 'realRatingMatrix' with 99392 ratings.
```

This data object contains multiple matrices which contain user ratings of different movies. You can look at the first distinct movies with:

```
> head(names(colCounts(MovieLense)))
[1] "Toy Story (1995)"                                    "GoldenEye
(1995)"
[3] "Four Rooms (1995)"                                  "Get Shorty
(1995)"
[5] "Copycat (1995)"                                       "Shanghai
Triad (Yao a yao yao dao waipo qiao) (1995)"
```

The movies are rated from the users with `1-5` and are rated with missing value if the user did not rate it so far. To build our collaborative model, we have to prepare the dataset and clean it. In our case, we want to remove the users that have only a few ratings and movies that have received too few ratings. Both types might be prone to outlier and impact the performance of our model negatively. To do so, we will set a threshold and restrict the training to users who have rated at least `10` movies or movies that have been rated at least `100` times.

```
ratings = MovieLense[rowCounts(MovieLense) > 10, colCounts(MovieLense)
> 100]
dim(ratings)
```

As discussed in the previous chapters, we then normalize the data so that the average rating given by each user is 0. This handles cases where a user consistently assigns higher or lower ratings to all movies compared to the average for all users. In other words, normalizing of data is done to remove the bias in each user's ratings.

```
ratings = normalize(ratings)
```

As in the previous chapter, we have to split our dataset into test and train sets. Because the data is in an user-movie format, we will use the `evaluationScheme()` function in `recommenderlab`. This function allows us to specify the splitting procedure with several parameters that are relevant for recommender systems:

```
ratings_sets = evaluationScheme(data=ratings, method="split",
                               train=0.8,
                               given=15,
                               goodRating=3,
                               k=1)
```

In the example, we split the dataset by specifying how many items are used for each user and the minimum value that indicates a good rating.

In our collaborative filtering model, we want to find users that are similar to a specific user. Building on the user sets we can identify the top-rated items by them. These items are then our recommendation for the given user.

In a next step, we now proceed to construct our collaborative filtering model utilizing the default settings of the `Recommender()` function. We apply this function to make predictions using the test set of our initial dataset. To evaluate the accuracy of our recommendations, we use the built-in functions from the `recommenderlab` package to compare them with the values present in the dataset.

```
recommender_ubcf = Recommender(data=getData(ratings_sets, "train"),
                               method="UBCF",
                               parameter=NULL)
```

Then we evaluate it with:

```
recommendations = predict(object=recommender_ubcf,
                          newdata=getData(ratings_sets, "known"),
                          n=10,
                          type="ratings")

eval_accuracy = calcPredictionAccuracy(x=recommendations,
                                       data=getData(ratings_sets,
"unknown"),
                                       byUser=TRUE)
head(eval_accuracy)
```

Congratulation! You build your first recommendation model. In business data analytics this kind of model is also called user-based collaborative filtering, because we made our recommendation based on users that are similar to the user.

Another approach to make recommendations is item-based collaborative filtering. Item-based collaborative filtering attempts to find, for a given user, items that are similar to items purchased by the user (Gorakala & Usuelli, 2015). Or in other words we now compare items based on how users have rated them. We then try to find items that are similar to each other and recommends those similar items to users.

Let us now create an item-based collaborative filtering model using the default settings of the `Recommender()` function:

```
recommender_ibcf = Recommender(data = getData(ratings_sets, "train"),
                               method = "IBCF",
                               parameter = NULL)
```

And evaluate it with:

```
recommendations = predict(object=recommender_ibcf,
                          newdata=getData(ratings_sets, "known"),
                          n=10,
                          type="ratings")
eval_accuracy = calcPredictionAccuracy(x=recommendations,
                                       data=getData(ratings_sets,
"unknown"),
                                       byUser=TRUE)
head(eval_accuracy)
```

That's it. You could implement this module into a web app with shiny and forward the user input in the `predict()` function to give user specific recommendations. If a user wants a recommendation, he or she can enter some information in your webapp and gets a recommendation. However, in most business scenarios, this is probably not usable. You have existing systems like an online shop or decision-support system and want to implement just the recommendation as a feature. In this case, you have to provide your recommendation service as API. Let us now have a look at it!

6.4.3 APIs with Plumber

As illustrated in the example before, there might be times you might want to share your R findings with others who have an existing app or information systems or don't use R. If they can't understand your work, they'll struggle to make use of it in their preferred software or programming language. However, if you make your results available through an API, anyone can access your results without the hassle of translation. APIs serve as a bridge between different applications and platforms.

For that purpose, their app has to send a request to your API. The API takes the request to the server and gets a response. The API then sends that response back to the app of the user with the request. The beauty of API responses is that they're universally understandable. Just as you use R to interact with the MovieLense API, others can utilize different tools like Python to access the MovieLense API.

For instance, imagine you're collaborating with a website developer who uses JavaScript. You've built a model in R, and you'd like to share its results. You can provide the developer with an API, allowing them to display those results on a website without having to recreate your model in another programming language. Moreover, the website can showcase real-time updates because it communicates directly with your API.

Setting up R APIs with Plumber is a straightforward process that involves a few key steps. First, you define the API endpoints by annotating your R functions with special comments. These comments describe the inputs, outputs, and behavior of each API endpoint. Next, you run the Plumber application, which automatically generates the API server and maps the defined endpoints to the corresponding R functions. Let us now have a look at this process.

First, please install the `plumber` package and load it with:

```
install.packages("plumber", dependencies = TRUE)
library(plumber)
```

Then you can create your first simple API. Best practice is to let RStudio setup everything for you by clicking "Plumber API" in the context menu.

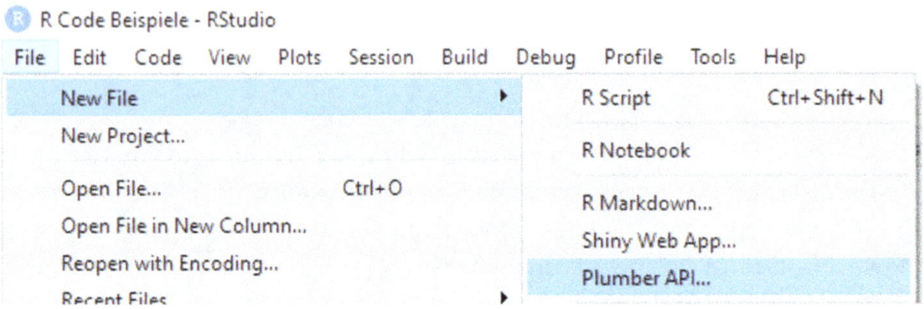

Figure 6-23 Generate an API easily in the toolbar.

Alternatively, you can make a new file. Name it "`my_api.R`" and copy-paste the following code:

```
library(plumber)

#* @apiTitle Plumber Example API

#* Echo back the input
#* @param msg The message to echo
#* @get /echo
function(msg = "") {
    list(msg = paste0("The message is: '", msg, "'"))
}

#* Plot a histogram
#* @png
#* @get /plot
function() {
    rand <- rnorm(100)
    hist(rand)
}

#* Return the sum of two numbers
#* @param a The first number to add
#* @param b The second number to add
#* @post /sum
function(a, b) {
    as.numeric(a) + as.numeric(b)
}
```

Now you should have a working API which can be host on an R-Connect server and will provide two possible endpoints. One is hosted at the path `/echo` and simply returns the message passed in. The other endpoint is hosted at the path `/plot` and returns an image.

Once you have Plumber installed, you can use the `pr()` function to translate this R file into a Plumber API named `root`.

```
root = pr("plumber.R")
```

Then you can test your API with:

```
root %>% pr_run()
```

You should see a message about your API running on your computer on port `8000`. The API will continue running in your R session until you press the `Esc` key. If you're running this code locally on your personal machine, you should be able to open *http://localhost:8000/echo* or *http://localhost:8000/plot* in a web browser to test your new API endpoints.

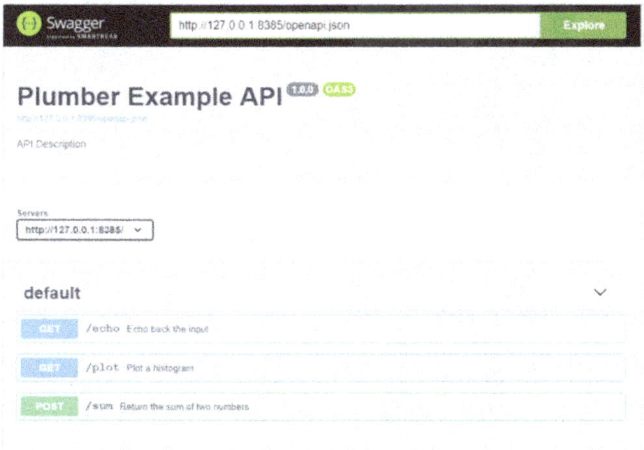

Figure 6-24 Your working API with the different endpoints.

Congratulations! You've just created your first Plumber API. You can call your different endpoints in the URL by adding them to the path or clicking on them in the web interface.

Now let us come back to the different endpoints. In Plumber you can define endpoints by annotating your R functions with special comments, like e.g.:

```
#* Return the sum of two numbers
#* @param a The first number to add
#* @param b The second number to add
#* @post /sum
function(a, b) {
    as.numeric(a) + as.numeric(b)
}
```

As you can see the annotations start with a `#*` symbol. Plumber parses your script to find these annotations. It uses them then to convert your function into an API endpoint. The first line of the annotation is a description of your function, then the parameters of your function are described with the @param annotation. Finally, you have to specify if the endpoint is a `GET` or `POST` request and how to access it. In this example we have a `POST` request (`@post`), which can be accessed by adding "`/sum`" to the URL of the API.

I recommend, that you can have a short look at the official documentation at:

https://www.rplumber.io/articles/rendering-output.html

before you start your API project. You can find there many working examples and further possible annotations that will help you to fine tune your API with `plumber`.

6.4.4 Analytics Systems Deployment

When it comes to publishing your business data analytics applications, you have several options depending on your specific needs and preferences. Best practice is to setup your own shiny server or even better use a RStudio Connect server. RStudio Connect is a publishing platform for data science content created in R and Python. It enables users to securely deploy and share Shiny applications, R Markdown reports, Plumber APIs, and Jupyter Notebooks. With RStudio Connect, business data analysts can easily publish their work from their development environment to a centralized web-based interface. The platform provides a secure and interactive environment for colleagues, stakeholders, and the public to access and explore the published content. RStudio Connect also offers collaboration and governance features, ensuring version control, access control, and scheduling capabilities for seamless team collaboration.

If you only want to test your app and when it does not contain company sensitive information, you can host it on one of the many public platforms for publishing and sharing your R Shiny applications:

- **Shinyapps.io**: Shinyapps.io is a cloud-based platform provided by RStudio specifically designed for hosting and deploying Shiny applications. It offers a user-friendly interface for deploying your apps with just a few clicks. You can choose from free or paid hosting plans based on your usage requirements.
- **GitHub Pages**: If your Shiny application is relatively simple and consists mainly of static content, you can host it on GitHub Pages. By pushing your application code to a GitHub repository and enabling GitHub Pages, you can make your application accessible via a custom URL.
- **Docker**: Docker is a popular platform for containerization, which allows you to package your Shiny application along with all its dependencies into a single container. You can then publish the container to platforms like Docker Hub, making it easy for others to deploy and run your application in their own environment.

Consider your specific requirements, such as scalability, cost, control, and collaboration need, when deciding on the best publishing method for your R Shiny applications. Each option has its own advantages and trade-offs, so choose the one that aligns with your goals and resources.

In the following example I will show you how to deploy your app with shinyapps.io, which is the most popular free-to-use service and has many possibilities to specify with different upgrades.

Before you can start, you have to register on the *shinyapps.io* website. If you registered successfully, you should see the following dashboard:

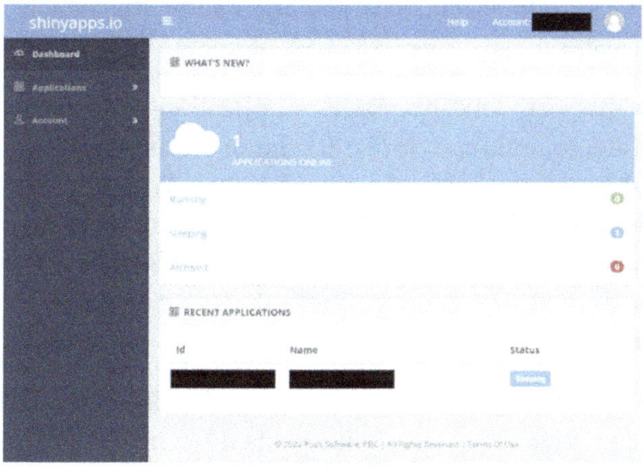

Figure 6-25 Dashboard with app overview and user information.

Where you can manage and upgrade your apps and services. Finally, you have to go back to RStudio and open the shiny app of your choice. Then you have to click on the "publish" command in the top right corner, next to the "Run App" command. You can also select "File" - „Publish" from the context menu.

Then the following dialogue will appear:

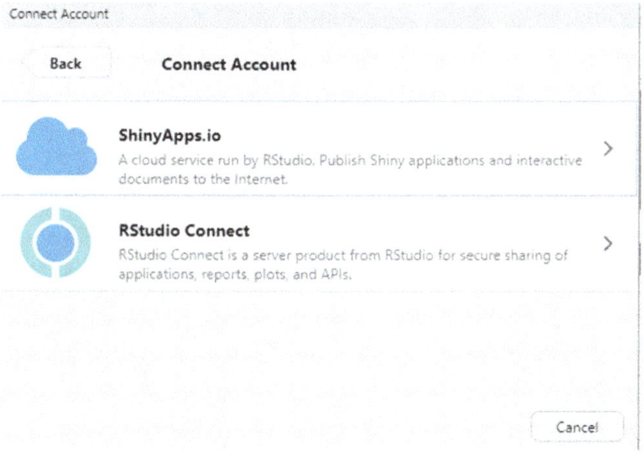

Figure 6-26 Dialog to select your connect account.

We select "ShinyApps.io", where you have to enter your credentials and account information. After this is done, you can select which files should be uploaded. Then your app will appear in the shinyapps.io dashboard where you can find the link to access it online.

6.5 Useful R Functions for Everyday App Development

As before, I would like to use this section to present some useful functions that might help you building and deploying your models into business data products. If you are reading this chapter the first time, I recommend that you skip that section and come back to it at a later time.

First of all, I want to mention the `cloudyr` project. This project is a collection of different initiatives and packages to make facilitate cloud computing with R. It contains many tools and functions to handle the most popular cloud computing platforms like Amazon Web Services (*https://aws.amazon.com*), Google Cloud Services (*https://cloud.google.com*) or Microsoft Azure (*https://azure.microsoft.com*). These platforms are the most common platforms to host apps and services in a professional business context and its very likely that your company is using them to host their existing systems and services.

The different packages are hosted in a `drat` repository in the cloudyr GitHub website: *https://cloudyr.github.io/drat*. The versions are updated daily and also pushed to CRAN. If you plan to use cloud services you can install the `cloudyr` packages directly from the repository:

```
if (!require("drat")) {
    install.packages("drat")
}
drat::addRepo("cloudyr", "https://cloudyr.github.io/drat")
install.packages("NameOfPackage")
```

6.6 Checklist

1. Phase: Business Understanding			
1.1	Determine business objectives	❯	• Background • Business objectives • Business success criteria
1.2	Assess situation	❯	• Inventory of resources and capabilities • Requirements, assumptions and constraints • Risks and contingencies • Terminology • Costs and benefits
1.3	Determine analytics goals	❯	• Business data analytics goals • Business data analytics success criteria
1.4	Produce project plan	❯	• Project plan • Initial assessment of tools and techniques
2. Phase: Business Data Understanding			
2.1	Collect initial data	❯	• Initial data collection report
2.2	Describe data	❯	• Data description report
2.3	Explore data	❯	• Data exploration report
2.4	Verify data quality	❯	• Data quality report
3. Phase: Business Data Preparation			
3.1	Select data	❯	• Rationale for inclusion/exclusion
3.2	Clean data	❯	• Data cleaning report
3.3	Construct data	❯	• Derived features • Generated records
3.4	Integrate data	❯	• Merged data

3.5	Format data	❯	• Reformatted data • Dataset • Dataset description
4. Phase: Modeling			
4.1	Select modeling techniques	❯	• Modeling technique • Modeling assumptions
4.2	Generate test design	❯	• Test design
4.3	Build model	❯	• Parameter settings • Models • Model descriptions
4.4	Assess model	❯	• Model assessment • Revised parameter settings
5. Phase: Evaluation			
5.1	Evaluate results	❯	• Assessment of results w.r.t. success criteria • Approved models
5.2	Review process	❯	• Review of process
5.3	Determine next steps	❯	• List of possible actions decisions
6. Phase: Deployment			
6.1	Plan Deployment	❯	• Deployment plan
6.2	Plan Monitoring and Maintenance	❯	• Monitoring and maintenance plan
6.3	Produce final report	❯	• Final report • Final presentation
6.4	Review project	❯	• Experience documentation

6.7 Business Case Exercise: A Dashboard for the Marketing Management

6.7.1 Scenario

In our ongoing collaboration with Lindsey Maegle to evaluate the different whisky characteristics, you have made significant strides in analyzing and clustering the diverse world of whiskies. You have successfully applied clustering algorithms to categorize whiskies based on various characteristics, creating valuable insights for her and her team. Consequently, Lindsey Maegle, sees great potential in the clustering results we have obtained. She envisions a dynamic and user-friendly tool that allows her to evaluate further characteristics more effectively. Additionally, she would like the flexibility to choose a specific variable, such as "body," and visualize the geographic distribution of whiskies with that attribute on an interactive map. As a consequence, you propose the development of a "Whisky Evaluation Dashboard", a reporting app tailored to Lindsey's needs. This app will provide her with the following key features:

- *The dashboard will display the results of our whisky clustering analysis in an intuitive and visually appealing manner. Lindsey will be able to explore different clusters and gain a deep understanding of the characteristics that define each one.*
- *To facilitate in-depth analysis, Lindsey can choose a variable of interest, such as "body" or any other relevant attribute. The dashboard will dynamically update to reflect the distribution of whiskies based on her selection.*
- *A powerful feature of the dashboard will be the interactive map. Lindsey can select a cluster or variable and view the geographical distribution of whiskies associated with her choice. This feature adds an unique dimension to her analysis, allowing her to explore the regional influence on whisky characteristics.*
- *The dashboard will be designed with user-friendliness in mind, ensuring that Lindsey can navigate through the data effortlessly. It will include intuitive controls and interactive elements for a seamless user experience.*

For that purpose, use your code and findings from the previous business case exercise and develop a notebook, dashboard or shiny app that helps Lindsey investigating the whisky dataset.

6.7.2 Task

Your task is to:

- Make a simple app with `flexdashboard` or `shiny` to visualize the variables `SMOKNESS` and `RICHNESS`. If that works out you can add further variables to the dataset.
- If you feel unsecure to build a whole webapp, try to start with a notebook. You can later transform it easily to a `flexdashboard` app.

- Implement the different requirements from Lindsey. Use the geographical data of the distillers to plot them dynamically on a map.
- You can reuse your code and findings from the previous exercise.

6.7.3 Remarks

Use the `whiskies.csv` file for your analysis.

6.7.4 Solutions

You find the solutions of this exercise on the course website.

7 Mastering Business Data Analytics

7.1 Introduction

After the last chapters, our journey in business data analytics comes to an end. We have tackled challenges, uncovered insights, and unlocked opportunities for the *Junglivet Whisky Company*, with you playing a central role in our success.

While you are reading this book, the demand for business data analytics experts is steadily increasing. In 2023, the World Economic Forum conducted an extensive investigation into the future of the workplace, highlighting the growing need for business data analysts and related professionals (World Economic Forum, 2023). They predict a 30% growth in the employment of (business) data analysts and scientists by 2027.

If you've found pleasure helping the different stakeholders at the *Junglivet Whisky Company* you might considering applying for a full-time business data analytics job in this positive market environment. If this is the case, this chapter is for you. By reading this book, you've already taken the first step. But if you take the tips and tricks in this chapter to heart, you will be well prepared to successfully advance your journey on this exciting career path.

In the following chapter, we'll delve into various facets of business data analytics careers, from roles and responsibilities to strategies for building a robust portfolio. We will also discuss navigating technical interviews, improving your business data analytics pitching and communication skills, and seizing opportunities to make a meaningful impact. So let us continue one last time to unravel the mysteries and possibilities that lie ahead as future business data analyst.

7.2 Start your Career as Business Data Analyst

When considering career paths in business data analytics, it becomes apparent that each individual's journey is unique. Some find their way into the role of business data analyst as career changers who have learnt the skills through practical experience. But typically, many university graduates, with a background in data-related studies such as information systems, business informatics, business administration, economics, statistics, or related disciplines, also join as entry-level business data analysts.

It's important to recognize that there is no one-size-fits-all approach for defining the "right" path in business data analytics. Instead, I encourage you to focus on what brings you joy and aligns with your current life circumstances. Whether you are driven by a passion for uncovering insights from data or motivated by the prospect of leveraging business data analytics to drive business growth, the key is to find opportunities that fit with your interests and goals.

For that purpose, I want to highlight the different job opportunities and roles, relevant companies, and possible career paths to guide you in your exploration in this field. In business data analytics, there is a diverse range of roles tailored to suit various skill sets and interests. Some individuals may thrive in roles such as senior business data analyst, where they focus on extracting insights from data and generating reports to support decision-making processes.

D. Jung, *The Modern Business Data Analyst*, https://doi.org/10.1007/978-3-031-59907-1_7

Others may find fulfillment in roles like data scientist, where they delve into advanced statistical analysis and machine learning to develop predictive models and algorithms.

Additionally, there are opportunities for specialization within specific industries or domains, such as finance, marketing, automotive or e-commerce. Within each industry, business data analysts may tackle unique challenges and work with domain-specific data sets to drive strategic initiatives and optimize business performance.

Moreover, the landscape of business data analytics is constantly evolving, with emerging technologies and methodologies shaping the development of the field. As such, there are opportunities for continuous learning and professional development, whether through further education, certifications, or hands-on experience with cutting-edge tools and techniques.

7.2.1 Related Job Roles

If you open your computer and enter the term "*Business Data Analytics*" in a job portal of your choice, you will probably be overwhelmed by a multitude of job descriptions and opportunities. Here is a list of possible job titles you might face that are related to business data analyst job profiles:

- (Business) data analyst
- (Business) data scientist
- (Business) data visualization expert
- Data architect
- Data engineer
- Data governance specialist
- Data steward
- Statistician
- Machine learning engineer
- …

All these job roles match different parts of the CRISP-DM process (Chapman et al., 1999) and business analytics project lifecycle you already know (see chapter 1). And if you start your career as a business data analyst, I promise you that you will face some of them in your everyday business. The bigger your employer will be, the more different roles and people will be in your projects.

Furthermore, you certainly already have likes and dislikes for the different sections you have seen in this book. And thus, it might be interesting for you to match typical job roles covering your favorite parts of the CRISP-DM as a specialization of your business data analyst profile. As I mentioned before, if you enjoyed the sections about investigating data and like to design sophisticated plots, then you can think about specializing in data visualization or alternatively the data visualization expert might be the job role of your choice. If you enjoy uncovering insights from data and exploring patterns and trends, a role as business data analysts with focus in statistical modelling may be a good fit for you. On the other hand, if you want to go further and leveraging advanced analytical techniques and machine learning algorithms to solve complex problems, consider a future career as a (business) data scientist or machine learning engineer. These roles will profit from a business data analyst background and involve developing predictive models, algorithms, and data-driven solutions to address business challenges, decisions and strategies.

As you explore different job opportunities, don't be afraid to tailor your applications and highlight relevant skills and experiences that demonstrate your fit for the role. For that purpose, let us now together demystify the job roles and titles and understand the concrete aspects behind the terms from the list above. As a starting point for this discussion, I designed an overarching plot matching the different job titles to the CRISP-DM framework in Figure 7-1.

As illustrated in the overview in Figure 7-1, over the whole lifecycle of a business data analytics project there are job roles covering the whole process, but also other job roles involved only in specific steps. All of them are typical roles that you will encounter in your projects, but also could be possible specializations or even future jobs for you based on your interest in the different topics and problems we faced in this book. Let us now have a short look at them.

The most generic roles in business data analytics projects are the **business data analyst** and **data scientist**. Both business data analysts and data scientists can play important roles in different phases of the CRISP-DM process, but their specific contributions may vary based on their skill sets and expertise.

While the business data analyst is a typical entrance job in the world of business analytics (see the following subsection 7.2.3 about career paths), business data scientists may have a data analyst background but gained significant modeling, coding and software engineering skills. Or as Hayashi et al. put it in a nutshell: you can think of a data scientist as a business data analyst with advanced computer science and statistical skills (Hayashi, 1998).

If you work as business data analyst together with a data scientist, you will divide the tasks across the project based on the individual skills and experiences. For instance, in the business understanding, the business data analysts will focus on gathering business requirements, define project goals, and identify key stakeholders, while the data scientists work closely with the business data analysts to understand the problem domain and align it with data science objectives and later focus on the modeling and statistical part of the project. However, the transition from business data analyst to data scientist is fluid and in small companies the business data scientists will do the job of business data analysts and vice-versa if they are not available.

After these two very generic job roles, let us now have a look at the specialized roles in a business analytics project.

While the business data analysts are mainly responsible for setting up and planning the project, sometimes it might be helpful to use the help of a **business data visualization expert** for your workshops and communication with the stakeholders. Business data visualization specialists create visual representations of data to facilitate understanding. They are relevant in the business understanding phase to communicate business insights and in the data understanding phase for exploring and presenting data. Sometimes they are also welcome partners in the deployment

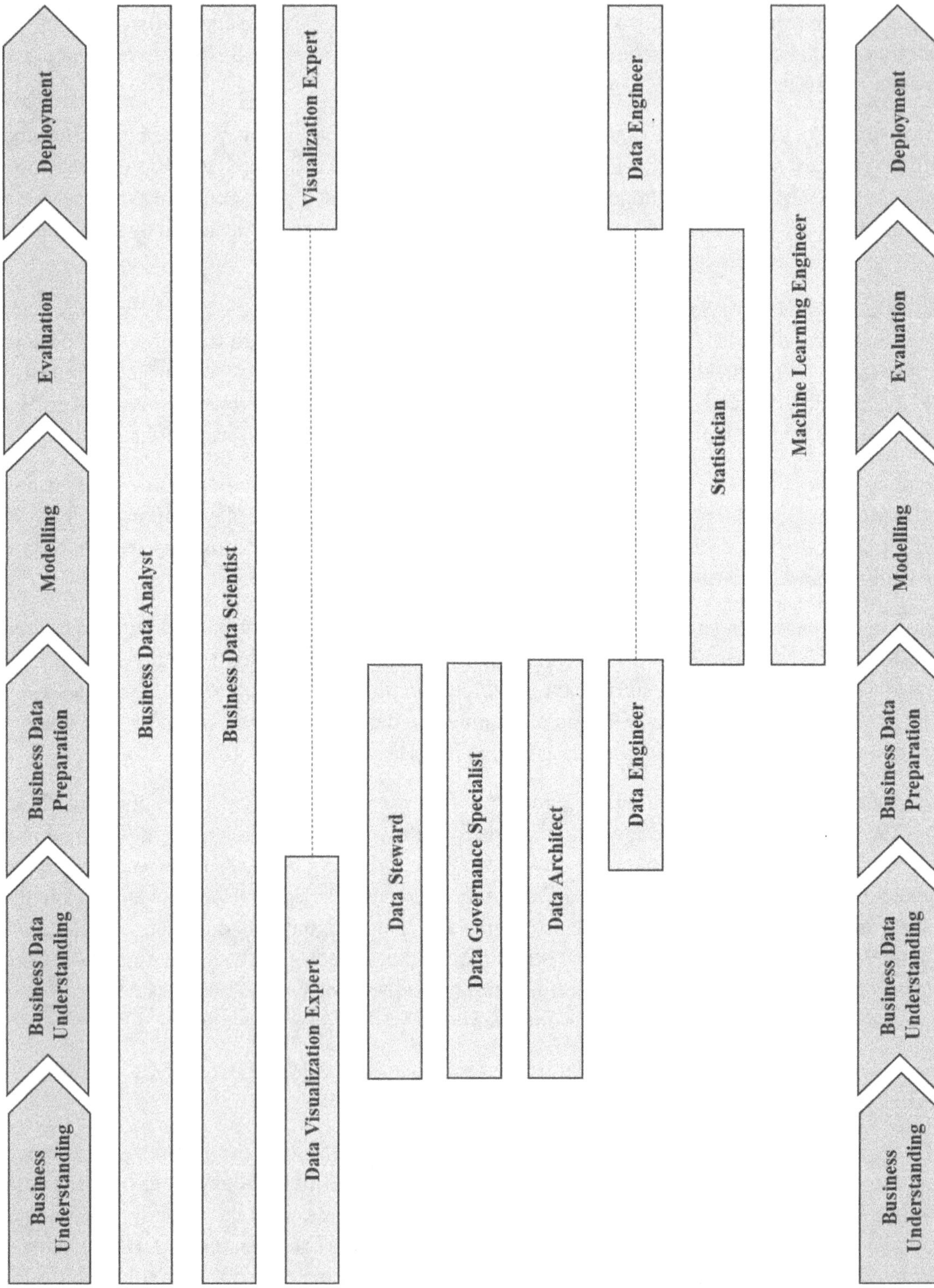

Figure 7-1 The different phases of the CRISP-DM framework and possible job roles.

phase if the business data analysts and data scientists prepare a management presentation for communicating results and findings.

In the next phase, the data understanding phase, the key players are the data steward, the data governance specialist, and the data architects. **Data stewards** are responsible for ensuring the quality and integrity of data. In the data understanding phase, they assist in identifying and profiling data sources. In the data preparation phase, they contribute to maintaining data quality and consistency.

In larger companies in particular, data is not simply freely available or accessible. This is where the **data governance specialist** comes into play. The data governance specialists focus on establishing and maintaining data governance policies and procedures. They play a crucial role in ensuring that data is managed effectively and complies with organizational standards.

After the feedback from the data steward and the data governance specialist you have to clarify the access and consumption of your data. **Data architects** design the structure and organization of your company's data systems. They contribute to the data understanding phase by identifying relevant data sources and in the data preparation phase by designing the data architecture.

Now that the most important questions of data understanding have been clarified, the data must be processed and prepared for analysis. In the data preparation phase business data analysts, data scientists and **data engineers** have to work together. The data engineer cleans, integrates, and transforms the data based on the feedback from data architects, data stewards and data governance specialists. The business data analysts ensures that the business data is in a suitable format for analysis, ensuring its quality and accessibility. While the data scientist collaborates with the data engineer to define the data preparation steps and validate their effectiveness by identifying patterns, relationships, and potential challenges. If the project is small or the data preprocessing is straight forward, this can be also done by a business data analyst or data scientist.

The next step is the modeling phase, where the data scientist selects and applies appropriate modeling techniques, building predictive or descriptive models based on the project objectives. Sometimes in big companies there are also one or multiple **statisticians** in the project. They use advanced statistical methods to analyze and interpret data. Together with data scientists and business data analysts they play a key role in the modeling phase by selecting appropriate statistical models and later in the evaluation phase by assessing model performance.

Furthermore, there might also be a **machine learning engineer** in the project. He or she implements and deploys the models, ensures scalability, efficiency, and accuracy if the final result has to be integrated in a bigger information system or the project is large, and many people are involved. If the modeling problem is out of the box, then these jobs are also done mostly by the business data analyst or data scientist.

In the evaluation, data scientists, business data analysts and domain experts work together. The analysts and the data scientists evaluate the model's performance, validate its effectiveness against predefined evaluation criteria in the business understanding. The domain experts are needed, because they provide domain-specific knowledge to assess the practicality and usefulness of the models' outcomes.

If the results are positive, there will be a deployment of a solution. In this phase data engineer and machine learning/software engineer must work together. The machine learning and/or sometimes even software engineers assist the business data analysts and data scientists in deploying the models, monitoring their performance, and ensuring proper maintenance. Machine learning specialists are mostly software developers and are essential in the modeling phase to integrate models into software applications. They also play a crucial role in deploying models into production environments, ensuring that they work seamlessly within existing systems. Together with the data engineer, they work on tasks such as data cleaning, transformation, and ensuring the scalability of data processing pipelines.

As we discussed before, sometimes data visualization experts are also relevant in the deployment phase, especially when the results of business data analytics models need to be presented to end-users in the form of a report or dashboard. They focus on creating user-friendly interfaces for visualizing and interacting with the data.

Finally, it's important to note that while all the mentioned job roles are commonly associated with specific steps, collaboration and cross-functional communication are crucial throughout the entire project. Effective teamwork and synergy between these roles contribute to the success of a business data analytics project. Also remember, these job roles may vary depending on the organization and project context, and individuals may have overlapping responsibilities (e.g., job focus of business data analysts vs. business data scientists).

7.2.2 Companies

There is no way to categorize companies offering jobs for business data analysts definitively, but as for software-engineering companies, they can generally be classified based on their industry and level of data maturity (Orosz, 2023).

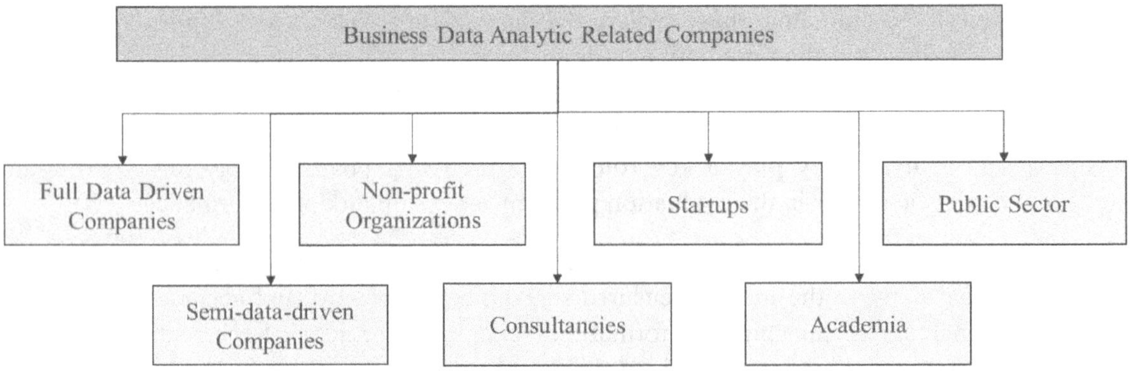

Figure 7-2 Business data analytics related companies.

Smaller companies like startups or academia may offer opportunities for hands-on experience and greater autonomy, while larger data-driven companies often have established business data analytics teams and infrastructure. Additionally, companies in data-intensive industries such as technology, finance, and e-commerce are more likely to prioritize data-driven decision-making which makes them a good fit for you, if you want to learn from more experienced business data analysts.

A **full data-driven company** is an organization that understand how to use data-driven insights and bases its decisions, strategies, and actions on business data analytics rather than on other factors (Henke & Jacques Bughin, 2016). Some data-driven companies have introduced data-driven business models that have taken entire industries by surprise.

Netflix as a popular example utilized business data analytics to make the successful "House of Cards" show by analyzing user viewing habits, preferences, and ratings to identify the element that would resonate with their audience (Carr, 2013). By leveraging this data-driven approach, Netflix was able to tailor the show's content, casting, and storyline to align with viewer interests, increasing the likelihood of its success and attracting subscribers to the platform.

Other full data-driven companies like Amazon and Google have revolutionized e-commerce and digital advertising respectively by leveraging data analytics to personalize recommendations for their users. Through business data analytics models analyzing user behavior, search queries, browsing history, and demographic information, these companies offer tailored product suggestions or ads and content recommendations.

If you look for data-driven companies, you will see that most of them are (big) tech companies. But there are also successful data-driven companies outside the tech sector. The German Allianz SE or the Chinese Ping An Insurance Group are some popular examples to name outside the US. Like the business models of banks are mainly data-driven and they rely on business data analytics in their key processes, e.g., to assess risk accurately, personalize services, and detect fraud. By leveraging data analytics, these companies optimized operational processes, enhanced productivity, and improved customer experience.

Tech and non-tech companies offer many interesting job opportunities for business data analysts because they recognize the critical role of (business) data analytics in driving their business success. You will have the opportunity to work with very talented coworkers and learn a lot. And you will have the chance to investigate vast amounts of business data from various sources, including customer interactions, market trends, and operational metrics. These company use state-of-the-art and advanced analytics platforms and techniques to extract actionable insights, identify opportunities for growth, and address business challenges which is also a unique opportunity for you as an entry-level business data analyst.

More traditional, **"semi-data-driven" companies** understand the need to transform to data-driven companies and have started to build business data analytics divisions or small teams to support the relevant business processes. These semi-data-driven companies, recognize the value of business data analytics in improving decision-making and gaining competitive advantages but may not fully integrate data-driven approaches across all aspects of their business.

Traditional examples are manufacturing companies such as General Electric, Toyota or VW Group are increasingly adopting data-driven approaches to enhance production processes, improve supply chain efficiency, and reduce operational costs. Similarly, big retail chains like Walmart in the US or the Schwarz Group in Germany have begun leveraging business data analytics to optimize inventory management, pricing strategies, and customer segmentation. A current example from Germany shows that this transformation must be taken seriously from the existing data-driven companies: Recently, the German Schwarz group, which owns the popular supermarket brands Lidl and Kaufland, has increasingly established itself as a serious player in the tech sector in Europe and competes with Amazon, Microsoft and SAP in the cloud and platform business in Europe.

Semi-data-driven companies typically have established business models that may rely on conventional strategies, industry experience, and intuition to make decisions. They have established dedicated business data analytics divisions or departments, tasked with collecting, analyzing, and interpreting data to provide insights and recommendations to other departments within the organization. While these teams may employ business data analysts related job roles to handle business data-related tasks, the integration of data-driven practices into the broader organizational culture and decision-making processes may still be evolving. And in the course of this natural development towards a "data-driven company", the example of the Schwarz Group shows that findings can be used very quickly to penetrate new markets and test new business models.

Consequently, semi-data-driven companies might offer interesting possibilities for business data analysts to contribute their expertise and drive the adoption of data-driven practices within the organization. In these companies, business data analysts play a crucial role in bridging the gap between traditional approaches and data-driven decision-making. By leveraging their analytical skills, domain knowledge, and understanding of data-driven methodologies, business data analysts can help these companies unlock the full potential of their data assets and drive strategic initiatives.

Startups offer business data analysts a dynamic and fast-paced work environment where they can make a significant impact and contribute to the company's growth from the ground up. In startups, business data analysts often have the opportunity to work on a wide range of projects, gain hands-on experience with cutting-edge technologies, and collaborate closely with cross-functional teams. Additionally, startups tend to have a culture of innovation and experimentation, providing business data analysts with opportunities for creativity, autonomy, and professional development.

Startups that provide equity to employees present both high-risk and high-reward opportunities. If the company succeeds and ultimately goes public or gets acquired, early employees who hold substantial amounts of stock can reap significant financial rewards. This setup incentivizes employees to contribute to the company's growth and success, as their personal financial outcomes are closely tied to the company's performance and eventual exit strategy.

Business data analytics plays a significant role in the **public sector**, contributing to better governance, improved service delivery, and enhanced decision-making. Across various domains, data analytics enables government agencies to leverage insights from large datasets to address societal challenges and meet the needs of citizens effectively.

One real-world example of the public sector using business data analytics is the COVID-19 Test and Trace program from the National Health Service (NHS) in the United Kingdom (Shelby et al., 2021). During the COVID-19 pandemic, the NHS implemented a comprehensive testing and contact tracing system to monitor and control the spread of the virus. The program utilized data analytics to track the spread of COVID-19, identify hotspots, and allocate testing resources efficiently. By analyzing various data sources, including test results, contact tracing information, and demographic data, the NHS could identify patterns and trends in virus transmission, assess the effectiveness of containment measures, and make data-driven decisions to prioritize testing and allocation of resources.

If you decide for a career in the public sector, you will have the opportunity to apply your skills and expertise to make a meaningful impact on society. Working in the public sector allows you to tackle complex problems, collaborate with diverse stakeholders, and drive positive change at a community or national level. Additionally, you would have the chance to work on projects that have a direct impact on public policy, social welfare, and the overall well-being of citizens, offering a rewarding and fulfilling career path.

Business data analytics plays a crucial role in **nonprofit organizations**, enabling them to operate more efficiently, effectively, and transparently in pursuit of their missions. Nonprofits rely on data analytics to inform decision-making, enhance donor engagement, measure impact, and optimize resource allocation.

Examples of non-profit organizations which offer interesting possibilities for you as a future business data analyst are the Wikimedia Foundation, World Health Organization (WHO) or the United Nations (UN).

One of the most popular show cases of business data analytics in the non-profit sector are the US presidential campaigns of Barack Obama in 2008 and 2012. Barack Obama and his team utilized business data analytics techniques to target voters effectively, mobilize supporters, and secure victory (Baumgartner et al., 2010). The campaign employed a data-driven strategy known as "microtargeting," which involved analyzing vast amounts of voter data to identify key demographics and tailor campaign messages accordingly. By leveraging business data analytics, the campaign was able to identify swing voters, prioritize resources, and craft personalized outreach efforts, such as door-to-door canvassing and targeted advertising.

Additionally, the campaign used predictive modeling to forecast voter behavior, optimize fundraising efforts, and allocate resources strategically across battleground states. Overall, Obama's successful use of data analytics revolutionized political campaigning and demonstrated the power of data-driven strategies in influencing electoral outcomes.

The presidential campaign of Barack Obama illustrates that non-profit organizations might be an interesting place for you to work as a business data analyst because you would have the opportunity to apply your analytical skills and expertise to support causes that align with your values and make a positive impact on society. Working in the nonprofit sector allows you to work closely with mission-driven organizations, collaborate with passionate individuals, and contribute to meaningful projects that address social, environmental, and humanitarian issues.

Business data analytics **consultancies** are professional service firms that specialize in providing expert guidance and solutions related to data analytics and business intelligence. These consultancies offer a wide range of services to businesses, organizations, and individuals seeking to leverage data-driven insights to improve decision-making, drive growth, and achieve their strategic objectives.

One of the primary roles of business data analytics consultancies is to help clients harness the power of data to gain actionable insights and make informed decisions. Consultants work closely with clients to understand their unique business challenges, objectives, and data-related needs. They then use advanced analytics tools and techniques to analyze large volumes of data, uncover hidden patterns, trends, and correlations, and derive meaningful insights that drive business value.

Examples of business data analytics consultancies are Accenture, Capgemini, or MHP.

As you will work on many different projects in many different companies due to your work in consulting, you will also gain a lot of valuable professional experience relatively quick. There is a well-known saying that one year in consulting is equivalent to three years of professional experience in another company. Whether this is true or not, working as a business data analytics consultant is certainly a good choice because it offers exposure to diverse industries, challenges, and methodologies.

Additionally, consultants often have the opportunity to work with cutting-edge technologies and collaborate with best-in-class coworkers from various backgrounds, fostering continuous learning and personal growth. Furthermore, the nature of consulting roles allows a high level of autonomy, responsibility, and impact, as consultants are often entrusted with driving critical projects and delivering tangible results for clients.

In data science and its little sibling business data analytics, there is a significant presence of individuals with doctoral degrees due to the complex and multidisciplinary nature of the subject matter. A doctoral degree provides you with in-depth knowledge and expertise in areas such as statistics, computer science, mathematics, and business, making you well-equipped to tackle sophisticated analytical challenges in the business world.

Consequently, **academia** can be a valuable avenue for gaining professional experience as a (business) data analyst. Academic institutions often engage in research projects and collaborations with industry partners, providing opportunities for analysts to work on real-world problems, access large datasets, and apply advanced analytics techniques in practical settings.

Additionally, academia offers a beneficial environment for continuous learning and skill development, with access to resources such as workshops, seminars, and conferences. Working in academia allows you as research data analysts to collaborate with leading experts in the field, contribute to cutting-edge research, and build a strong professional network, ultimately enhancing your credibility and marketability as future business data analysts.

7.2.3 Career Paths

Within a large company (compared to academia or non-profit organizations) the career paths of business data analysts are very straight forward. At the end, you have generally spoken three

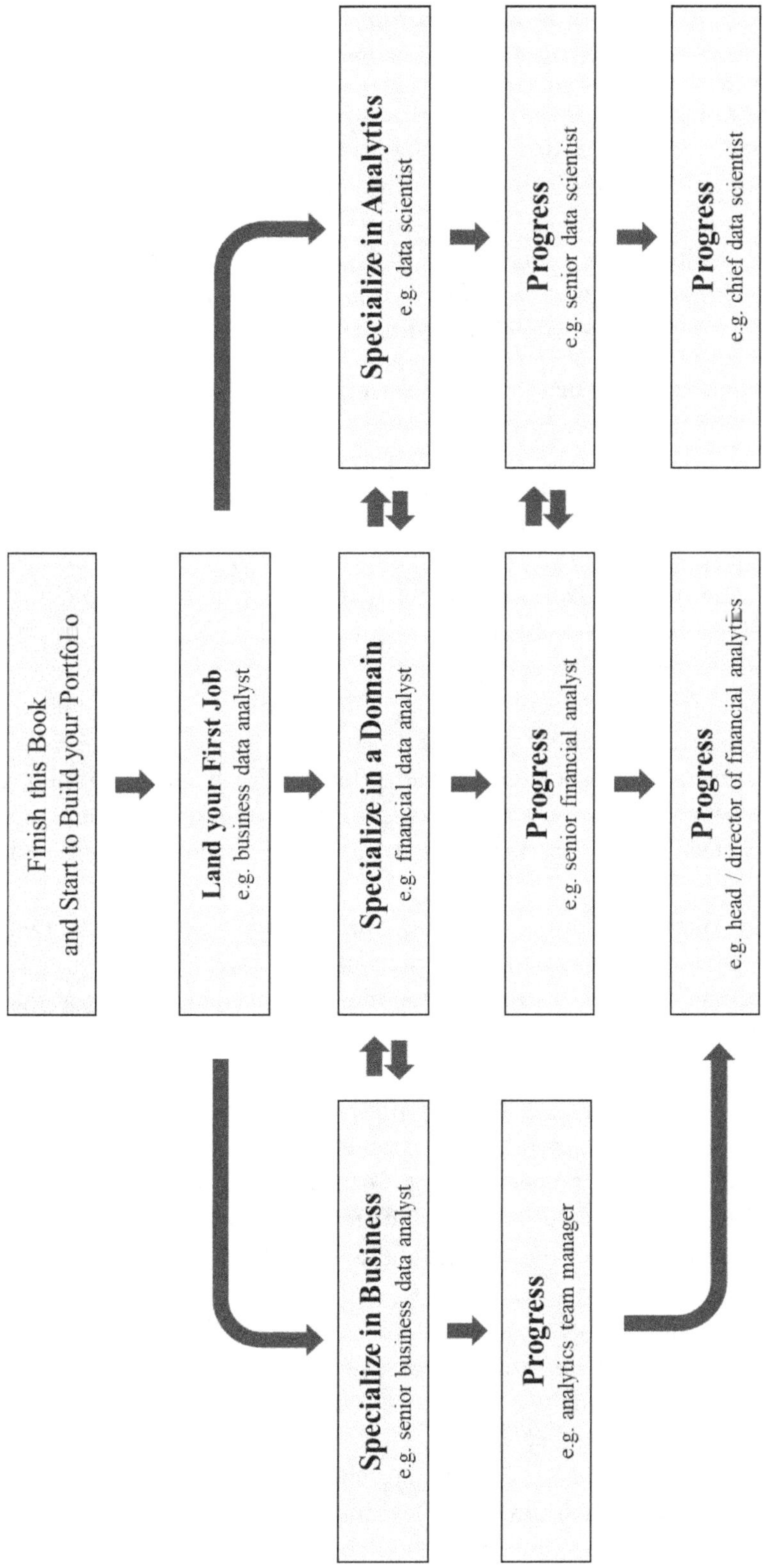

Figure 7-3 Different career paths in business data analytics.

options: specialize in your job **as business data analyst**, specialize in the **domain** of your company or specialize in **analytics and data science** in general.

Specializing in your role involves becoming a business data analytics expert in specific business data analytics tools, techniques, or methodologies, such as dashboarding, data visualization or predictive modeling. Focusing on the domain of your company, means you will focus gaining deep knowledge and understanding of the industry or sector in which your company operates, such as finance, healthcare, or retail. Finally, specializing in analytics in general involves broadening your skills and expertise across various domains (and industries), allowing you to work on diverse analytics projects and adapt to different business contexts.

Each path offers unique opportunities for professional growth and advancement within the field of business data analytics and the following new job roles:

- **Business data analyst** (Entry-Level): Many business data analysts start their careers in entry-level positions, such as business data analysts or data analyst. In these roles, they are responsible for collecting, analyzing, and interpreting business data to provide insights and support decision-making within their organization. Entry-level business data analysts typically focus on tasks such as data cleaning, data visualization, basic statistical analysis and reporting.
- **Senior business data analyst**: After gaining some experience in entry-level positions, business data analysts may progress to senior or lead analyst roles. In these positions, they take on more responsibility and often work on more complex projects. Senior business data analysts may be involved in designing and implementing data analytics solutions, leading project teams, and providing mentorship and guidance to junior analysts.
- **Specialist roles** (Domain): As business data analysts gain expertise within a particular field or domain, such as finance, they may transfer into specialized roles, such as financial data analysts. In these roles, they further enhance their analytical skills by focusing on domain-specific methods or techniques, such as time series analysis or econometrics.
- **Specialist roles** (Data Science): Alternatively, business data analysts can also expand their methodological expertise in computer science and machine learning or statistics and pursue a career as a data scientist. Data scientists not only analyze data as business data analyst but also experiment with different models and algorithms, and often build systems for data collection, processing, and deployment by themselves, which makes them a valuable expert (the Harvard Business Review titled that this makes them the sexiest job of the 21st century (Davenport & Patil, 2012)).
- **Managerial roles**: With further experience and demonstrated leadership capabilities, senior business data analysts, specialized data analysts or data scientists may advance into managerial or supervisory roles. In these positions, they are responsible for overseeing teams of analysts, managing projects and resources, and providing strategic direction for data analytics initiatives within their organization. Managerial roles require strong leadership, communication, and decision-making skills.

An important aspect of your career path is, of course, the salary and culture of the team or company you're considering. Therefore, I recommend investigating your potential employer on Glassdoor (*www.glassdoor.com*) and Levels.fyi (*www.levels.fyi*). Glassdoor provides information regarding company culture, employee reviews, and insights into the workplace environment. If you read the reviews of your potential employer, you can gain a first insight into the values, work-life balance, and overall employee satisfaction. On the other hand, Levels.fyi is an excellent resource to find average salaries for business data analyst positions. This platform offers detailed salary data based on many factors like experience, location, and company size, allowing you to benchmark your compensation expectations against industry standards.

In addition to hard facts such as salary and employee evaluations, there are of course numerous other factors that can have a significant impact on your career path. I'm thinking of factors such as location, work-life balance and work location (remote, hybrid or on-site). It's important not to overlook these factors when evaluating job offers. Prioritizing elements such as a supportive work environment, flexibility and alignment with personal values can ultimately contribute to long-term career satisfaction and fulfilment than a title or high pay.

7.3 Prepare for your Business Data Analyst Job

In business data analytics, technical job interviews are the gateway to start at big data-driven and market leading companies like Amazon, Google, Microsoft, BlackRock or data-driven consultants like Boston Consulting Group or McKinsey. To stand out as a candidate and secure an invitation for such interviews, a compelling portfolio of projects and showcases, coupled with strong academic credentials, is essential. Practical experience gained through internships, working student roles, or notable awards can further increase your success rates.

However, gaining an invitation is just the first step. The on-site job interviews pose additional challenges, requiring a mix of technical expertise, problem-solving prowess, and effective communication skills. These interviews typically incorporate case studies or problem-solving exercises to assess candidates' analytical abilities, alongside a barrage of technical questions. Additionally, expect inquiries about your practical experience and achievements.

In the subsequent section, we will delve deeper into these areas, providing insights on how to effectively prepare yourself to secure the job of your dreams.

7.3.1 Building a Portfolio

One strategy to build an individual portfolio that highlights your experience and skills as a business data analytics expert is establishing an online presence through a personal website or blog, where you can highlight your business data analytics projects, achievements, and capabilities. Alternatively, leveraging social networks or platforms like LinkedIn (*www.linkedin.com*) offer a valuable opportunity to build a comprehensive portfolio, allowing you to showcase your skills, accomplishments, and projects through detailed profiles, endorsements, recommendations, and original content.

Ensure that your online portfolio reflects your background, education, career goals, and contact information to maximize visibility to potential employers or collaborators. If you have noteworthy projects to showcase, provide links to your code, datasets, analysis results, and other relevant resources e.g., on GitHub (*www.github.com*). Additionally, consider enhancing your visibility and credibility by contributing blog articles to platforms like Medium (*www.medium.com*), R-bloggers (*www.r-bloggers.com*), or Kaggle (*www.kaggle.com*), or by publishing in scientific outlets such as conferences or journals.

If you have the time and skills, then expanding your personal website by producing video tutorials for platforms like YouTube is also a very good way to show your expertise. By diversifying your online presence and actively sharing your expertise in these platforms, you can effectively position yourself as a standout candidate.

7.3.2 Coding Challenges and Hackathons

Hackathons and coding challenges are a relatively new but increasingly popular form of solving a business case. Many of these competitions are announced online and have gained worldwide fame, such as Netflix's "One Million Dollar Challenge" which was an open competition to increase the user rating model for Netflix films (Bennett et al., 2007). In the meantime, there are also more and more competitions or live coding events where you compete on-site in front of others.

Meanwhile, many companies use hackathons to recruit new talent and generate new ideas. One of the best-known ones for students in Europe is the *Data Mining Cup*, in which I took part in a team as a student myself. In this competition, students have several weeks to develop a model for a real business problem with a real partner on anonymized data. There are even several cash prizes and internships to be won[2]. But many companies also organize such events themselves to develop talent and get people interested in exciting topics. During my time at industry, there was an annual competitive capstone event with my company and selected universities in which students were invited to participate and could win prices.

Here is a short but certainly incomplete list of popular business data analytics challenges and events that might be interesting:

- **Data-Mining Cup**: the competition for students has been organized annually since 2000 by GK Software SE from Chemnitz (Germany). Every year, mainly universities from the USA and Europe take part, but also some universities from the rest of the world. There is high prize money and internships to be won. More information at: *www.data-mining-cup.com*
- **European Big Data Hackathon**: every two years, the European Commission (Eurostat) organizes the European Big Data Hackathon, bringing teams together from all over Europe. They compete to find the best ways to solve challenging statistical problems. These teams come up with new and creative methods, applications, and data products by combining official statistics and big data. These solutions aim to help answer important questions related to EU policies and statistics: *www.cros-legacy.ec.europa.eu*

[2] I would like to point out that my classmates and I at university won second place and spent our prize money on a legendary dinner at an Italian restaurant in Karlsruhe (and donated the rest to charity).

- **Tianchi Big Data Science Competition**: this competition is primarily aimed at researchers or data science experts and is considered as one of the most important AI competitions in the world. It is organized by the Alibaba Group and aims to stimulate and support the development of big data science, artificial intelligence and machine learning: *www.tianchi.aliyun.com*

In addition to these events, there is also a very popular platform called Kaggle (*www.kaggle.com*). Kaggle is an online platform that hosts data science competitions, provides datasets for analysis, and offers a collaborative community for data scientists, machine learning practitioners and researchers. It was founded in 2010 and has become one of the largest and most popular platforms in the data science community.

On Kaggle, future business data analysts can participate in competitions organized by companies, organizations, or individuals, where they are given access to real datasets and challenged to develop the best predictive models or solutions. These competitions cover various fields, such as healthcare, finance, image recognition, natural language processing and more. Participants submit their models, which are scored against a predefined metric, and winners receive prizes and recognition.

In addition to the competitions, Kaggle offers an extensive collection of datasets for users to explore and analyze. These datasets come from a wide range of sources and cover a variety of topics. They serve as valuable resources for data analysis, developing machine learning models and conducting research.

Kaggle also provides a collaborative environment through forums where users can ask questions, share insights, and discuss topics related to data science. This encourages knowledge sharing and community learning and allows users to learn from experienced practitioners and engage in discussions with other data enthusiasts.

7.3.3 Technical Interview

In this part, let us have a short discussion about the technical interview and typical technical interview questions. For that purpose, I prepared 100 example questions you probably will have somehow to answer if you take part at a technical interview for a business data analytics position. The questions in this book focus on business data analytics in general with the purpose to assess your technical proficiency in R. You can find my selection of 100 technical interview questions I ask during technical job interviews in the appendix of this book or as flashcards on the course website.

If you want, you can also use the questions as a starting point to prepare yourself for the technical interview. For that purpose, I also made a digital flashcard deck containing all my technical interview questions, which you can use to recapitulate this book and prepare you for your first real interview (you can download it from the book website). I also recommend strongly that you create and add your own flashcards with topics and questions you have learnt in your business data analytics journey so far. Regularly testing your knowledge with flashcards will help improve your problem-solving abilities, analytical reasoning, and decision-making skills, ensuring you excel in technical interviews.

The free-to-use software Anki is my choice for practicing flashcards everywhere on my phone. Anki is a smart flashcard software that utilizes spaced repetition, a technique proven to enhance long-term memory retention. By creating personalized flashcards, you can focus on key concepts, formulas, algorithms, or statistical techniques that are frequently tested in business data analytics interviews. Anki's algorithm schedules flashcards based on your performance, ensuring that you review difficult concepts more frequently while reinforcing your understanding of previously learned material.

From my experience, interviews can be nerve-wracking, especially when facing unfamiliar or challenging questions. Learning these flashcards with Anki can boost your confidence simulating common technical interview scenarios. Regularly reviewing these flashcards will help you become more comfortable with the interview process, refine your responses, and develop a fluent and confident communication style.

7.3.4 Business Cases

This book is part of a lecture, and I always motivate my students to solve business cases in teams. As we have seen in Figure 7-1, real-world business data analytics projects often require working in teams and collaborating with stakeholders from different functional areas. This requires communication skills, the ability to work in diverse groups, and an understanding of how to effectively present and communicate analytical insights to non-technical audiences. These skills are invaluable for modern business data analysts as they often need to work with cross-functional teams and communicate their findings to stakeholders.

In the appendix I collected therefore business problems and case studies that I found interesting. The scenario of the *Junglivet Whisky Company* and the other data used in this case studies are complete fictional. However, they are based on real-world problems I encountered and I think they will help you to deepen your skills you learnt in the chapters before.

Explaining the business background and domain knowledge would be out of scope of this book. I simplified the cases as much as possible so you can solve them without domain knowledge. The goal is to show pitfalls and aspects of business data analytics problems without distracting you.

You can find the data of the following examples in the git repository accompanying this book, or in the R package `dstools`. Furthermore, I plan to publish further business cases on the course website in near future you can use to train your skills even further.

7.3.5 How to Continue

Besides building a strong portfolio, it is crucial to continually equip yourself with the right skills and knowledge in business data analytics. One way to go to expand your skills and learn state-of-the-art techniques and methods is through online courses, workshops, and certification programs offered by popular institutions and platforms such as:

- **Coursera**: offers a wide range of courses and specializations in data science, machine learning and data analytics, taught by industry experts and top universities, providing comprehensive learning resources and practical projects to enhance your skills. I made myself some courses and liked it: *www.coursera.com*

- **Udacity**: provides nanodegree programs, the most popular certified and structured courses at the moment. The content is designed to equip you with hands-on examples, personalized mentorship and feedback, and real-world projects for practical application: *www.udacity.com*
- **DataCamp**: specializes in interactive, hands-on learning courses, offering many tutorials, exercises, and projects that allow you to deepen your coding and analytics skills easily: *www.datacamp.com*

These resources provide opportunities to deepen your understanding of key concepts, learn new tools and techniques, and stay current with industry best practices.

Additionally, networking with peers, attending industry conferences, and participating in online communities and forums can offer valuable insights and opportunities for collaboration and knowledge sharing. Engaging in mentorship relationships with experienced professionals can also provide guidance and support as you navigate your career in business data analytics.

Since I can never cover the wealth of topics in business analytics in a single book, I would like to take this opportunity to point out some other really outstanding books that are well worth reading in the next step.

Here are some suggestions how to continue your career:

- **Berthold & Hand (2007)**: A detailed overview and guide to the analysis of data in general, with in-depth mathematical explanations and examples. Not for the business data analyst beginner, but a good read to follow up on after reading and understanding the concepts in this book.
- **Field et al., (2012)**: Has been one of my first books about R and is often recommended. Andy Field is a super approachable person and taught me an incredible amount about statistics with R. I highly recommend this book if you want to improve your statistics and modelling skills.
- **James et al. (2013)**: In this book we were unfortunately only able to cover the topic of business data analytics modelling very briefly. If you are looking for a good book that teaches you the theoretical background with in-depth mathematical explanations as well as applications in R, you should take a look at this bestseller. Gareth James et al. teaches you everything you need to know about modelling as a business data analyst and business data scientist.
- **Wickham et al. (2023)**: Hadley Wickham is (my personal) superstar of the current R analytics world. He has done incredible work, is the author of the most important packages, co-developer of RStudio and many other tools without which modern analytics with R would not be possible. His latest textbook on data science with R is therefore a must-read.

7.4 Business Data Analyst Best Practices

Getting business analytics projects and tasks done is far from trivial. In this section, I will show you some best-practices and practical strategies for effectively managing and executing business analytics projects. By mastering these techniques, you can enhance your productivity, drive project success, and deliver actionable insights to your stakeholders.

7.4.1 Getting Things Done

Well-known software engineer coach and author Gergerly Orosz[3] says that as an software-engineer, it is very important to know how to get things done (Orosz, 2023). Because most business data analyst will write a significant number lines of code during their lifetime, I promise you that Orosz's tips are also useful for you.

If you master this skill of "getting things done", you build a reputation for being able to do that, you are the person getting the exciting projects in the future. Which in turn gives you the potential to learn and grow. Exciting projects are also projects that have the potential to accelerate your career because they generate visibility at the management or are very well received in your individual portfolio if you are looking for examples in the technical interview.

In order to successfully implement projects, Gergerly Orosz suggests the following seven rules illustrated in Figure 7-4 to get things done (Orosz, 2023), which I adjusted for business data analysts and suggest you to consider.

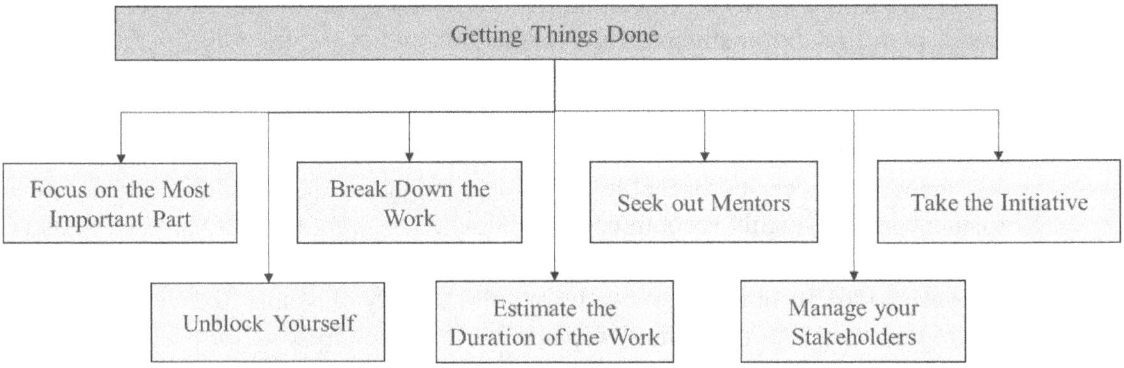

Figure 7-4 Different techniques to get your business data analytics project done, based on Gergely (2023)

Identifying the most important task at any point in the project is critical when analyzing business data, as projects often involve numerous data sources, analysis techniques and deliverables. For example, if you are tasked with improving marketing ROI, you may prioritize identifying key performance indicators and analyzing customer segmentation data before moving on to more detailed analysis. Throughout the project, the aim is always to focus on the most important task. Even if that means not completing other tasks, skipping meetings, or putting other things aside.

---

[3] Gergerly Orosz is a software engineer and author of the world-famous blog *pragmatic engineer*, which provides a look behind the scenes of the tech industry and offers many valuable tips and tricks. It's worth a visit: *www.blog.pragmaticengineer.com*.

In most projects you will encounter blockers such as missing data, technical limitations, or unclear project requirements. To unblock yourself, you must learn how to collaborate with the different roles we discussed in the previous section, or you have to say no to other tasks. For instance, if encountering difficulties in accessing a specific dataset, reaching out to the data engineering team for assistance can help resolve the issue efficiently.

Breaking down complex analyses into smaller, manageable tasks is essential for maintaining clarity and progress. For instance, when you are tasked with building a regression model for customer churn, you can use the CRISP-DM framework to break down the work into different steps like data preprocessing, feature engineering, model selection, and validation. Using frameworks and dividing your tasks into subtasks reduces complexity, enables collaboration, and facilitates tracking of progress at each stage.

Another very important skill is accurate time estimation. I promise you that one of the main reasons why analytics projects fail is because they break the timelines and overstretch resources. It is critical in business data analytics to ensure that there are presentable deliverables at a certain point in the project. For example, if you have difficulty estimating the duration of data cleansing and preparation tasks, the complexity of the data set can provide realistic estimates. Alternatively, you can be prepared for the fact that, as a rule of thumb, data preparation and data understanding will normally account for around 60-80% of the project duration and effort (Press, 2016).

From my time at university until today, mentors or more experienced experts have played an important role for me. Budding data analysts in particular benefit enormously from the collaboration and advice of more experienced analysts. So be sure to seek the advice and support of experienced colleagues. Alternatively, as mentioned in the previous section (*7.3.5 - How to Continue*), there are also very good specialized books by experts on a wide range of topics.

The next tip is a more generic one and aims to emphasize that building positive relationships and maintaining effective communication with stakeholders, colleagues, and team members from the kick-off until the end of the project is critical to project success. I have seen so many projects, where the project team forgot about stakeholder management and to communicate relevant information with the other stakeholders. Stakeholder management is essential to prevent misunderstandings and to manage expectations on both sides. Actively seeking feedback on the results, you are planning to present is important to make sure that you are on track. Nothing is worse than a meeting where you realize that you have been working around the topic or problem for weeks.

The most experienced business data analysts I met during my career were able to drive project success by proactively identifying opportunities for improvement or optimization within their scope of work. They were very fast in identifying inefficiencies in processes and workflows, and proposed automation solutions to streamline many complicated processes. Their ability to think critically and strategically allowed them to not only address immediate project challenges but also anticipate future needs and trends.

7.4.2 Cracking Problems

The cases discussed in this book cover many of the common problem characteristics and solutions that you will need in most of your business data analysis projects. However, as a newcomer to business data analytics, you will inevitably encounter new types of problems that I haven't been able to prepare you for in this book.

While the examples discussed in the case studies and exercises provide a solid foundation, real-world scenarios can often be diverse and unpredictable. To help you navigate these challenges, let's explore some general techniques and principles tailored to business data analytics (Spraul, 2013):

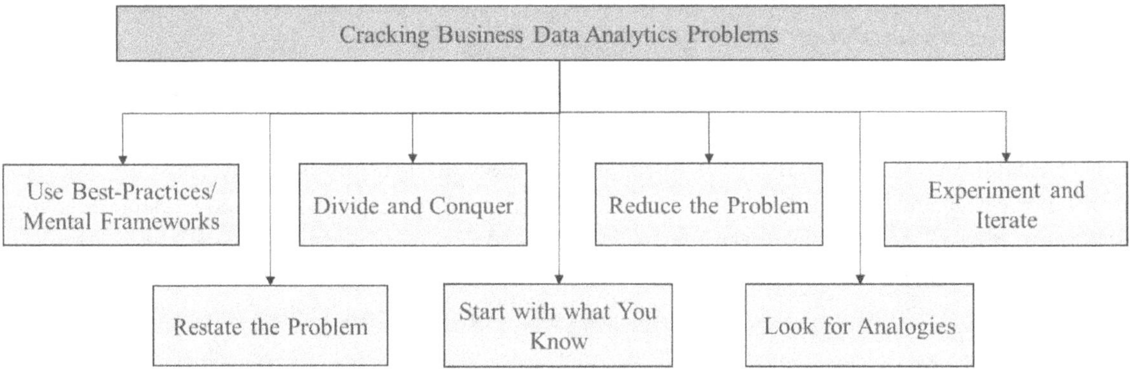

Figure 7-5 Best-practices to crack new types of business data analytics problems, based on Spraul (2013).

First, when facing a new problem, it's crucial to have a structured plan to guide your analysis. The CRISP-DM framework we used in this book is particularly useful for this. Even if the problem or business case you are facing, the framework encapsulates the knowledge from over 200 members of the CRISP-DM special interest group that build the guideline (Chapman et al., 1999). It is designed to help you if you are not familiar with the problem you are facing. Following the framework and activities in this book ensures that you address all necessary aspects of the problem systematically and efficiently. So, try to orientate yourself very strongly on this framework, especially with new problems, and you will see how many things will solve themselves.

Secondly, I always use abstraction as a technique if I face a new type of problem. For instance, you could abstract the problem of customer churn prediction, which is new for you, by drawing analogies between customer churn prediction and predicting the aging process of whisky. Just as certain flavor notes (e.g., hints of oak, honey) indicate the maturity of whisky, certain customer behaviors (e.g., decreased usage, decreased satisfaction) may indicate the likelihood of churn. By abstracting the problem in this way, I can simplify the complexity of customer churn prediction and focus on identifying actionable steps or insights in my mind.

The special thing about business data analytics is that it doesn't really matter what business you work in or what kind of business data you analyze. Predicting the number of whisky bottles sold based on orders from the last few quarters is no more or less difficult than predicting the number of fridges or hair ties sold. Let us assume the scenario from chapter 5 where the *Junglivet Whisky*

Company wants to improve customer segmentation for targeted marketing campaigns. As business data analyst you can use abstraction to break down the process of customer segmentation into simpler components. For example, focus on identifying common characteristics or patterns among customers in a first step to find relevant features and datasets, such as their purchasing history, frequency of visits, or product preferences. By abstracting the concept of customer segmentation into these fundamental elements, you can better understand the underlying factors driving customer behavior and tailor marketing strategies accordingly. And as you see, it doesn't matter what kind of product the customers are consuming!

It is also a good idea to always start with the aspects of a problem or task that you already know or are familiar with. The idea is to warm up to the problem cognitively and achieve as much as possible using what you already know. Identify and try any relevant data sources, information or techniques that you can apply. For instance, if you have experience with customer segmentation techniques, you might apply them to a new marketing campaign analysis project to identify target customer groups.

As discussed in the previous section, it is crucial to identify the most important aspect when solving problems. Try to simplify or reduce the complexity of the problem by focusing on the most critical aspects or variables. In business data analytics, it's often essential to prioritize actionable insights over exhaustive analysis. For example, when analyzing sales data, you might focus on key product categories or geographical regions rather than attempting to analyze every individual transaction.

Since it is basically not so important what type of data you analyze, it can be worthwhile to derive relationships or insights from other projects using analogies. Always try to find parallels to similar problems in your industry or domain. Drawing connections between different business functions or processes can provide valuable insights and guide your approach. For instance, if you're analyzing sales data for the *Junglivet Whisky Company*, you might find parallels in the marketing strategies used by other companies like luxury watch brands or sport cars. While the products may differ vastly, both industries rely heavily on brand image, customer perception, and targeted marketing to drive sales. By examining successful marketing tactics employed in the watch industry, such as influencer collaborations or experiential marketing events, you can gain inspiration for innovative marketing approaches within the whisky industry.

Don't hesitate to experiment with different analytical approaches and solutions. Iterate on your analysis, refining your methods based on feedback and insights gained along the way. In business data analytics, the iterative process is essential for uncovering hidden patterns, refining models, and validating assumptions. For example, if you're building a model to predict whisky sales, start by exploring various techniques we discussed in this book such as linear regression, decision trees, or neural networks. Test each model using different sets of features and parameters to see which ones has the most accurate predictions. Then, iterate on your models by incorporating new data sources or refining existing features to improve accuracy further. Additionally, consider using techniques such as A/B testing to evaluate the effectiveness of different business strategies or marketing campaigns. By comparing the performance of two or more variations in a controlled experiment, you can determine which approach is the most promising.

Remember that encountering new problems is a natural part of the learning process in business data analytics. Stay patient, persistent, and resilient in the face of challenges. Each problem you encounter presents an opportunity to expand your skills, deepen your understanding, and deliver value to your organization. Or as Hadley Wickham, chief scientist at RStudio, said in an interview (Waggoner, 2018): "It's easy when you start out programming to get really frustrated and think, oh it's me, I'm really stupid, or, I'm not made out to program. But that is absolutely not the case. Everyone gets frustrated. I still get frustrated occasionally when writing R code. It's just a natural part of programming. So, it happens to everyone and gets less and less over time. Don't blame yourself. Just take a break, do something fun, and then come back and try again later".

7.4.3 Pitching your Project

In business data analytics capstones, hackathons and case studies it is often required to pitch the results to a committee or jury. In practice, it is also common to pitch or present the results and the status of an important business data analytics project to the management on a regular basis. So, there are many reasons for considering what makes a good business data analytics pitch or presentation.

One reason why case study pitches are so challenging is that there is no right or wrong answer in most situations. Instead, case studies require you to create a hypothesis for an analytical question and then produce data to support or validate your hypothesis. In other words, it's not just about your technical skills, but also your creative problem solving and ability to communicate with stakeholders.

Consequently, pitching a business data analytics case study requires careful preparation and an effective presentation strategy. But do not worry! That's why we used business cases all across the book to help you to think this way from the beginning. Furthermore, here are some tips from my experience to help you pitch a business data analytics case study successfully:

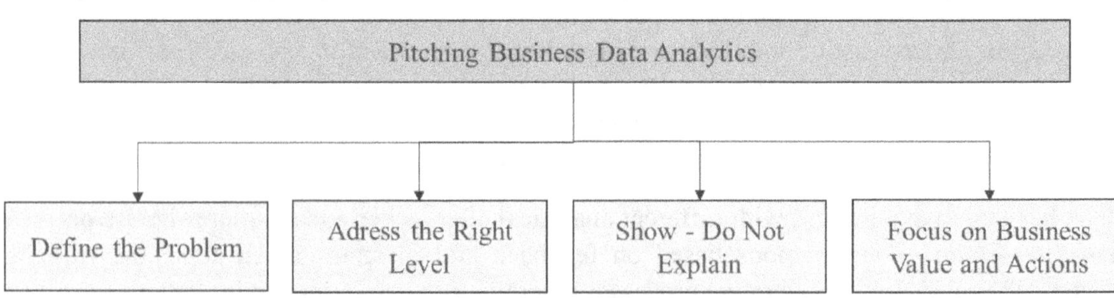

Figure 7-6 Key strategies to pitch a business data analytics project.

First, before you prepare your pitch, or your PowerPoint presentation define the problem and objective. And yes, this is exactly the point where the attentive students in my courses always say: "Thanks for this tip but shouldn't we have done this already in the business understanding phase?". Yes, normally, this should have been already done in the business understanding phase, but let me say you, sometimes the requirements change during the project, after your first findings or the first evaluation cycle.

You should also clearly articulate the problem or challenge addressed in the case study in your presentation. Explain why it is important and how data analytics can provide insights or solutions. State the objective of the case study, such as improving operational efficiency, optimizing marketing campaigns, or enhancing customer experience. And finally, make clear which metric or KPI is used to measure the success of the project. In my life, I have often judged the performance of many student teams as a judge at hackathons or capstones and many students with good modeling results still don't do well at the end presentation because they don't do exactly that. It is elementary not only to solve a problem but also to clearly emphasize the problem and its importance for the business. If you do not know how to do this, I recommend that you have a look at the business data analytics project slide deck I prepared for you on the book website.

Second, present the methodology considering the level of technical knowledge of your audience. If a presentation is very technical and detailed, then it may be perfect for colleagues in the analytics team, but not for managers who must make a lot of very complicated decisions every day and need a different abstraction level. Describe the data analytics methodology used in the case study in the wording of your audience, e.g., the business owner or management. For that purpose, use a clear and concise communication style. Focus on the interesting parts and not on the procedure (everyone knows that you worked hard, you must now show that you are smart and can generate business value). Explain the data collection process if it is relevant or contributes to your message, otherwise skip it. The same holds for data cleaning and preprocessing techniques, and the analytical methods applied. Highlight any innovative or advanced techniques employed to solve the problem or interesting finding or implication for the business but do not get too technical. The management trust you that you have the technical knowledge, but they want to understand and assess the business impact and implications.

Third, show do not explain! Many presentations by inexperienced business data analysts consist of very extensive slides with lots of bullet points. They often explain to me in detail what they have done and read the key points out loud from the slides. Please do not do this. Present your findings and outcomes of the case study with a compelling and engaging presentation. Use visualizations, charts, and graphs to illustrate the insights gained from the data analysis and not long lists of bullet points. If you lack information, then make realistic assumptions. This also shows your understanding of the problem, the company, or the business. I also recommend that you try to quantify the impact of the analytics solution in terms of key metrics, such as cost savings, revenue growth, or improved customer satisfaction. Demonstrate confidence in your findings and be prepared to answer questions or address concerns from the audience.

Finally, always highlight the business value and give actionable recommendations. Many presentations end suddenly and leave the audience with a so-what attitude. It is often completely unclear how to proceed after the project or case and what managers or those affected should do now. It is your job as a business data analyst to make suggestions based on your analysis. During the presentation emphasize the business value and impact of the case study. Demonstrate how the data analytics solution positively influenced decision-making, strategic planning, or operational efficiency. Connect the findings to specific business outcomes and demonstrate the return on investment or other relevant benefits. And finally, based on the findings, offer actionable recommendations for the audience. For instance, you can highlight opportunities for improvement or future areas of focus that can benefit from business data analytics or even suggest further business data analytics projects, questions, or objectives. Show that the case study is not just a retrospective analysis but also a valuable source of insights for future decision-making.

7.5 Some Last Words

Thank you for reading this book and joining me on my journey through the field of business data analytics. I hope this book has been a valuable resource for you to learn about R and analytics in an enjoyable and understandable way. And that it might have made you smile at one point or another.

If you like, I invite you to connect with me on LinkedIn to stay in touch (*www.linkedin.com/in/dominikjung42*), share insights and keep up to date with the ever-evolving field of analytics.

Should you have any comments, find any errors or inaccuracies: Your engagement and feedback mean a lot to me. Please feel free to write to me!

I wish you continued success and exciting discoveries in your data-driven endeavors.

8 Appendix

8.1 100 Technical Interview Questions

In the following, you find my favorite 100 questions for you, you can use to test your business data analytics knowledge. As I mentioned before, they are also available as flashcards on the book website.

8.1.1 Business Understanding (15)

1. What are the five relevant aspects of a business data analysts' mindset?
A business data analyst needs curiosity and inquisitiveness while exploring business data, a natural problem-solving mindset while breaking down problems into manageable parts, must see himself as an advocate for data-driven decision-making, be-detail oriented to ensure accuracy and needs to understand the need for continuous and customer-centric communication.

2. What is the SEMMA framework?
The framework is an alternative to the CRISP-DM framework and organizes business data analytics in the phases sampling, exploring, modifying, modeling, and assessment

3. Please name two further business data analytics frameworks besides CRISP-DM!
SEMMA, Microsoft TDSP, DASC-PM

4. What does the abbreviation CRISP-DM mean?
Cross-industry standard process for data mining

5. How are business data analytics projects organized according to CRISP-DM?
The CRISP-DM framework divides analytic projects into six phases that are related to each other. The six phases are business understanding, business data understanding, business data preparation, business data analytics, evaluation and deployment.

6. Please name a common trade-off when deciding for an analytics solution!
Performance/complexity vs. understandability

7. What are the relevant activities in the business understanding phase?
Determine business objective, assess the situation, determine project and analytics goals, and project planning.

8. How do business data analysts determine business objectives?
They gather background information about the business and business problem, define primary or business objectives and business success criteria.

9. How do business data analysts assess the situation in the business understanding phase?
Make an inventory of resources, gather requirements, assumptions and constrains, evaluate risk and contingencies, and conduct a cost and benefit analysis.

10. What are possible methods to determine business objectives and assess the situation in business understanding?
Cognitive maps, glossaries, documentations, user stories and mockups

D. Jung, *The Modern Business Data Analyst*, https://doi.org/10.1007/978-3-031-59907-1_8

11. What should you keep in mind when inviting participants for a cognitive map workshop?
Gather a diverse group of participants of the business domain where you plan to do your project. It is important that they have relevant knowledge and expertise related to the chosen topic. Do not invite people just for fun or because they like the cookies you offer in the workshop and ask you if they can join. Including the wrong individuals is not helpful, but including the right individuals from different backgrounds can offer unique perspectives and enrich the cognitive map.

12. What is the format of a user story? Give an example!
As a [user role], I want [goal] so that [benefit].

13. What are mockups?
Mockups are a visual representations or prototypes of a product, often used in the early stages of design to demonstrate its layout, structure, and user interface.

14. Please give a possible business analytics objective for the business objective "Enhance online sales to current customers"!
The objective could be "Forecast the number of whiskies a customer will purchase, considering their past three years of buying history, demographic details (age, salary, city, etc.), and item price.".

15. What are the minimal requirements for a business data analytics project plan?
A typical project plan in business analytics consists at least of a written definition of the business objectives, a short project roadmap and a list with the project team or stakeholder.

8.1.2 Business Data Analytics Toolbox (17)

16. What do you understand by a workspace in R programming language?
The current R working environment of a user that has user defined objects like variables or functions is referred to as workspace in R language.

17. How do you recognize R script files?
They end with ".R"

18. What is RStudio?
It is an integrated development environment (IDE) to write R code.

19. What is the output of the following code?
```
> string = "Hello World!"
> nchar(string)
[1] 8
```

20. How can you check if a variable is numeric?
With `is.numeric()`

21. How can you cast a variable as integer?
With `as.integer()`

22. How can you delete an object or variable?
You have to set it `NULL`, *e.g.* `my_vector = NULL`

23. How are missing values represented in R?
`NA` (Not Available) is used to represent missing values.

24. Please name a 2-dimensional data structure in R!
Matrix or data frame

25. How would you import an excel file named "productionlog_sample.xlsx" into R?
You can run the following command:
```
dataset = read_excel("productionlog_sample.xlsx")
```
Or use the RStudio Import function.

26. How do you write comments in R?
With the `#`*-command, e.g.*
```
# test123
```

27. How can you print the names of a data frame named `df`?
```
names(df)
```

28. Write a for loop to print the numbers 1 to 3!
```
for(i in 1:3){
  print(i)
}
```

29. What is CRAN?
R's central software repository, supported by the R Foundation

30. How can you install the package tidyverse?
```
install.packages("tidyverse")
```

31. What are R-projects? What are the benefits to work with projects?
R projects provide a separate directory, package library, and workspace for each project, making it easier to keep work organized, isolate project-specific settings, and ensure reproducibility.

32. Where would you look for information if you encountered a problem while programming?
Google, Stackoverflow, github, Chatgpt

8.1.3 Business Data Understanding (17)

33. What are the different steps of the business data understanding phase?
Collect initial business data, describe business data, explore business data, verify business data quality.

34. What is a common package for business data manipulation with R?
dplyr

35. What is the split-apply-combine paradigm? Please explain it!
The split-apply-combine paradigm is like making a sandwich. First, you split your ingredients into groups like bread, cheese, and veggies. Then, you apply an action to each group, like spreading mayo on the bread. Finally, you combine the groups back together to make the

finished sandwich. In data analysis, it means dividing data into groups, performing operations on each group, and then putting the results back together.

36. How can you pick rows in a data frame with `dplyr` based on their values?
With filter()

37. How can you count the distinct elements in a group?
n()

38. Please name the three elements of a plot with `ggplot2`!
data, geometric objects, aesthetic mappings

39. How can you create separate subplots of your data?
With facets

40. What does the line in the middle of a boxplot stand for?
The middle line of the boxplot is the median of your dataset.

41. What is a Sankey diagram?
A Sankey diagram is like a flowchart that shows how things move from one place to another. It's often used to visualize the flow of energy, money, or materials in a system. The width of the arrows in the diagram represents the amount being transferred, making it easy to see where most of the flow is going.

42. What are dummy variables?
Dummy (or binary) variables are commonly used in analysis and statistics. A dummy column is a column that has a value of `TRUE` *or* `1` *if a categorical event or value occurs and a value of* `FALSE` *or* `0` *if it does not. In most cases, a dummy variable is a characteristic of the event or object being described. For example, if the dummy variable was for being a Scottish whisky the dummy variable gets a value of* `TRUE` *if it is a Scottish whisky, if it is not, it gets a value of* `FALSE`*.*

43. How can you export your visualizations in R?
With the wizard in RStudio or the command `ggsave()`

44. How can you check the frequency distribution of a categorical variable?
You can make a density plot or use the `table()` *function*

45. Please name some commands to compute descriptive statistics in R!
`min()`, `max()`, `quantile()`, `mean()`, `median()`, `sd()`, or `range()`

46. How can you find the unique values of a variable in R?
`unique()`

47. How can you compute summary statistics?
`summary()`

48. What is the idea of the GIGO principle in business data analytics?
The GIGO principle in business data analytics means "Garbage In, Garbage Out." It reminds us that if we put low-quality or incorrect data into our analysis, we'll get unreliable or incorrect results. In other words, the quality of the output depends on the quality of the input data.

49. How can you show the last rows of a dataset?
`tail()`

8.1.4 Business Data Preparation (17)

50. In which phase of the CRISP-DM framework do you clean the business data?
Business data preparation

51. What do you understand under data integration?
All related activities to merge data from different sources or include external datasets.

52. Can you name the most relevant business data sources in business data analytics?
Flat files, web data, APIs, databases

53. Why are CSV files no more used in modern business data analytic projects?
CSV files are used less and less in modern projects. The reason is that the comma (",") separator is not very usable because the comma has become a very common character often used in the business data value itself.

54. What is an API in the context of business data preparation?
API stands for application programming interface and means that the data you want to access is provided directly for usage by a data provider.

55. How can you fix a corrupted value?
Ignore them or replace the corrupted value with `NA` *or even a meaningful value like* `0` *or apply methods you use to handling missing values.*

56. How can you remove all rows that contain `NA` values?
`na.omit()`

57. How can you check if an element is present in a vector?
Use `%in%` *which returns a Boolean value if the element is in the vector.*

58. What are the most popular approaches to handle different scales of your variables?
Min-max normalization, z-score standardization

59. What does the curse of dimensionality mean?
If the number of features is higher than the number of rows, many algorithms will fail to produce reliable results. This problem is called the curse of dimensionality.

60. Why can we use entropy to find "good" features?
Entropy measures the level of disorder or randomness in the values of a feature. High entropy indicates greater disorder, while low entropy signifies more order and predictability. Or in other words high entropy in features is associated with high variance and thus we can assume that the feature contains a lot of information.

61. What is PCA, and why do you need it in business data analytics?
Principal component analysis (PCA) is used to reduce the number of features in datasets by reducing the number of features while computing new features preserving the essential information of the original features.

62. How can you read all xlsx files in a directory?
With the `read_dir()` *function from the* `dstools` package.

63. Which function is used to save an object in the `.RData` format?
`save()`

64. How can you list all files in your working directory?
`list.files()`

65. Which functions are used to merge different strings into one single string?
`Paste()` *or paste0()*

66. What is a popular issue when working with big numbers in R? How can you avoid it?
The numbers are saved in scientific notation. For instance, a number like 12345678987654321, becomes this in R: `1.23456e+16`. *If you want to avoid this you can turn off scientific notation for numbers using* `options(scipen = 999)`.

8.1.5 Modeling (17)

67. What are the relevant steps in the modeling phase?
Select modeling techniques, generate test design, build model, assess model.

68. What are common modeling techniques in business data analytics if you want to predict categorial variables? Name two.
Decision trees, logistic regression

69. Give an example business problem or question which is typical for association analysis!
Which products in my online shop are normally bought together?

70. Which modeling technique is suitable if you want to predict which products are normally bought together in a supermarket?
Association analysis

71. What is the difference between a training and a test set?
A training set is used to train your model. It's the portion of data used to teach the model patterns and relationships within the data. A test set, on the other hand, is a separate portion of data that is not used during training. It's reserved for evaluating the model's performance and assessing how well it can make predictions on unseen data. Essentially, the training set helps build the model, while the test set helps assess its accuracy and generalization to new, unseen data.

72. Which modeling technique and function is used to find out whether the means of two groups are equal to each other or not?
T-tests and it's implemented in the `t.tests()` *function*

73. What is the basic idea behind an A/B testing?
The basic idea of A/B testing is to systematically prepare a data sample measuring the outcome of two competing business decisions and test these two different alternatives against each other.

74. What is a very common error in cluster analysis?
A very common error in clustering is that business data analysts forget to scale their data.

75. What does the `k` in "k-nearest-neighbor" stand for?
The letter "k" refers to the number of nearest neighbors considered in the algorithm. When a new data point (e.g. customer) is presented, the k-NN algorithm identifies the k closest customers in the feature space and classifies the new point based on the majority class among these neighbors.

76. What is the difference between Type 1 Error and Type 2 Error in classification tasks?
In classification tasks, Type 1 Error and Type 2 Error are two different types of mistakes made by a classification model. Type 1 Error would be classifying a customer of the Junglivet online shop as a high-value customer (e.g., offering them premium services) when they're not, which could lead to unnecessary costs. Type 2 Error would be failing to identify a high-value customer when they actually are one, missing out on the opportunity to provide them with targeted, loyalty-enhancing offers.

77. What is a good error measure for classification tasks. Name one and explain it.
Accuracy, measures how many predictions the model got correct out of all predictions. It's the mean error rate across all observations calculated as (true positives + true negatives) / (true positives + true negatives + false positives + false negatives).

78. What is the difference between correlation and causation. Why does the first not imply the other one?
Correlation is a statistical relationship between variables but doesn't imply causation, which means that a change in one variable directly causes a change in another. For instance, if sales and profits of the Junglivet Whisky Company are correlated, it doesn't mean that increasing sales will automatically increase profits. Other factors may be involved.

79. What is a decision tree? How do you interpret it?
You can imagine a decision tree as a flowchart that helps you make decisions by following the path that best fits your data. It has a tree-like structure and consists of nodes, branches, and leaves. Each node represents a decision or a test on an attribute, each branch represents an outcome of that test, and each leaf node represents a class label. To interpret a decision tree, you start at the root node and follow the branches based on the conditions tested at each node until you reach a leaf node, which provides the classification or decision.

80. How will you measure the probability of a binary response variable in R language?
Logistic regression can be used for this and the function `glm()` *provides this functionality.*

81. What is the difference of time series and regression analysis in business data understanding? When do you use which method?
In general, you would use time series forecasting when you are interested in making predictions about future values of a time series, and regression modeling when you are interested in understanding the relationship between variables.

82. What are the three conceptual components of the Prophet algorithm?
The Prophet algorithm uses a decomposable time-series model with three main model components: Trend, Seasonality and Holidays.

83. What is the use of the `set.seed()` function?
The `set.seed()` *function is used by business data analysts to set a seed for the random number generator. This means that the function will allow business data analysts to reproduce the same results as observed by someone who uses the same R scripts.*

8.1.6 Business Data Products (17)

84. What are the three steps of the evaluation phase? Please explain them briefly!

- *Evaluate results: Assess your outputs with respect to the business success criteria you defined in the beginning.*
- *Review process: Review the project and process*
- *Determine next steps: Make a list of possible actions and decisions for or against an analytics solution deployment*

85. What are the three relevant types of data products in business data analytics?
Insights, reports and systems

86. Please name some reporting data products in business data analytics
Reports, posters, notebooks and dashboards

87. What is the `officer` package? For what do you use it?
The `officer` *package in R is a package used for creating and manipulating Microsoft Office documents. It provides functions and tools to generate and format files from R, making it particularly useful for creating reports, documents, or presentations that require integration with Microsoft Office.*

88. What is a notebook in business data analytics?
Interactive notebooks in R refer to documents that combine code, text, and visualizations in an interactive environment.

89. Given the following scenario: The marketing team wants to monitor the sales in the online shop of the Junglivet whisky company. How would you build a dashboard for them. What are the most relevant components?
To build a dashboard for the marketing team to monitor sales at the Junglivet whisky company, you can use the shiny or flexdashboard package. Here are some possible components:

- *Sales Trends: Include line charts or area graphs that show sales trends over time. You can break this down by day, week, month, or other relevant time intervals.*

- *Product Performance: Use bar charts or tables to compare sales of different whisky products. Include metrics like top-selling products, products with declining sales, and products with the highest growth.*
- *Customer Segmentation: Analyze customer demographics and preferences. Create segments such as new customers, loyal customers, or high-value customers, and show how they contribute to sales.*
- *Conversion Rate: Monitor the conversion rate for the online shop, tracking the percentage of visitors who make a purchase.*
- *Marketing Campaign Performance: Include a section that tracks the performance of various marketing campaigns. This can help the marketing team understand which campaigns are driving sales.*

90. How can you implement an interactive plot in shiny?
In Shiny, you can implement an interactive plot using the `renderPlot()` *function. You specify a reactive expression that generates the plot and then you use* `plotOutput` *in the UI to display the plot.*

91. Please name some popular business analytics systems!
Decision support systems, recommender systems, APIs, or pipelines

92. Please give a short draft of the file structure in shiny projects.

```
name_of_your_project
├── app.R
├── data
│   └── e.g. onlineshop.csv
└── www
    ├── e.g. junglivet.jpg
    └── e.g. helper.R
```

It is best practice to split your shiny project in two folders. The `data` *folder is used to store your data that is used by the app. This is necessary if you do not work with live data connections. And the* `www` *folder contains all files that are otherwise used by the app like images (*`junglivet.jpg`*) or other R functions (*`helper.R`*).*

93. What is the difference between user-based collaborative filtering and item-based collaborative filtering?
User-based collaborative filtering recommends items based on the behavior of users who are similar to the target user. It identifies users with similar preferences and suggests items they have liked. In contrast, item-based collaborative filtering recommends items by comparing the similarity between items, suggesting products similar to those the user has interacted with. The choice between these methods depends on the nature of the available data and the specific requirements of the recommendation system.

94. Why are APIs used in business data analytics?
APIs in business data analytics serve as a bridge for connecting different systems, allowing them to interact and share insights or data efficiently. For instance, you can provide your model as a service for other systems in your company without integrating it in every other system.

95. What is RStudio Connect?
RStudio Connect is a publishing platform for data science content created in R and Python. It enables users to securely deploy and share Shiny applications, R Markdown reports, Plumber APIs, and Jupyter Notebooks.

96. Have you heart of shinyapps.io? What is it good for?
Shinyapps.io is a cloud-based platform provided by RStudio specifically designed for hosting and deploying Shiny applications.

97. Please name other possibilities to host your own shiny app besides shinyapps.io!
For instance, you can host your app on your own shiny server. RStudio offers this product for hosting and managing applications on your own servers. Or you can containerize your app using Docker and deploy it to various cloud platforms like Amazon Web Services or your own server.

98. What is the `cloudyr` project?
This project is a collection of different initiatives and packages *to make facilitate cloud computing with R. It contains many tools and functions to handle the most popular cloud computing platforms like Amazon Web Services, Google Cloud Services or Microsoft Azure.*

99. What is the last phase of the CRISP-DM framework?
Deployment

100. What are the relevant steps in the deployment phase?
Plan deployment, plan monitoring and maintenance, produce final report and review project.

8.2 Business Case Study 1: Analyzing Transactions in the Junglivet Online Shop

8.2.1 Scenario

You are on your way to your office when you see Ted Gumble standing in front of a large quantity of whisky at the warehouse. He swears and shoos his colleagues Carl and Alice across the yard while cursing under his breath: "Bloody hell, why are they sending us their whiskies back now of all times?" You hesitate for a moment and then turn to Ted and ask him what it's all about. He explains that the online store is seeing an increase in returns and that he has already asked Lindsey Maegle several times to get to the bottom of it. You offer to take a look at the sales in the online store and help him get the situation under control.

After taking a quick look at the data and having a further exchange with Ted, the problem (business understanding) has become relatively clear to you. You are convinced that there are two things to do here. Firstly, a model for predicting returns could certainly help to recognize returns in advance in order to initiate countermeasures. For example, you could have the order confirmed again separately or offer different shipping costs. Secondly, a short report would be useful to understand whether certain products or customer types are frequently returning goods. Cheerfully you get straight to work.

8.2.2 Task

Your task is to analyze the `onlineshop` dataset from the `dstools` package in context of a CRISP-DM project. In particular, you want to understand and model the returns of the customer. For that purpose, implement the following tasks:

- Make some descriptive statistics to understand the dataset and the target variable. Generate visualizations you can use in the final management pitch.
- Derive at least five hypotheses to explain the return shipments. Use them later to derive action items.
- Build multiple models to predict if the transaction is a return or not. Use multiple evaluation criteria and discuss the results with your team mates.
- Design a final report to pitch your result at the Junglivet management. The report should investigate the reasons for returns, the specific customer types, the returns in general and possible action items to reduce the return shipments.

8.2.3 Remarks

Use the `onlineshop` dataset for this case study.

8.3 Business Case Study 2: Developing a Car Maintenance System

8.3.1 Scenario

Your long experience as business analytics expert has led you to a new project in collaboration with the aftersales department of a leading automotive giant, Luxe Motors. Leveraging their extensive reservoir of vehicle service history and diagnostic data, the aftersales team is eager to test a predictive analytics model aimed at foreseeing upcoming workshop visits for car maintenance.

After the first business understanding workshop, the team has provided you with a full sample of a vehicle analysis log file encapsulating diverse facets of customer vehicle data. This kind of dataset contains many critical parameters, including but not limited to car mileage, service history, engine diagnostics, previous maintenance procedures, and component replacements.

In this project you set yourself the objective to build a predictive model to predict if a customer's vehicle is likely to necessitate another workshop visit within a few weeks following their previous maintenance appointment. The model's successful implementation promises to revolutionize Luxe Motors' aftersales service by significantly curtailing unwarranted workshop visits and notably enhancing overall service quality. Moreover, it aims to empower Luxe Motors' customers with proactive and efficient vehicle maintenance. Furthermore, you will have to setup a reporting solution visualizing the vehicle analysis logfiles for the aftersales team. This enables the team to better recognize quality problems in the long term and rectify them promptly.

8.3.2 Task

Your task is to analyze the `car_maintenance` dataset from the `dstools` package in context of a CRISP-DM project. In particular, you want to understand and model the likeliness that a customer has a serious issue and has to return again to the Luxe Motors repair shops. For that purpose, implement the following tasks:

- Try to find further information about the features (data understanding). Make some descriptive statistics to understand the dataset and the target variable.
- Derive at some hypotheses which features might be connected to the issue and which features are irrelevant.
- Build multiple models which allow the Luxe Motors team to setup a predictive maintenance process. Use different evaluation criteria and discuss them.
- Design a final report to pitch your result at the Luxe Motors management. The report should illustrate your process, descriptive information about the dataset, the reasons for the quality problems and present possible action items to increase the service quality.
- Your pitch should end with the presentation of a live demo of a reporting solution (e.g. dashboard) that can be used from the aftersales team to investigate their vehicle analysis logfiles.

8.3.3 Remarks

Use the `car_maintenance` dataset for this case study.

8.4 Business Case Study 3: Take Part in an Analytics Competition

8.4.1 Scenario

An important building block for a good CV of a business data analytics expert is to be able to demonstrate practical experience. Hackathons and competitions in particular offer young professionals and newcomers the opportunity to demonstrate their skills and get in touch with potential employers. They also challenge individuals to be innovative and solve problems efficiently and purposefully.

The following business case is therefore about entering one of the most famous analytics competitions in the world: The Titanic Challenge. The challenge is hosted by the data science platform Kaggle and allows you to playfully learn how to participate in a real competition and improve your skills without time pressure.

Go to the platform's website and find out more about the challenge. Identify on your own how to sign up for this challenge, how to download the data sets and then submit your results. Try to achieve at least 90% accuracy.

8.4.2 Task

Your task is to analyze the `Titanic - Machine Learning from Disaster` challenge hosted by the Kaggle platform. The challenge is based on the historical event of the RMS Titanic's sinking in 1912. The challenge involves using analytic techniques to predict which passengers aboard the Titanic were likely to survive based on various attributes such as age, gender, ticket class, family relationships, and more. To solve this business case, implement the following tasks:

- Find the challenge, and read the description, accept the competition rules and gain access to the competition dataset.
- Download the datasets, build analytic models on it locally and generate a prediction file to submit at the platform. Do this building on the CRISP-DM process we used in the other business cases.
- Upload your prediction as a submission on Kaggle and receive an accuracy score. Your final score should be about 90%. Do not worry, if you are not good enough at the moment, you can submit a better solution later.
- Learn how to benchmark: See how your model ranks against other Kagglers on our leaderboard. And check out the discussion forum to find lots of tutorials and insights from other competitors. Try to use this information to improve your solution.

8.4.3 Remarks

Use the `train.csv` and `test.csv` dataset for this case study. You can download it at: *www.kaggle.com/c/titanic/overview*.

References

Allahyari, H., & Lavesson, N. (2011). User-oriented assessment of classification model understandability. *11th Scandinavian conference on Artificial intelligence*.

Aronson, J. E., Liang, T.-P., & MacCarthy, R. V. (2005). *Decision support systems and intelligent systems* (Bd. 4). Pearson Prentice-Hall Upper Saddle River, NJ, USA:

Baumgartner, J. C., Mackay, J. B., Morris, J. S., Otenyo, E. E., Powell, L., Smith, M. M., Snow, N., Solop, F. I., & Waite, B. C. (2010). *Communicator-in-chief: How Barack Obama used new media technology to win the White House*. Lexington Books.

Bennett, J., Lanning, S., & others. (2007). The Netflix prize. *Proceedings of KDD Cup and Workshop, 2007*, 35.

Berthold, M. R., & Hand, D. J. (2007). *Intelligent data analysis: An introduction*. Springer.

Breiman, L. (2001). Statistical modeling: The two cultures (with comments and a rejoinder by the author). *Statistical Science*, *16*(3), 199–231.

Brown, M. S. (2015). What IT needs to know about the data mining process. *Published by Forbes, 29*.

Carr, D. (2013, Februar 24). The Media Equation—Giving Viewers What They Want. *The New York Times*. https://www.nytimes.com/2013/02/25/business/media/for-house-of-cards-using-big-data-to-guarantee-its-popularity.html

Chapman, P., Clinton, J., Kerber, R., Khabaza, T., Reinartz, T., Shearer, C., & Wirth, R. (1999). The CRISP-DM user guide. *4th CRISP-DM SIG Workshop in Brussels in March, 1999*.

Davenport, T. H., & Patil, D. (2012). Data scientist. *Harvard business review*, *90*(5), 70–76.

Fayyad, U. M., Piatetsky-Shapiro, G., Smyth, P., & others. (1996). Knowledge Discovery and Data Mining: Towards a Unifying Framework. *KDD*, *96*, 82–88.

Field, Z., Miles, J., & Field, A. (2012). Discovering statistics using R. *Discovering Statistics Using R*, 1–992.

Fisher, R. A. (1936). The use of multiple measurements in taxonomic problems. *Annals of eugenics*, *7*(2), 179–188.

Fisher, R. A. (1956). Mathematics of a lady tasting tea. *The world of mathematics*, *3*(part 8), 1514–1521.

Geman, S., Bienenstock, E., & Doursat, R. (1992). Neural networks and the bias/variance dilemma. *Neural computation*, *4*(1), 1–58.

Gorakala, S. K., & Usuelli, M. (2015). *Building a recommendation system with R*. Packt Publishing Ltd.

Hayashi, C. (1998). What is data science? Fundamental concepts and a heuristic example. *Data Science, Classification, and Related Methods: Proceedings of the Fifth Conference of the International Federation of Classification Societies (IFCS-96), Kobe, Japan, March 27–30, 1996*, 40–51.

Henke, N., & Jacques Bughin, L. (2016). *The age of analytics: Competing in a data-driven world*.

D. Jung, *The Modern Business Data Analyst*, https://doi.org/10.1007/978-3-031-59907-1

Inbar, O., Tractinsky, N., & Meyer, J. (2007). Minimalism in information visualization: Attitudes towards maximizing the data-ink ratio. *Proceedings of the 14th European conference on Cognitive ergonomics: invent! explore!*, 185–188.

James, G., Witten, D., Hastie, T., Tibshirani, R., & others. (2013). *An introduction to statistical learning* (Bd. 112). Springer.

Jung, D., Dorner, V., Glaser, F., & Morana, S. (2018). Robo-advisory: Digitalization and automation of financial advisory. *Business & Information Systems Engineering, 60*, 81–86.

Kahneman, D. (2011). *Thinking, fast and slow*. Macmillan.

Keen, P. G. (1980). Decision support systems: A research perspective. *Decision support systems: Issues and challenges: Proceedings of an international task force meeting*, 23–44.

Knuth, D. E. (1984). Literate programming. *The computer journal*, *27*(2), 97–111.

MacQueen, J. (1967). Classification and analysis of multivariate observations. *5th Berkeley Symp. Math. Statist. Probability*, 281–297.

Michalczyk, S., Nadj, M., Beier, H., & Maedche, A. (2021). Designing a self-service analytics system for supply base optimization. *Intelligent Information Systems: CAiSE Forum 2021, Melbourne, VIC, Australia, June 28–July 2, 2021, Proceedings*, 154–161.

Microsoft Corporation. (2020). *The Team Data Science Process lifecycle*. https://docs.microsoft.com/en-us/azure/architecture/data-science-process/lifecycle

Orosz, G. (2023). *The Software Engineer's Guidebook*. Pragmatic Engineer B.V.

Patil, D. (2012). *Data jujitsu: The art of turning data into data product. Radar*. O'Reilly. http://radar. oreilly. com/2012/07/data-jujitsu.html

Piatetsky, G. (2014). CRISP-DM, still the top methodology for analytics, data mining, or data science projects. *KDD News*.

Press, G. (2016). Cleaning big data: Most time-consuming, least enjoyable data science task, survey says. *Forbes, March, 23*, 15.

Quinlan, J. R. (2014). *C4.5: Programs for machine learning*. Elsevier.

Resnick, P., & Varian, H. R. (1997). Recommender systems. *Communications of the ACM*, *40*(3), 56–58.

SAS Institute. (2017). *Introduction to SEMMA*. https://documentation.sas.com/doc/en/emref/14.3/n061bzurmej4j3n1jnj8bbjjm1a2.htm

Schulz, M., Neuhaus, U., Kaufmann, J., Badura, D., Kuehnel, S., Badewitz, W., Dann, D., Kloker, S., Alekozai, E. M., & Lanquillon, C. (2020). *Introducing DASC-PM: A Data Science Process Model*.

Shannon, C. E. (1948). A mathematical theory of communication. *The Bell system technical journal*, *27*(3), 379–423.

Shelby, T., Schenck, C., Weeks, B., Goodwin, J., Hennein, R., Zhou, X., Spiegelman, D., Grau, L. E., Niccolai, L., Bond, M., & others. (2021). Lessons learned from COVID-19 contact tracing during a public health emergency: A prospective implementation study. *Frontiers in public health*, *9*, 721952.

Spraul, V. A. (2013). *Think like a programmer*. MITP-Verlags GmbH & Co. KG.

Sturges, H. A. (1926). The choice of a class interval. *Journal of the american statistical association*, *21*(153), 65–66.

Taylor, S. J., & Letham, B. (2018). Forecasting at scale. *The American Statistician*, *72*(1), 37–45.

Verleysen, M., & François, D. (2005). The curse of dimensionality in data mining and time series prediction. *International work-conference on artificial neural networks*, 758–770.

Voigt, P., & Von dem Bussche, A. (2017). The eu general data protection regulation (gdpr). *A Practical Guide, 1st Ed., Cham: Springer International Publishing*, *10*, 3152676.

Waggoner, P. (2018). Advice to Young (and Old) Programmers: A Conversation with Hadley Wickham. *R-Bloggers.com (org. R-Posts.com)*. https://www.r-bloggers.com/2018/08/advice-to-young-and-old-programmers-a-conversation-with-hadley-wickham/

Wickham, H. (2010). A layered grammar of graphics. *Journal of Computational and Graphical Statistics*, *19*(1), 3–28.

Wickham, H. (2011). The split-apply-combine strategy for data analysis. *Journal of statistical software*, *40*, 1–29.

Wickham, H. (2014). Tidy Data. *Journal of Statistical Software*, *59*(10), 1–23. https://doi.org/10.18637/jss.v059.i10

Wilkinson, L., Anand, A., & Grossman, R. (2005). Graph-theoretic scagnostics. *Information Visualization, IEEE Symposium on*, 21–21.

World Economic Forum (Hrsg.). (2023). *The Future of Jobs Report*. https://www.weforum.org/publications/the-future-of-jobs-report-2023/

Printed in the USA
CPSIA information can be obtained
at www.ICGtesting.com
CBHW080318150724
11591CB00007B/232

9 783031 599064